T0235101

Communications in Computer and Information Science   778

Commenced Publication in 2007
Founding and Former Series Editors:
Alfredo Cuzzocrea, Xiaoyong Du, Orhun Kara, Ting Liu, Dominik Ślęzak,
and Xiaokang Yang

Editorial Board

Simone Diniz Junqueira Barbosa
Pontifical Catholic University of Rio de Janeiro (PUC-Rio),
Rio de Janeiro, Brazil
Phoebe Chen
La Trobe University, Melbourne, Australia
Joaquim Filipe
Polytechnic Institute of Setúbal, Setúbal, Portugal
Igor Kotenko
St. Petersburg Institute for Informatics and Automation of the Russian
Academy of Sciences, St. Petersburg, Russia
Krishna M. Sivalingam
Indian Institute of Technology Madras, Chennai, India
Takashi Washio
Osaka University, Osaka, Japan
Junsong Yuan
Nanyang Technological University, Singapore
Lizhu Zhou
Tsinghua University, Beijing, China

More information about this series at http://www.springer.com/series/7899

Dimitar Trajanov · Verica Bakeva (Eds.)

ICT Innovations 2017

Data-Driven Innovation

9th International Conference, ICT Innovations 2017
Skopje, Macedonia, September 18–23, 2017
Proceedings

 Springer

Editors
Dimitar Trajanov ⓘ
Computer Science and Engineering
Ss. Cyril and Methodius University
Skopje
Macedonia

Verica Bakeva ⓘ
Computer Science and Engineering
Ss. Cyril and Methodius University
Skopje
Macedonia

ISSN 1865-0929   ISSN 1865-0937 (electronic)
Communications in Computer and Information Science
ISBN 978-3-319-67596-1   ISBN 978-3-319-67597-8 (eBook)
DOI 10.1007/978-3-319-67597-8

Library of Congress Control Number: 2017952857

© Springer International Publishing AG 2017
This work is subject to copyright. All rights are reserved by the Publisher, whether the whole or part of the material is concerned, specifically the rights of translation, reprinting, reuse of illustrations, recitation, broadcasting, reproduction on microfilms or in any other physical way, and transmission or information storage and retrieval, electronic adaptation, computer software, or by similar or dissimilar methodology now known or hereafter developed.
The use of general descriptive names, registered names, trademarks, service marks, etc. in this publication does not imply, even in the absence of a specific statement, that such names are exempt from the relevant protective laws and regulations and therefore free for general use.
The publisher, the authors and the editors are safe to assume that the advice and information in this book are believed to be true and accurate at the date of publication. Neither the publisher nor the authors or the editors give a warranty, express or implied, with respect to the material contained herein or for any errors or omissions that may have been made. The publisher remains neutral with regard to jurisdictional claims in published maps and institutional affiliations.

Printed on acid-free paper

This Springer imprint is published by Springer Nature
The registered company is Springer International Publishing AG
The registered company address is: Gewerbestrasse 11, 6330 Cham, Switzerland

Preface

The ICT Innovations 2017 conference created and managed a vibrant environment, where participants shared the latest discoveries and best practices, and learned about data-driven innovation. The conference promoted the development of models, methods, and instruments of data science and provided a unique environment for the presentation and discussion of new approaches and prototypes in the field.

ICT Innovations conferences are organized by the Association for Information and Communication Technologies (ICT-ACT), whose mission is the advancement of ICT technologies. The main co-organizer and supporter of the 9th International ICT Innovations conference was the Faculty of Computer Science and Engineering and Ss. Cyril and Methodius University in Skopje, Macedonia.

The special conference topic was "data-driven innovation". Data-driven innovation forms a key pillar in the 21st-century sources of growth. The confluence of several trends, including the increasing migration of socio-economic activities to the Internet and the decline in the cost of data collection, storage, and processing, are leading to the generation and use of huge volumes of data. These large datasets are becoming a core asset in research and economics, fostering new discoveries, new industries, new processes, and new products. Data innovation is happening everywhere. It is transforming how and what people purchase, and how they communicate and collaborate. In response, data and analytics are being used to motivate radical new business models and disrupt traditional industry structures. Outperforming organizations will be those that make data and analytics central to their innovation processes, as well as to the innovation itself. Data promises to provide the input for creative endeavors and new ideas. But the importance of data and analytics will transcend ideation and inspiration. The conference also focused on a variety of ICT fields like: big data analytics, cloud computing, data mining, digital signal processing, e-health, embedded systems, emerging mobile technologies, multimedia, Internet of Things, machine learning, software engineering, security and cryptography, coding theory, wearable technologies, wireless communication, sensor networks, and other related topics.

The ICT Innovations 2017 received 90 submissions from 232 authors coming from 22 different countries. All these submissions were peer reviewed by the ICT Innovations 2017 Program Committee consisting of 260 high-quality researchers coming from 56 different countries. In order to assure a high-quality and thoughtful review process, we assigned each paper to more than five reviewers, resulting in 5.1 average assigned reviewers per paper, and at the end of the review process, there were 4.26 reviews per paper. Based on the results from the peer reviews, 26 full papers were accepted, resulting in a 28.8% acceptance rate, which was within our goal of less than 30%.

ICT Innovations 2017 was held in held in Skopje at two locations, the Hotel Aleksandar Palace and the Faculty of Computer Science and Engineering, during September 18–23, 2017.

We would like to express sincere gratitude to the invited speakers for their inspirational talks, to the authors for submitting their work to this conference, and the reviewers for sharing their experience during the selection process. Special thanks to Aleksandra Popovska Mitrovik, Vladimir Zdraveski, and Milos Jovanovik for their technical support during the conference and their help during the preparation of the conference proceedings.

September 2017   Dimitar Trajanov
                                                             Verica Bakeva

Organization

ICT Innovations 2017 was organized by the Association for Information and Communication Technologies (ICT-ACT).

Conference and Program Chairs

Dimitar Trajanov	Ss. Cyril and Methodius University, Skopje, Macedonia
Verica Bakeva	Ss. Cyril and Methodius University, Skopje, Macedonia

Program Committee

Jugoslav Achkoski	Military Academy "General Mihailo Apostolski", Macedonia
Nevena Ackovska	University Ss.Cyril and Methodius, Macedonia
Syed Ahsan	Technische Universität Graz, Austria
Marco Aiello	University of Groningen, Netherlands
Zahid Akhtar	University of Udine, Italy
Dietrich Albert	University of Graz, Austria
Azir Aliu	Southeastern European University of Macedonia, Macedonia
Giner Alor Hernandez	Hernandez Instituto Tecnologico de Orizaba, Mexico
Adel Alti	University of Setif, Algeria
Luis Alvarez Sabucedo	Universidade de Vigo, Spain
Hani Alzaid	King Abdulaziz City for Science and Technology, Saudi Arabia
Ljupcho Antovski	University Ss.Cyril and Methodius, Macedonia
Goce Armenski	University Ss.Cyril and Methodius, Macedonia
Hrachya Astsatryan	National Academy of Sciences of Armenia, Armenia
Tsonka Baicheva	Bulgarian Academy of Science, Bulgaria
Verica Bakeva	University Ss.Cyril and Methodius, Macedonia
Ekaterina Balandina	FRUCT, Academia-to-Industry Competence Incubator, Finland
Valentina Emilia Balas	Aurel Vlaicu University of Arad, Romania
Antun Balaz	Institute of Physics Belgrade, Serbia
Angel Barriga	IMSE/University of Seville, Spain
Lasko Basnarkov	University Ss.Cyril and Methodius, Macedonia
Hrvoje Belani	Croatian Health Insurance Fund, Croatia
Marta Beltran	Rey Juan Carlos University, Spain
Gennaro Boggia	DEI - Politecnico di Bari, Italy

Slobodan Bojanic	Universidad Politécnica de Madrid, Spain
Mirjana Bortnar Kljaji	University of Maribor, Slovenia
Dragan Bosnacki	Eindhoven University of Technology, Netherlands
An Braeken	EHB, Belgium
Zaki Brahmi	RIADI-Lab, Tunisia
Torsten Braun	Universität Bern, Switzerland
Andrej Brodnik	University of Ljubljana, Slovenia
Serhat Burmaoglu	Izmir Katip Celebi University, Turkey
Francesc Burrull	Universidad Politécnica de Cartagena, Spain
David Camacho	Universidad Autonoma de Madrid, Spain
Nick Cavalcanti	UFPE, UK
Somchai Chatvichienchai	University of Nagasaki, Japan
Jenhui Chen	Chang Gung University, Taiwan
Ljubomir Chitkushev	Boston University, USA
Ivan Chorbev	University Ss.Cyril and Methodius, Macedonia
Ioanna Chouvarda	Aristotle University of Thessaloniki, Greece
Ping-Tsai Chung	Long Island University, New York, USA
Betim Cico	Southeastern European University of Macedonia, Macedonia
Emmanuel Conchon	Institut de Recherche en Informatique de Toulouse, France
Marilia Curado	University of Coimbra, Portugal
Bozidara Cvetkovic	Jozef Stefan Institute, Slovenia
Robertas Damasevicius	Kaunas University of Technology, Lithuania
Pasqua D'Ambra	ICAR-CNR, Italy
Danco Davcev	University Ss.Cyril and Methodius, Macedonia
Antonio De Nicola	ENEA, Italy
Boris Delibasic	University of Belgrade, Serbia
Goran Devedzic	University of Kragijevac, Serbia
Vesna Dimitrievska Ristovska	University Ss.Cyril and Methodius, Macedonia
Vesna Dimitrova	University Ss.Cyril and Methodius, Macedonia
Ivica Dimitrovski	University Ss.Cyril and Methodius, Macedonia
Salvatore Distefano	University of Messina, Italy
Milena Djukanovic	University of Montenegro, Montenegro
Ciprian Dobre	University Politehnica of Bucharest, Romania
Martin Drlik	Constantine the Philosopher University in Nitra, Slovakia
Kristina Drusany Staric	University Medical Centre Ljubljana, Slovenia
Saso Dzeroski	Jozef Stefan Institute, Ljubljana, Slovenia
Joshua Ellul	University of Malta, Malta
Suliman Mohamed Fati	Universiti Sains Malaysia, Malaysia
Deborah Fels	Ryerson University, Canada
Majlinda Fetaji	Southeastern European University of Macedonia, Macedonia
Sonja Filiposka	University Ss.Cyril and Methodius, Macedonia

Predrag Filipovikj	Mälardalen University, Sweden
Neki Frasheri	Polytechnic University of Tirana, Albania
Kaori Fujinami	Tokyo University of Agriculture and Technology, Japan
Slavko Gajin	University of Belgrade, Serbia
Joao Gama	University of Porto, Portugal
Ivan Ganchev	University of Limerick, Ireland/University of Plovdiv Paisii Hilendarski, Bulgaria
Todor Ganchev	Technical University of Varna, Bulgaria
Nuno Garcia	Universidade da Beira Interior, Portugal
Andrey Gavrilov	Laboratory Hybrid Intelligent Systems, Russia
Amjad Gawanmeh	Khalifa University, United Arab Emirates
John Gialelis	University of Patras, Greece
Sonja Gievska	University Ss.Cyril and Methodius, Macedonia
Dejan Gjorgjevikj	University Ss.Cyril and Methodius, Macedonia
Danilo Gligoroski	Norwegian University of Science and Technology, Norway
Rossitza Goleva	Technical University of Sofia, Bulgaria
Abel Gomes	Univeristy of Beira Interior, Portugal
Saso Gramatikov	University Ss.Cyril and Methodius, Macedonia
George Gravvanis	Democritus University of Thrace, Greece
Andrej Grguric	Ericsson Nikola Tesla - Research and Innovations Unit, Croatia
Daniel Grosu	Wayne State University, USA
David Guralnick	Teachers College, Columbia University, USA
Marjan Gushev	University Ss.Cyril and Methodius, Macedonia
Yoram Haddad	Jerusalem College of Technology, Israel
Elena Hadzieva	University of Information Science and Technology (UIST) St. Paul the Apostle, Macedonia
Tianyong Hao	Guangdong University of Foreign Studies, China
Natasa Hoic-Bozic	University of Rijeka, Croatia
Violeta Holmes	University of Huddersfield, UK
Fu-Shiung Hsieh	University of Technology, Taiwan
Yin-Fu Huang	University of Science and Technology, Taiwan
Ladislav Huraj	University of SS. Cyril and Methodius, Slovakia
Hieu Trung Huynh	Industrial University of Ho Chi Minh City, Vietnam
Barna Laszlo Iantovics	Petru Maior University of Tg. Mures, Romania
Vacius Jusas	Kaunas University of Technology, Lithuania
Sergio Ilarri	University of Zaragoza, Spain
Natasha Ilievska	University Ss.Cyril and Methodius, Macedonia
Minna Isomursu	VTT Technical Research Centre, Finland
Mirjana Ivanovic	University of Novi Sad, Serbia
Boro Jakimovska	University Ss.Cyril and Methodius, Macedonia
Smilka Janevska-Sarkanjac	University Ss.Cyril and Methodius, Macedonia
Yichuan Jiang	Southeast University, China
Mile Jovanov	University Ss.Cyril and Methodius, Macedonia

Milosh Jovanovikj	University Ss.Cyril and Methodius, Macedonia
Slobodan Kalajdziski	University Ss.Cyril and Methodius, Macedonia
Alexey Kalinov	Cadence Design Systems, Russia
Kalinka Kaloyanova	University of Sofia - FMI, Bulgaria
Aneta Karaivanova	Bulgarian Academy of Sciences, Bulgaria
Takahiro Kawamura	The University of Electro-Communications, Japan
Richard Knepper	Indiana University, USA
Ljupcho Kocarev	University Ss.Cyril and Methodius, Macedonia
Natasa Koceska	University Goce Delcev, Macedonia
Saso Koceski	University Goce Delcev, Macedonia
Dragi Kocev	Jozef Stefan Institute, Slovenia
Peter Kokol	University of Maribor, Slovenia
Margita Kon-Popovska	University Ss.Cyril and Methodius, Macedonia
Magdalena Kostoska	University Ss.Cyril and Methodius, Macedonia
Ivan Kraljevski	VoiceINTERconnect GmbH, Germany
Aleksandar Krapezh	Serbian Academy of Sciences and Arts, Serbia
Andrea Kulakov	University Ss.Cyril and Methodius, Macedonia
Siddhivinayak Kulkarni	Federation University, Australia
Ashok Kumar Das	International Institute of Information Technology, India
Brajesh Kumar Singh	Faculty of Engineering and Technology, RBS College, India
Anirban Kundu	Kuang-Chi Institute of Advanced Technology, Singapore
Minoru Kuribayashi	Kobe University, Japan
Eugenijus Kurilovas	Vilnius University, Lithuania
Arianit Kurti	Linnaeus University, Sweden
Jan Kwiatkowski	Wroclaw University of Technology, Poland
David Lamas	Tallinn University, Estonia
Alexey Lastovetsky	University College Dublin, Ireland
Sanja Lazarova-Molnar	University of Southern Denmark, Denmark
Nhien An Le Khac	University College Dublin, Ireland
Rita Yi Man Li	Hong Kong Shue Yan University, Hong Kong
Hwee-San Lim	Universiti Sains Malaysia, Malaysia
Thomas Lindh	KTH, Sweden
Igor Ljubi	Croatian Institute for Health Insurance, Croatia
Suzana Loshkovska	University Ss.Cyril and Methodius, Macedonia
Jos Machado Da Silva	FEUP, Portugal
Ana Madevska Bogdanova	University Ss.Cyril and Methodius, Macedonia
Gjorgji Madjarov	University Ss.Cyril and Methodius, Macedonia
Piero Malcovati	University of Pavia, Italy
Augostino Marengo	Università degli Studi di Bari Aldo Moro, Italy
Ninoslav Marina	University St. Paul the Apostole, Macedonia
Smile Markovski	University Ss.Cyril and Methodius, Macedonia
Cveta Martinovska	University Goce Delcev, Macedonia
Fulvio Mastrogiovanni	University of Genoa, Italy
Darko Matovski	University of Southampton, UK

Marcin Michalak	Silesian University of Technology, Poland
Hristina Mihajloska	University Ss.Cyril and Methodius, Macedonia
Aleksandra Mileva	University Goce Delcev, Macedonia
Biljana Mileva Boshkoska	Faculty of Information Studies in Novo Mesto, Slovenia
Miroslav Mirchev	University Ss.Cyril and Methodius, Macedonia
Georgina Mircheva	University Ss.Cyril and Methodius, Macedonia
Anastas Mishev	University Ss.Cyril and Methodius, Macedonia
Igor Mishkovski	University Ss.Cyril and Methodius, Macedonia
Kosta Mitreski	University Ss.Cyril and Methodius, Macedonia
Pece Mitrevski	University St. Kliment Ohridski, Macedonia
Irina Mocanu	University Politehnica of Bucharest, Romania
Anne Moen	University of Oslo, Norway
Ammar Mohammed	Cairo University, Egypt
Radouane Mrabet	Mohammed V - Souissi University, Morocco
Irena Nancovska Serbec	University of Ljubljana, Slovenia
Andreja Naumoski	University Ss.Cyril and Methodius, Macedonia
Viorel Nicolau	Dunarea de Jos University of Galati, Romania
Alexandru Nicolin	Horia Hulubei National Institute of Physics and Nuclear Engineering, Romania
Manuel Noguera	Universidad de Granada, Spain
Anthony Norcio	University of Maryland, Baltimore County, USA
Novica Nosovic	University of Sarajevo, Bosnia and Herzegovina
Ivana Ognjanovi	Univerzitet Donja Gorica, Montenegro
Pance Panov	Jozef Stefan Institute, Slovenia
Eleonora Pantano	University of Calabria, Italy
Marcin Paprzycki	IBS PAN and WSM, Poland
Peter Parycek	Danube-University Krems, Austria
Matteo Pastorino	Life Supporting Technologies - UPM, Spain
Shushma Patel	London South Bank University, UK
Christian Fischer Pedersen	Aarhus University, Denmark
Maritta Perälä-Heape	Centre for Health and Technology, Finland
Dana Petcu	West University of Timisoara, Romania
Predrag Petkovic	University of Ni, Serbia
Antonio Pinheiro	Universidade da Beira Interior, Portugal
Niels Pinkwart	Humboldt Universität zu Berlin, Germany
Matus Pleva	Technical University of Koice, Slovakia
Vedran Podobnik	University of Zagreb, Croatia
Florin Pop	University Politehnica of Bucharest, Romania
Zaneta Popeska	University Ss.Cyril and Methodius, Macedonia
Aleksandra Popovska-Mitrovikj	University Ss.Cyril and Methodius, Macedonia
Marco Porta	University of Pavia, Italy
Rodica Potolea	Technical Univeristy of Cluj-Napoca, Romania
Ustijana Rechkoska Shikoska	UINT, Macedonia

Manjeet Rege	University of St. Thomas, USA
Kalinka Regina Castelo Branco	Institute of Mathematics and Computer Sciences, Brazil
Miriam Reiner	Technion – Israel Institute of Technology, Israel
Blagoj Ristevski	University St. Kliment Ohridski, Macedonia
Sasko Ristov	University Ss.Cyril and Methodius, Macedonia
Philippe Roose	LIUPPA, France
Jatinderkumar Saini	Narmada College of Computer Application, India
Simona Samardziska	University Ss.Cyril and Methodius, Macedonia
Snezana Savovska	University Ss.Cyril and Methodius, Macedonia
Wolfgang Schreiner	Research Institute for Symbolic Computation (RISC), Austria
Loren Schwiebert	Wayne State University, USA
Bryan Scotney	University of Ulster, UK
Ramo Endelj	Univerzitet Donja Gorica, Montenegro
Vladimr Sildi	Matej Bel University, Slovakia
Josep Silva	Universitat Politècnica de València, Spain
Manuel Silva	Instituto Superior de Engenharia do Porto, Portugal
Roel Smolders	VITO, Belgium
Ana Sokolova	University of Salzburg, Austria
Michael Sonntag	Johannes Kepler University Linz, Austria
Dejan Spasov	University Ss.Cyril and Methodius, Macedonia
Susanna Spinsante	Università Politecnica delle Marche, Italy
Georgi Stojanov	The American University of Paris, France
Igor Stojanovic	University Goce Delcev, Macedonia
Biljana Stojkoska	University Ss.Cyril and Methodius, Macedonia
Stanimir Stoyanov	University Paisii Hilendarski, Bulgaria
Ariel Stulman	The Jerusalem College of Technology, Israel
Chandrasekaran Subramaniam	Kumaraguru College of Technology, Coimbatore, India
Chang-Ai Sun	University of Science and Technology Beijing, China
Ousmane Thiare	Gaston Berger University, Senegal
Biljana Tojtovska	University Ss.Cyril and Methodius, Macedonia
Dimitar Trajanov	University Ss.Cyril and Methodius, Macedonia
Ljiljana Trajkovic	Simon Fraser University, Canada
Vladimir Trajkovik	University Ss.Cyril and Methodius, Macedonia
Denis Trcek	University of Ljubljana, Slovenia
Kire Trivodaliev	University Ss.Cyril and Methodius, Macedonia
Yuh-Min Tseng	National Changhua University of Education, Taiwan
Marek Tudruj	Polish Academy of Sciences, Poland
Carlos Valderrama	University of Mons, Belgium
Zlatko Varbanov	Veliko Tarnovo University, Bulgaria
Sergio, A. Velastin	Universidad de Santiago de Chile, Chile
Goran Velinov	University Ss.Cyril and Methodius, Macedonia
Elena Vlahu-Georgievska	University St. Kliment Ohridski, Macedonia
Irena Vodenska	Boston University, USA

Boris Vrdoljak	University of Zagreb, Croatia
Katarzyna Wac	University of Geneva, Switzerland
Santoso Wibowo	Central Queensland University, Australia
Michal Wozniak	Wroclaw University of Technology, Poland
Lai Xu	Bournemouth University, UK
Shuxiang Xu	University of Tasmania, Australia
Wuyi Yue	Konan University, Japan
Filip Zavoral	Charles University Prague, Czech Republic
Zoran Zdravev	University Goce Delcev Macedonia
Katerina Zdravkova	University Ss.Cyril and Methodius, Macedonia
Xiangyan Zeng	Fort Valley State University, USA

Organizing Committee

Cveta Martinovska	University Goce Delcev, Macedonia
Gjorgji Madjarov	University Ss.Cyril and Methodius, Macedonia
Elena Vlahu-Georgievska	University St. Kliment Ohridski, Macedonia
Azir Aliu	Southeastern European University of Macedonia, Macedonia
Ustijana Reckoska Shikoska	UINT, Macedonia

Technical Committee

Aleksandra Popovska Mitrovik	University Ss.Cyril and Methodius, Macedonia
Milos Jovanovik	University Ss.Cyril and Methodius, Macedonia
Vladimir Zdraveski	University Ss.Cyril and Methodius, Macedonia

Contents

Invited Keynote Paper

Proceeding Papers

Invited Keynote Paper

Video Pandemics: Worldwide Viral Spreading of Psy's Gangnam Style Video

Zsófia Kallus[1,2], Dániel Kondor[1,3], József Stéger[1], István Csabai[1],
Eszter Bokányi[1], and Gábor Vattay[1(✉)]

[1] Department of Physics of Complex Systems,
Eötvös University, Pázmány P. s. 1/A, Budapest 1117, Hungary
vattay@elte.hu
[2] Ericsson Research, Budapest, Hungary
[3] Senseable City Laboratory, Massachusetts Institute of Technology,
Cambridge, USA

Abstract. Viral videos can reach global penetration traveling through international channels of communication similarly to real diseases starting from a well-localized source. In past centuries, disease fronts propagated in a concentric spatial fashion from the source of the outbreak via the short range human contact network. The emergence of long-distance air-travel changed these ancient patterns. However, recently, Brockmann and Helbing have shown that concentric propagation waves can be reinstated if propagation time and distance is measured in the flight-time and travel volume weighted underlying air-travel network. Here, we adopt this method for the analysis of viral meme propagation in Twitter messages, and define a similar weighted network distance in the communication network connecting countries and states of the World. We recover a wave-like behavior on average and assess the randomizing effect of non-locality of spreading. We show that similar result can be recovered from Google Trends data as well.

Keywords: Geo-social networks · Meme dynamics · Online news propagation · Graph embedding

1 Introduction

According to Wikipedia, the music video of 'Gangnam Style' by recording artist Psy reached the unprecedented milestone of one billion YouTube views on December 21, 2012. It was directed by Cho Soo-Hyun and the video premiered on July 15, 2012. What makes this viral video unique from the point of view of social network research is that it was spreading mostly via human-to-human social network links before its first public appearance in the United States on August 20, 2012 and before Katy Perry shared it with her 25 million followers on Twitter on August 21.

Other viral videos spread typically via news media and reach worldwide audiences quickly, within 1–3 days. Our assumption is that only those online

© Springer International Publishing AG 2017
D. Trajanov and V. Bakeva (Eds.): ICT Innovations 2017, CCIS 778, pp. 3–12, 2017.
DOI: 10.1007/978-3-319-67597-8_1

viral phenomena can show similarities to global pandemics that were originally constrained to a well localized, limited region and then, after an outbreak period, reached a worldwide level of penetration. In 2012, the record breaking 'Gangnam Style' [1] marked the appearance of a new type of online meme, reaching unprecedented level of fame despite its originally small local audience. From the sub-culture of k-pop fans it reached an increasingly wide range of users of online media - including academics - from around the World (Refs. [2–5]). We approximately reconstructed its spreading process by filtering geo-tagged messages containing the words 'Gangnam' and 'style'. In Fig. 1 we show the location of geo-tagged posts containing the expression 'Gangnam Style' from our collection of the public Twitter feed, as of September 2012. For this purpose we used our historical Twitter dataset and collection of follower relations connecting 5.8M active Twitter users who enabled access to their location information while posting messages to their public accounts (see Ref. [6] for details). When tracing videos on the Twitter social platform we have access only to the public posts, and only look at geo-tagged messages. Location information allows us to record the approximate arrival time of a certain news to a specific geo-political region. In the real space this process looks indeed random, but the 'local to global' transition is also apparent as the messages cover a progressively larger territory. We collected the approximate first arrival time of the video in different geo-political regions of the World. In order to study the viral spreading of the appearances of the video in the Twitter datastream first we coarse-grained the World map into large homogeneous geo-political regions. We used regions of countries and states of the World as the cells (i.e. the nodes) and aggregated the individual links connecting them. By performing an aggregation into geo-political regions we construct a weighted graph connecting 261 super nodes. Thus the ratio of edge weights can be interpreted as an approximation of the relative strength of communication between pairs of the connected regions. This high-level graph is thus naturally embedded into the geographic space giving a natural length to its

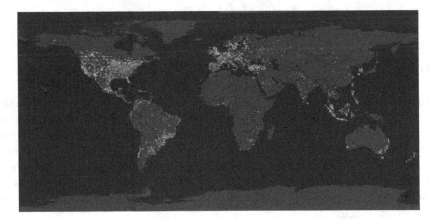

Fig. 1. Geo-locations of Twitter messages containing 'Gangnam Style'.

edges. We then connect individuals in the spatial social network and then recreate a high-level aggregated weighted graph between regions by querying a large database containing the collection of historic, freely available Twitter messages [6–8]. In Fig. 2 we show the resulting weights in graphical form for Twitter users in California.

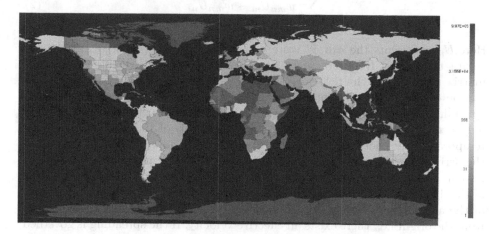

Fig. 2. Social connection weights between large geo-political regions of the World. The map shows our 261 geo-political regions and the number of friendships (mutual Twitter followers) between users in California and the rest of the World. Colour codes the number of friendships with users in California in our database. Deep red means that Californians have $\sim 10^5$ friendships within California and deep blue indicates that $\sim 10^0$ friendship connects them for example to certain regions of Africa. (Color figure online)

2 Weighting the Speed of Spreading in the Network

Brockmann and Helbing in Ref. [9] worked out a method by which they were able to predict the effective spreading time of a real disease between two nodes of air-traffic based on the network and the number of passengers travelling between them in unit time and introduced an effective measure of distance of the two nodes. Here, we repeat their derivation, except, we use the number of mutual social contacts (mutual Twitter followers called friends) between geo-located Twitter users in the geo-political regions. Each user can be in one of two possible states: *susceptible* for the viral video (never seen the video before) or *infected* (already seen the video) and affects the state of others when contact occurs between them via sharing the video in their social network feeds. The model that we adopt is based on the meta-population model [10,11]. After dividing the world map into geo-political regions the dynamics within the nth spatial unit is modelled by the *SI equations* [12–15] with disease specific parameters:

$$\partial_t S_n = -\alpha I_n S_n / N_n,$$
$$\partial_t I_n = -\beta I_n + \alpha I_n S_n / N_n .$$

(1)

This local dynamics is complemented by the weights that connect the separated cells as described by

$$\sum_{m \neq n} w_{mn} U_m - w_{mn} U_n.$$

(2)

Here U_n represents the nth SI state variable, and $w_{mn} = F_{nm}/N_m$ is the per capita traffic flux from site m to site n, F_{nm} being the weighted adjacency matrix representing the network. This means that the human contact network can be effectively divided into interconnected layers and the spatial reach of a spreading process is determined mainly by the weighted network of regions. We only need this high-level information and the details of the large and complex temporal contact network can be neglected. Once we have a weighted graph and the arrival times of the video at each of the nodes the embedding of the graph into an abstract space can be performed. Its goal is to uncover the wave pattern of the dynamics. This means finding a source region from which the dependence of the arrival times at a region is linear on the effective distance of the region from the origin i.e., there exists an effective velocity. If the spreading is governed by the Eqs. (1–2) the effective distance (spreading time) between nodes m and n can be defined as follows:

$$d_{mn} = 1 - log(P_{mn}) \leq 1,$$

(3)

where $P_{mn} = F_{mn}/\sum_m F_{mn}$ is the flux fraction from m to n i.e., the probability to choose destination m if one is in the region n. This way a minimal distance means maximal probability, and additivity of the distances is ensured by the logarithmic function.

3   From Geographic to Network-Based Embedding

The embedding comprises of the following steps. After calculating the connection weighted adjacency matrix, each link weight is transformed into the *effective quasi distance* defined by Eq. 3. Starting from each node a *shortest path tree* can be constructed from all nodes reachable from the selected one. These shortest paths correspond by definition to the most probable active routes that the equations of epidemics and the weights would predict. In such a tree the distance of each node from the origin is the length of the shortest path connecting them. The original graph and one of its shortest path trees is shown in Fig. 3. If the node selected as origin corresponds to the most effective source, the video will arrive first at the closest nodes and then propagate towards the periphery. If the arrival times show a linear dependence on the effective distance, it is likely that we found the source of the spreading. Comparing different trees by measuring the goodness of a linear fit allows us to find the most probable and most effective source node on the graph. For the 'Gangnam Style' video it is the region of the *Philippines*. While the obvious center should be Seoul/South Korea, where the video has been created, it seems that the most intensive source of social network spreading was in the Philippines. This is probably due to the fact that we are not able to measure the very short time it took the video to spread from South Korea to the Philippines and more importantly, the Philippines is much more connected socially to the rest of the world than South Korea. This is partially due to the English language use and a well spread diaspora of the Philippines.

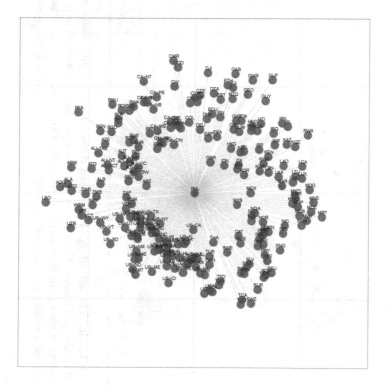

Fig. 3. Embedded shortest path tree. The countries of the World are partitioned to 261 large administrative regions. An aggregated version of the weighted and directed graph has been used from the individual-level follower relations. Effective distances are represented as the radial distance from the node at the origin in arbitrary units, and the angular coordinate of the nodes is arbitrary as well.

The linear fit, on the other hand, is not perfect. Uncertainty is introduced into our analysis by multiple factors. These are the heterogeneous nature of the use of the social platforms around the world and the variability of activities over time; the partial measurements obtained from the publicly available geo-tagged sample of tweets; and the external effect of other media propagating the same news. These circumstances all add to the uncertainty of the first arrival times in our dataset. In order to recover the average wave form we had to use smoothing in

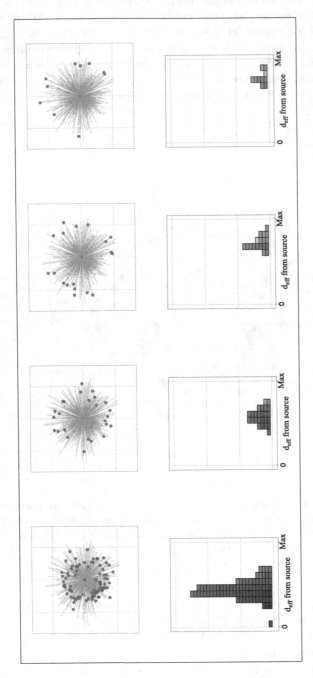

Fig. 4. Progressive stages of the pandemic. The spreading of the wave is shown in four progressive stages of the propagation. Each stage is defined by separate time slice of equal length. The nodes where the news has just arrived in that slice are first shown on the shortest path tree. Second, a corresponding histogram is created based on effective distances. Each rectangle represents one of the regional nodes and a common logarithmic color scale represents the number of users of the nodes (color scale of Fig. 5 is used). (Color figure online)

space and in time as well. The spatial averaging achieved by regional aggregation is also justified by the assumption that users within the same region are likely to be more connected to each other than to the rest of the world [8]. We used a moving window over the arrival times and used the linear fitting to the average distances and the average times of the windows. The noise can be effectively reduced by choosing a window of two weeks to a month. As shown on Fig. 4 by the colour scale, the smaller the number of Twitter geo-users in a region, the less reliable the analysis becomes. Moreover these small regions form the peripheral ring as seen from the most probable source region's point of view.

4 Comparison of Twitter and Google Trends Distances

Once the site of origin is selected, the embedding is straightforward. The effective distances between nodes and the source node are equivalent to the shortest path distances on the embedded graph. Figure 5 shows how the underlying order can be uncovered by this transformation. The seemingly randomized left panel – representing the arrival of the video at various *geographic* distances – becomes structured, and a linear trend emerges as a function of the *effective* distance. Linearity breaks down only at large distances, where remote peripheral regions are left waiting for the video to arrive at last. We also measured arrival times by looking at the Google search engine records through Google Trends analytics service. In Fig. 6 we show the results. Its sparsity is coming from a lower temporal resolution of Google Trends compared to our Twitter based dataset.

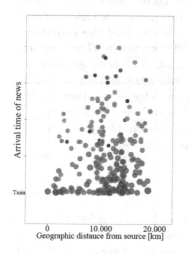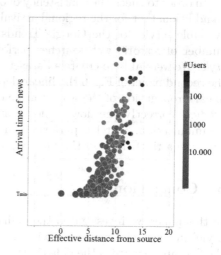

Fig. 5. Geographic distance vs. effective distance on Twitter. Here each dot represents a state, colored according to the number of users, using the same logarithmic color scale as before. The horizontal axis represents the distance from the source node of the Philippines, while the vertical axis is the arrival time of the video at that regional node. The clear order is uncovered in the second panel, where geographic distance is replaced by the effective distance. (Color figure online)

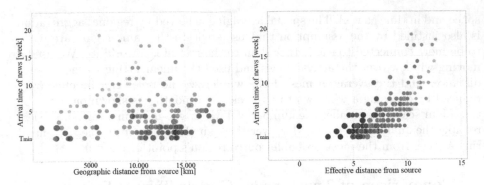

Fig. 6. Geographic distance vs. effective distance in Google Trends. Here each dot represents a state, colored according to the number of Users (same color scale as Fig. 5), using the same logarithmic color scale as before. The horizontal axis represents the distance from the source node of the Philippines, while the vertical axis is the arrival time of the video at that regional node. The clear order is uncovered in the second panel, where geographic distance is replaced by the effective distance. (Color figure online)

It is, however, very remarkable that the network embedding we found for the Twitter data works also for the Google Trends dataset and creates the proper reordering of the nodes. This result further underlines the effectiveness of our Twitter measurements.

In order to check the consistency of our results with other sources we analyzed an additional set for the regional arrival times. For this purpose we used publicly available service of the Google Trends platform, and gathered the cumulative number of historic web searches performed with the keywords *Gangnam* and *style*. We were able to create a dataset with a resolution of 2 weeks. As shown on the second panel of Fig. 6 the linear dependence is once again recovered, and the linear propagation of the wave can be effectively seen. This proves that Twitter predicts correctly the flow of news and information on the global network of communication. Searches performed first in each region seem to be synchronized with the arrival times of the first tweets coming from Twitter.

5   Conclusions

In this paper we investigated the online spreading of the viral video 'Gangnam Style' in the social network Twitter. Using geo-localized tweets we determined the first appearance of the video in the 261 major geo-political areas of the World. We adapted the tools developed in Ref. [9] for the spreading of infectious diseases over the air-travel network for the present problem. We showed that the number of friendships (mutual following) of Twitter users between geo-political regions is analogous to the passenger traffic volume of air-traffic from the point of view of online information spreading. Using our historic public tweet database we could calculate the relative weights of information traffic between the regions. We then reproduced the shortest path effective distance of geo-political regions from the epicenter of the social outbreak of the 'Gangnam Style' video pandemic, which seems to be in the Philippines, not far from South Korea, where it has been produced. We managed to verify the spreading pattern independently, based on the search results in Google Trends in the same time period. The synchrony between first appearances in Twitter and Google suggest that a universal pattern of social information flow (information highways) exist between geo-political regions, which is inherently non-technological, determined by the strength of social ties between different countries, cultures and languages. Further research is necessary to understand the main features of this global network.

Acknowledgement. The authors would like to thank János Szüle, László Dobos, Tamás Hanyecz, and Tamás Sebők for the maintenance of the Twitter database at Eötvös University. Authors thank the Hungarian National Research, Development and Innovation Office under Grant No. 125280 and the grant of Ericsson Ltd.

References

1. Gruger, W.: Psy's 'Gangnam Style' Video Hits 1 Billion Views, Unprecedented Milestone. Billboard com (2012)
2. Jung, S., Shim, D.: Social distribution: K-pop fan practices in Indonesia and the 'Gangnam Style' phenomenon. Int. J. Cult. Stud. 1367877913505173 (2013)
3. Weng, L., Menczer, F., Ahn, Y.Y.: Virality prediction and community structure in social networks. Sci. Rep. **3** (2013)
4. Keung Jung, S.: Global Audience Participation in the Production and Consumption of Gangnam Style. ScholarWorks@ Georgia State University (2014)
5. Wiggins, B.E., Bowers, G.B.: Memes as genre: a structurational analysis of the memescape. New Media Soc. **17**(11), 1886–1906 (2015)
6. Dobos, L., Szüle, J., Bodnar, T., Hanyecz, T., Sebők, T., Kondor, D., et al.: A multi-terabyte relational database for geo-tagged social network data. In: 2013 IEEE 4th International Conference on Cognitive Infocommunications (CogInfoCom), pp. 289–294. IEEE (2013)
7. Szüle, J., Kondor, D., Dobos, L., Csabai, I., Vattay, G.: Lost in the city: revisiting Milgram's experiment in the age of social networks. PLoS ONE **9**(11), e111973 (2014)
8. Kallus, Z., Barankai, N., Szüle, J., Vattay, G.: Spatial fingerprints of community structure in human interaction network for an extensive set of large-scale regions. PLoS ONE **10**(5), e0126713 (2015)
9. Brockmann, D., Helbing, D.: The hidden geometry of complex, network-driven contagion phenomena. Science **342**(6164), 1337–1342 (2013)
10. Balcan, D., Gonçalves, B., Hu, H., Ramasco, J.J., Colizza, V., Vespignani, A.: Modeling the spatial spread of infectious diseases: the GLobal Epidemic and Mobility computational model. J. Comput. Sci. **1**(3), 132–145 (2010). http://www.sciencedirect.com/science/article/pii/S1877750310000438
11. Vespignani, A.: Modelling dynamical processes in complex socio-technical systems. Nat. Phys. **8**(1), 32–39 (2012)

12. Dietz, K.: Epidemics and rumours: a survey. J. R. Stat. Soc. Ser. A (General), 505–528 (1967)
13. Newman, M.E.J.: Spread of epidemic disease on networks. Phys. Rev. E **66**, 016128 (2002). http://link.aps.org/doi/10.1103/PhysRevE.66.016128
14. Barthelemy, M., Barrat, A., Vespignani, A.: Dynamical Processes on Complex Networks. Cambridge University Press, Cambridge (2008)
15. Isham, V., Harden, S., Nekovee, M.: Stochastic epidemics and rumours on finite random networks. Phys. A **389**(3), 561–576 (2010)

Proceeding Papers

A Secure Discharging Protocol for Plug in Electric Vehicle (SDP-V2G) in Smart Grid

Khaled Shuaib[✉], Juhar Ahmed Abdella, Ezedin Barka, and Farag Sallabi

College of Information Technology, The United Arab Emirates University,
P.O Box 15551 Al-Ain, UAE
K.shuaib@uaeu.ac.ae

Abstract. Penetration of Plug in electric vehicles (PEVs) is expected to rise in the next few years especially in areas with new deployed smart power grid systems. Charging and discharging of PEVs will introduce several challenges related to load stabilization and information security. In this paper, we discuss a secure discharging protocol where users can be protected from possible information security and privacy attacks. The protocol also incorporates required remote authorization and payment transaction mechanisms. Our protocol is developed based on the use of encryption mechanisms and the dual signature approach. Using the security protocol verification tool, Automatic Verification and Analysis of Internet Security Protocols (AVISPA), the security aspects of the proposed protocol are verified. Our approach is robust against misuse of electric vehicles and unfair payment issues as it allows for user-based authentication in addition to the authentication of associated electric vehicles.

Keywords: Smart grid · Plug in electric vehicles · Information security · Protocol · Vehicle to grid

1 Introduction

One of the important components of Smart Grid (SG) systems is the Vehicle-to-Grid (V2G) network. The V2G network describes a network of power systems in which plug-in electric vehicles (PEVs) are connected to the SG as mobile distributed energy resources. V2G systems are capturing the attention of both the electricity providers and end users due to the various advantages gained by their deployment. On one hand, power suppliers benefit from utilizing PEVs to better manage demand response services and ancillary services (e.g. spinning reserves, reactive power support, frequency and voltage regulation) to stabilize the power system. On the other hand, users can get incentives from power providers by providing the aforementioned services. PEVs can store energy by charging their batteries during off-peak hours when the power supply from the grid or renewable energy resources is more than the demand. During peak hours when the energy demand exceeds the energy supply, PEVs can sell power back to the SG by discharging their batteries. The other advantage of V2G systems is that PEVs promote environmental benefits by reducing the CO_2 emissions. V2G networks are based on a SG system that supports a bi-directional flow of electricity and data communication [1].

© Springer International Publishing AG 2017
D. Trajanov and V. Bakeva (Eds.): ICT Innovations 2017, CCIS 778, pp. 15–26, 2017.
DOI: 10.1007/978-3-319-67597-8_2

The dependency of V2G networks on a two-way data communication allows efficient information exchange between different parties and provides a secure, flexible, responsive, and reliable payment system [2]. However, the reliance of V2G networks on two-way data communication gives rise to different kinds of security and privacy problems related to the confidentiality, integrity and availability of the system [3–7]. Some of the potential security attacks in V2G networks include but not limited to eavesdropping, DoS attacks, replay attacks and repudiation attacks. Moreover, the privacy of PEV owners could be violated by involved entities during discharging. This can take place when users' or vehicle based sensitive information such as the real user identity and vehicle identity submitted to the supplier for authentication and billing purposes. Furthermore, PEVs in V2G networks can be misused by adversaries for financial benefits by discharging PEVs of others [8, 9]. Therefore, charging protocols used between the SG and PEVs for charging/discharging should be equipped with end-to-end security and privacy preservation techniques. One of the challenging behaviors of V2G networks is that one-way authentication, where only the power company authenticates the PEV user, is not sufficient. In V2G networks, there is a requirement for mutual authentication. PEV users need to be able to sell power back to suppliers anywhere while getting credited for it by their contracted home suppliers. For this reason, to avoid impersonation attacks, a PEV user needs to be protected against dealing with illegitimate aggregators used as intermediators between PEVs and power suppliers. Protection is needed against any repudiation attacks by charging stations or aggregators. Therefore, any used charging station or aggregator needs to be properly authenticated. On the other hand, users of PEVs need to be authenticated to guarantee that only legitimate users can discharge their PEVs. By doing so, misuse of PEVs and any payment disputes will be avoided when multiple users are allowed to use a single PEV.

While there are several studies conducted on V2G networks [11–15], only few of them discuss PEV discharging protocols [11, 12]. The authors in [11] proposed an anonymous authentication protocol for V2G networks based on group signature and identity based restrictive partially blind signature technique to provide security and user privacy-preserving. The approach allows a charging station/aggregator to authenticate PEVs anonymously and to manage them dynamically. In addition, their system supports aggregation to reduce the communication overhead that may be caused by multiple PEVs communicating with the aggregator simultaneously. A mutual authentication scheme is suggested by [12] to avoid redirection and impersonation attacks that may exist in unilateral authentication. The system also supports anonymous authentication based on pseudonym IDs to protect users' privacy. However, there are some major issues which were not addressed by these two previous approaches. Both of these approaches do not include a payment mechanism and do not support user-based authentication but rely on vehicle-based authentication which may lead to misuse.of PEVs and unfair payment issues. Moreover, the approach proposed in [11] does not support mutual authentication. In addition, both approaches achieve anonymous authentication by using methods such as pseudonym IDs and group signature which have their own drawbacks. According to [21], group signature based authentication is not suitable for V2G networks due to the dynamic nature of PEVs which will lead to spatial and temporal uncertainties. The pitfall of Pseudonym ID based authentication is that its management is difficult for large number of vehicles as it usually requires frequent replacement of Pseudonym IDs [11]. In this paper, we propose a secure and privacy-aware PEV discharging protocol. The protocol supports anonymous mutual authentication and an anonymous payment mechanism achieved through the utilization of encryption mechanisms and a dual signature (DS) approach as used by the well-known Secure Electronic Transaction (SET) protocol [25]. Using the dual signature, we achieve anonymity without using complex techniques such as group based signature, Pseudonym ID or blind signature. Moreover, our approach is robust against misuse of electric vehicles and unfair payment problems as it allows for user-based authentication.

The remainder of this paper is organized as follows: We introduce the V2G discharging architecture in Sect. 2. Section 3 describes the proposed discharging protocol. Security analysis of the proposed protocol is presented in Sect. 4. Section 5 concludes the paper.

2 V2G Discharging Architecture

In this section, we describe the architecture of V2G networks. Figure 1 depicts a V2G architecture that shows the various entities involved in the discharging process and their interconnections. There are seven entities involved in the discharging process. A Service Provider (SP) is a utility company that provides electricity to end users who have established contracts with it. Aggregators (AGR) do not exist in the traditional power system architecture. Aggregators come into existence because of some new requirements imposed on the SG system as a result of integrating PEVs. When PEV users want to sell power back to the grid operators by discharging their batteries, the power discharged from a single PEV is not sufficient enough to provide ancillary service to the grid as PEVs have a limited battery capacity ranges from 10 kw to 40 kw. A certain minimum amount of power is required to become eligible for providing ancillary service. For example, the minimum amount of power that is required to provide ancillary service in the UK is 3 MW [4]. Hence, a new entity called Aggregator is introduced to act as an intermediary between PEVs and grid operators to accumulate the power discharged from distributed electric vehicle batteries into a single load or source and provide it to the power grid system [16–19]. Moreover, since PEV users visit charging stations randomly, uncontrolled PEV charging can cause unpredicted overload to the distribution system [20]. Therefore, Aggregators are also responsible for stabilizing, optimizing and controlling the charging process to protect the reliability of the power grid system. Aggregators need to frequently communicate with Distribution System Operators (DSO) to fulfill their objective. DSOs in turn have to communicate with the Transmission System Operators (TSO) on a regular basis to exchange supply/demand information. Aggregators usually sign contracts with suppliers and provide charging/discharging services for end customers. As can be seen from Fig. 1, PEV users can charge at different charging/discharging locations such as home charging point, offices or public charging stations (CS). A given charging location consists of a Smart Meter (SM) and one or more Electric Vehicle Supply Equipment (EVSEs). A SM is an electronic device that continuously records electric energy consumption and sends it to the supplier at some pre-defined time intervals. The EVSE is an intelligent device that is used as a charging point connecting the PEV to the smart grid system. Charging locations could be connected to an aggregator that is located in one of the three locations: An aggregator situated in the user's home area referred to as Home Aggregator (HAG), outside the user's home area but inside its supplier network called Visiting Aggregator (VAG) and outside the user's supplier network known as External Aggregator (EAG). In the second case, the user is roaming but internally. This scenario is referred to as Internal Roaming Charging (IRC). The roaming in the third case is called External Roaming Charging (ERC) as the user is roaming in an external supplier network.

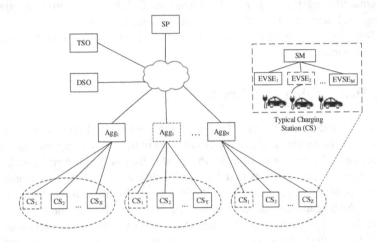

Fig. 1. V2G discharging architecture.

The discharging architecture and protocol presented in this paper supports only the first two cases. The ERC case will be incorporated as part of future work due to page limitation. Communications in a V2G architecture is based on real time communication between the various entities and can take place over different kinds of communication networks [24]. In our architecture, we assume that the PEV and EVSE are connected via Power Line Communication (PLC) and hence the communications between them is secure with no need for additional security configuration. EVSE and SM can be connected using wireless communication technologies such as ZigBee, Wi-Fi and or Wired technologies such as Ethernet. All other communications are assumed to be done through long haul wireless/wired networks such as 4G/LTE or fiber optics.

The V2G architecture demands a one-time system initialization in order to exchange information between system entities. Suppliers need to first obtain certificates from a Certificate Authority (CA) before they can issue certificates for users, AGRs, SMs and EVSEs under their territory. User registration involves generating a unique user ID (U_{ID}) for the user, issuing a smart card (SC) and registering the electric vehicles that the user is allowed to charge/discharge. The smart card provided to the user contains the public/ private key pairs of the user, the U_{ID} and the public key of the supplier. The private key of the user will be used by the SC to sign charging/discharging messages on behalf of the user and it is stored encrypted using a PIN number known only to the user. The PIN number is set by the user during the registration phase with the supplier. PEVs are identified by a unique Vehicle ID (V_{ID}) that is provided by the manufacturer during production. Suppliers register PEVs using this ID. We assume that the V_{ID} is also embedded into the PEV's firmware so that it can be used by the SC during charging. System initialization also includes installing the public key of the SM on the EVSE and vice versa. Moreover, the public key of the AGR is also installed on the SM. Aggregators establish an agreement with suppliers to provide charging/discharging service to end users. During the contract agreement, aggregators obtain the needed certificates from suppliers. Aggregators also get the list of public keys of smart meters in the area they are serving.

Suppliers hold a table of access control list (ACL) that associates users, PEVs and permissions to avoid misuse of PEVs and promote fair payment between multiple users of a single PEV. Let U and V represent the set of all users and PEVs registered by the supplier respectively such that $U = \{U_1, U_2, \ldots U_n\}$ and $V = \{V_1, V_2, \ldots V_m\}$. There are two kinds of permissions associated with PEVs, charging and discharging. Let P represents the set containing these two permissions i.e. $P = \{C, D\}$ where C represents charging and D discharging. Therefore, the elements of an ACL can be represented as: $ACL_i = \{U_j, V_k, P_l\}$ where $U_j \in U$, $V_k \in V$, $P_l \in P$ and ACL_i is the ith element of ACL. For example, the set $\{U_1, V_3, C\}$ indicates that user U_1 is allowed to charge vehicle V_1 while $\{U_2, V_4, D\}$ shows that user U_2 is allowed to discharge vehicle V_4.

3   Discharging Protocol

In this section we discuss the proposed discharging protocol. The protocol consists of three steps: discharging request, mutual authentication and payment capture. The following specific scenario will be used to explain the protocol: User U_1 who is driving vehicle V_1 visits a certain charging point to discharge his PEV's battery. We assume that various information related to charging/discharging is available on a display screen attached to the EVSE to help the user decide on whether to be served or not. The information displayed to the user on the screen includes: the available power type (Level1, Level2, Level3...), charging rate (CR), discharging rate (DR), maximum available amount of energy etc. CR and DR are the electricity price over some time period as it might change based on the dynamics in supply and demand. The CR and DR are pre-calculated by the AGR and communicated to the EVSE on a regular basis. If user U_1 decides to discharge his PEV battery selling back power to the grid based on the information available on the display screen, he can start the process by connecting his PEV to the EVSE and inserting his SC into the card reader (CRD). The next three subsections show the details of the steps taken to complete the discharging process securely.

3.1   Discharging Request

(a) The SC prompts the user for a password/PIN to verify that the user holding the SC is the legitimate user and to invoke the use of the user's private key. If successful, the user will be directed to a screen that allows him to select the type of service he is interested in (Charging or Discharging) and the amount of power in KW (power to be charged or power to be discharged). For our example, user U_1 selects discharging (D) and the amount of power to be discharged (PD). The user can only select a PD amount that is less than or equal to the maximum available battery power (MABP) of the PEV which is displayed to the user during the selection process. The MABP is calculated by the EVSE using the connection to the PEV. Once completed, a discharging request will be initiated between the user's SC (on behalf of the user) and the EVSE. An initial message (InMess) is sent from the user's SC to the EVSE which can be expressed as: SC → EVSE: = InMess where InMess = D || PD where || represents the concatenation operator.

(b) Upon receiving the initial message, the EVSE prepares an initial response message (InResMess) by concatenating the InMess with a unique transaction ID (TID), DR and the payment, P, the user will be credited for based on the PD and the discharge rate.
The actual power discharged and the actual payment the user will receive may be different from PD and P as the user may decide to stop discharging in the middle before the maximum requested power is reached. This is represented as: EVSE → SC: = InResMess where InResMess = D || PD || TID || DR || P.

(c) When the SC receives the initial response message, it prepares the discharging request (DReQ) using dual signature. The dual signature is made up of the User Related Information (URI) and Power Related Information (PRI). This is represented as: $DS = E_{KR_{U1}}[h(h(PRI) || h(URI))]$ where URI = D || TID || V_1 || U_1 || P, PRI = D || TID || PD || DR || P and h(x) is the hash of x. The process of generating the dual signature is shown in Fig. 2.

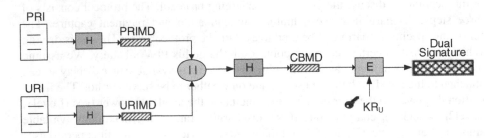

Fig. 2. Dual signature generation

(d) The SC then prepares the discharging request (DReQ) message based on the generated DS. The DReQ is composed of two messages: a message targeted to the aggregator (AggM) and another one intended for the SP (SpM). The SpM is encrypted using SP's public key so that the AGR is not able to see its content. Hence, DReQ will be expressed as: DReQ = AggM || SpM || Ts_1 where AggM = PRI || DS || h(URI), SpM = $E_{KU_{SP}}$[URI || DS || h(PRI)], and Ts_1 is a time stamp.

The SC sends the DReQ to EVSE. The message is then delivered from EVSE to SM, then from SM to AGR and finally to SP. Starting from EVSE, the message is encrypted using the public key of the receiver and signed by the private key of the sender after hashing. At the receiver end, the receiver verifies the integrity and source authenticity of the message using the public key of the sender. For example, the DReQ as it travels from EVSE to SM can be represented as: EVSE → SM: = Sig(EVSE, DReQ) ‖ $E_{KU_{SM}}$ [DReQ], Sig(X, M) is the signature of entity X over message M and is equal to E_{KR_X} (h(M)) where h(M) is the hash of M and E_{KR_X} (h(M)) is the encryption of h(M) with the private key of entity X. The verification of DReQ at the SM is performed as Ver(DReQ, KU_{EVSE}) where Ver(M, KU_Y) => D_{KU_Y} (Sig(Y, M)) = h(M) and reads as the verification of message M using the public key of entity Y.

(e) Upon receiving the DReQ message from the SM, the AGR saves the AggM for future use (e.g. payment disputes with the SP) and sends the message SpM ‖ Ts_2 to the SP. The interaction diagram for the discharging request is shown in Fig. 3.

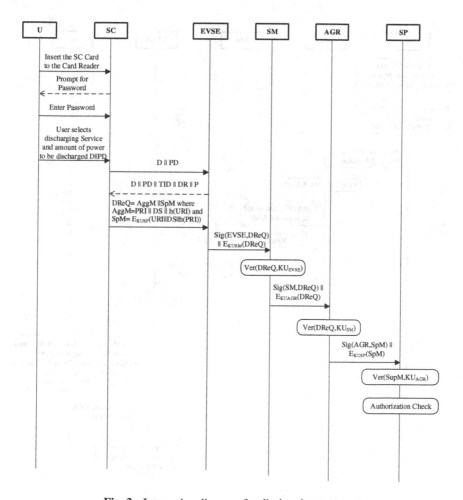

Fig. 3. Interaction diagram for discharging request

3.2 Mutual Authentication

In V2G networks, both the SP and the user should be mutually authenticated to verify that the user is legitimate and proper payments are done. After receiving the SpM ∥ Ts₂ from AGR, the SP decrypts it and gets the URI part which includes the user ID (U₁), the vehicle ID (V₁), and the requested service (D). The SP checks if there exists an entry {U₁, V₁, D} in the ACL table to authenticate the user and verify his access rights. An authorization response (AuthRes) will be sent back which takes the following format: AuthRes $= E_{KR_{SP}}$[DEC ∥ D ∥ TID] where DEC stands for decision and takes two values, Allow or Deny. For example, the AuthRes = Allow ∥ D ∥ TID conveys the meaning "Allow Discharging for the transaction with ID of TID". The AuthRes is delivered to the user (SC) as shown on the interaction diagram in Fig. 4. When the AuthRes reaches the SC, the SC first verifies that the AuthRes was generated by the expected SP by using the SP's public key. SC also makes sure that the TID in the AuthRes is the same as the

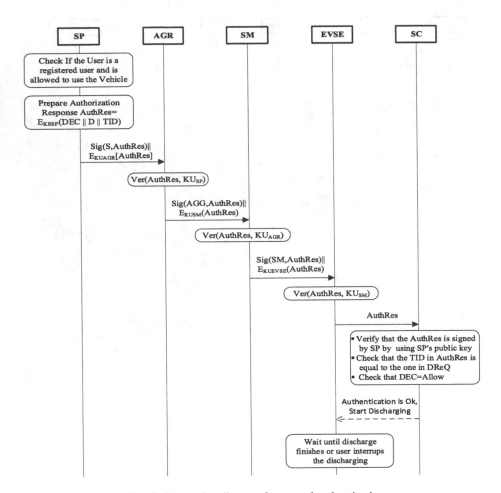

Fig. 4. Interaction diagram for mutual authentication

one sent in the DReQ. The SC then informs the EVSE of the decision (allow or Deny) with respect to the discharging request.

3.3 Payment Capture

The EVSE records the actual power discharged (APD) to the grid by the PEV. The EVSE will stop the discharging process if the PD amount mentioned in the DReQ is reached or the user deliberately interrupts the process. When discharging is completed, the EVSE prepares a power discharging report (PDR) containing the TID, the APD and the actual payment (AP) calculated based on the APD. EVSE then encrypts it with its private key and is sent to the SP as: $PDR = E_{KR_{EVSE}}[TID\|APD\|AP]$. The message is sent to the SP in a similar fashion as before (encrypting with the public key of the receiver and signing with private key of the sender). The EVSE also provides the user with a report containing the APD, the TID, and the AP.

4 Security Analysis

One of the security issues associated with PEVs is the lack of access control mechanisms for charging and discharging. The usage of PEVs by multiple drivers, such as in the cases of fleet management or in car renting companies, can lead to the misuse of the vehicles and can result in unfair payments if access to PEVs is not properly controlled/ managed. Adversaries may try to charge/discharge the PEV to gain financial benefits taking advantage of the price changes due to supply and demand. For example, a dishonest employee using a company's vehicle will conduct multiple charging/ discharging at different charging locations and times at the expense of the PEV owner for personal benefits. Therefore, there is a need to authenticate users who are allowed to use a PEV while controlling their access rights on the PEV i.e. charging/discharging. In addition, messages exchanged between all entities need to be protected against any possible passive or active attacks such as traffic analysis, message content modification, intentional delays and others while not violating the privacy of users and the confidentiality of exchanged information. As was seen in the above description of the proposed protocol, this can be achieved through the utilization of proper encryption techniques, the use of time stamps and hash functions.

4.1 Formal Verification Using AVISPA

One of the important criteria for security protocols is their robustness against various security attacks. In this section, we present the formal verification of the proposed discharging protocol to show that it is safe from several security attacks. The formal verification of our protocol is performed by using the well-known security protocol verification tool known as Automatic Verification and Analysis of Internet Security Protocols (AVISPA) [22]. AVISPA verification works under the assumption that the intruder has full control over the communication channels. To analyze a protocol using

AVISPA, it has to be described in a language called High Level Protocols Specification Language (HLPSL) [23]. AVISPA is composed of four back end servers for verification (OFMC, CL-AtSe, SATMC and TA4SP). Verification of a protocol can be performed by using any one of the four back-end servers. We verified our protocol based on OFMC and CL-AtSe and the test results show that the protocol is safe from attacks such as confidentiality breaches, message modification, nonrepudiation, source authentication, and replay attacks as shown in Table 1.

Table 1. Attacks which were tested for using AVISPA

Attack type	Safe
Message secrecy attacks	✓
Message integrity attacks	✓
Impersonation	✓
Replay attacks	✓
Repudiation	✓

5　Conclusions

Charging and Discharging of PEVs in a smart grid environment where two-way communication is needed implies the need for additional information security measures to be implemented to ensure confidentiality, integrity, availability and accountability. In this paper, we have introduced a secure protocol which can be used to guarantee these security features when PEVs discharge their batteries selling power back to the grid. The protocol was based on the use of the dual signature mechanism and validated using AVISPA to show that it is safe from certain possible information based security attacks.

Acknowledgment. This research was funded by a United Arab Emirates University, research grant, UPAR, number 31T060.

References

1. Chan, A.C.F., Zhou, J.: A secure, intelligent electric vehicle ecosystem for safe integration with the smart grid. IEEE Trans. Intell. Transp. Syst. **16**(6), 3367–3376 (2015)
2. Fan, Z., et al.: Smart grid communications: overview of research challenges, solutions, and standardization activities. IEEE Commun. Surv. Tutorials **15**(1), 21–38 (2013)
3. Chaudhry, H., Bohn, T.: Security concerns of a plug-in vehicle. In: 2012 IEEE PES Innovative Smart Grid Technologies (ISGT), Washington, DC, pp. 1–6 (2012)
4. Mustafa, M.A., Zhang, N., Kalogridis, G., Fan, Z.: Smart electric vehicle charging: security analysis. In: 2013 IEEE PES Innovative Smart Grid Technologies (ISGT), Washington, DC (2013)
5. Carryl, C., Ilyas, M., Mahgoub, I., Rathod, M.: The PEV security challenges to the smart grid: analysis of threats and mitigation strategies. In: 2013 International Conference on Connected Vehicles and Expo (ICCVE), Las Vegas, NV, pp. 300–305 (2013)

6. Aloula, F., Al-Alia, A.R., Al-Dalkya, R., Al-Mardinia, M., El-Hajj, W.: Smart grid security: threats, vulnerabilities and solutions. Int. J. Smart Grid Clean Energy **1**(1) (2012)
7. Han, W., Xiao, Y.: Privacy preservation for V2G networks in smart grid: a survey. Comput. Commun. **91–92**(1), 17–28 (2016)
8. Mustafa, M.A., Zhang, N., Kalogridis, G., Fan, Z.: Roaming electric vehicle charging and billing: An anonymous multi-user protocol. In: Smart 2014 IEEE International Conference on Grid Communications, Venice, pp. 939–945 (2014)
9. Shuaib, K., Barka, E., Ahmed Abdella, J., Sallabi, F.: Secure charging and payment protocol (SCPP) for roaming plug-in electric vehicles. In: Proceeding of the 4th International Conference on Control, Decision and Information Technologies (CoDIT 2017), Barcelona, Spain, 5–7 April 2017
10. Liu, H., Ning, H., Zhang, Y., Xiong, Q., Yang, L.T.: Role-dependent privacy preservation for secure V2G networks in the smart grid. IEEE Trans. Inf. Forensics Secur. **9**(2), 208–220 (2014)
11. Chen, J., Zhang, Y., Su, W.: An anonymous authentication scheme for plug-in electric vehicles joining to charging/discharging station in vehicle-to-Grid (V2G) networks. China Commun. **12**(3), 9–19 (2015)
12. Saxena, N., Choi, B.J.: Authentication scheme for flexible charging and discharging of mobile vehicles in the V2G networks. IEEE Trans. Inf. Forensics Secur. **11**(7), 1438–1452 (2016)
13. Liu, H., Ning, H., Zhang, Y., Guizani, M.: Battery status-aware authentication scheme for v2 g networks in smart grid. IEEE Trans. Smart Grid **4**(1), 99–110 (2013)
14. He, M., Zhang, K., Shen, X.: PMQC: a privacy-preserving multi-quality charging scheme in v2g network. In: 2014 IEEE (GLOBECOM 2014), Austin, USA, pp. 675–680 (2014)
15. Hoang, D.T., Wang, P., Niyato, D., Hossain, E.: Charging and discharging of plug-in electric vehicles (PEVs) in vehicle-to-grid (V2G) systems: a cyber insurance-based model. IEEE Access **5**, 732–754 (2017)
16. García-Villalobos, J., Zamora, I., San Martín, J.I., Asensio, F.J., Aperribay, V.: Plug-in electric vehicles in electric distribution networks: a review of smart charging approaches. Renew. Sustain. Energy Rev. **38**, 717–731 (2014)
17. San Román, T.G., Momber, I., Abbad, M.R., Miralles, Á.S.: Regulatory framework and business models for charging plug-in electric vehicles: infrastructure, agents, and commercial relationships. Energy Policy **39**(10), 6360–6375 (2011)
18. Guille, C., Gross, G.: Design of a conceptual framework for the V2G implementation. In: 2008 IEEE Energy 2030 Conference, Atlanta, GA (2008)
19. Bessa, R.J., Matos, M.A.: The role of an aggregator agent for EV in the electricity market. In: The Seventh Mediterranean Conference and Exhibition on Power Generation, Transmission, Distribution and Energy Conversion, (MedPower 2010). IET, AgiaNapa, Cyprus 2010, pp. 123–131 (2010)
20. Shuaib, K., Sallabi, F., Al Hussien, N., Abdel-Hafez, M.: Simulation of PEV service admission control (PEVSAC) model for smart grid using MATLAB. SoutheastCon 2016, Norfolk, VA (2016)
21. Hajy, S., Zargar, M., Yaghmaee, M.: Privacy preserving via group signature in smart grid. http://confbank.um.ac.ir/modules/conf_display/conferences/eiac2013/207_2.pdf. Last accessed 2 Feb 2017
22. AVISPA -automated validation of internet security protocols and applications (2006). http://www.avispa-project.org/. Last accessed 23 Jan 2017
23. AVISPA. Deliverable 2.1: The High-Level Protocol Specification Language (2003). http://www.avispa-project.org/. Last accessed 23 Jan 2017

24. Shuaib, K., Barka, E., Al Hussien, N., Abdel-Hafez, M., Alahmad, M.: Cognitive radio for smart grid with security considerations. Computers **5**(2) (2016)
25. Stalling, W.: Cryptography and Network Security: Principles and Practice, 6th edn. Prentice Hall Inc, Upper Saddle River (2014). ISBN:13:978-0133354690

ECGalert: A Heart Attack Alerting System

Marjan Gusev[1]($\boxtimes$), Aleksandar Stojmenski[1], and Ana Guseva[2]

[1] FCSE, Ss. Cyril and Methodius University, Skopje, Macedonia
{marjan.gushev,aleksandar.stojmenski}@finki.ukim.mk
[2] Innovation Dooel, Skopje, Macedonia
ana.guseva@innovation.com.mk

Abstract. This article presents a system for early detection and alerting of the onset of a heart attack. The system consists of a wireless and mobile ECG biosensor, a data center, smartphone and web applications, and a remote 24 h health care. The scientific basis of this system is founded on the fact that a heart attack can be detected at least two hours before its onset, and that a timely medical attention can dramatically reduce the risk of death or serious tissue damage.

So far, there are no commercial products matching the goals and functionalities proposed by this system, even though there are a number of proof-of-concept studies, and a number of similar products on the market. For the greater part, these currently offered solutions are specifically intended for conducting stress tests in modern hospitals, or as personal fitness devices. Most of them have limited battery power, do not use algorithms for heart attack detection, and/or require constant supervision by medical personnel.

Keywords: Heart monitoring system · ECG wearable sensor

1 Introduction

The motivation for realization of such a product arises from the latest statistics claiming that more than half of mortalities are caused by cardio vascular diseases. When it comes to matters of the heart, the statistics are dire: Cardiovascular disease is the top cause of death in the United States and around the globe. Americans spend more than \$300 billion annually on the costs of heart disease, stroke, and other cardiovascular diseases [17]. Statistics on the mortality rate in EU and wider show that more than 40% of deaths are due to cardiovascular diseases [12]. However, 80% of premature heart disease and stroke is preventable [21]. Additionally, because of dynamic and stressful lifestyle, recently the age limit for incidences of cardiac arrests has shifted from people aged 65 to people aged 40 or older.

It's no wonder, then, that many observers are making alarming predictions about the future of heart health. Unless current trends are halted or reversed, a World Health Organization report noted, over a billion people will die from cardiovascular diseases in the first half of the 21st century.

© Springer International Publishing AG 2017
D. Trajanov and V. Bakeva (Eds.): ICT Innovations 2017, CCIS 778, pp. 27–36, 2017.
DOI: 10.1007/978-3-319-67597-8_3

Within the cardiovascular diseases, a heart attack is a serious medical emergency, which occurs when blood flow (and oxygen supply) stops to a part of the heart. If impaired blood flow to the heart lasts long enough, it triggers a process called an ischemic cascade - the heart cells die (mainly by necrosis followed by apoptosis) and do not grow back. A timely medical attention, however, may save the patient of premature death, and drastically reduce the risk of serious damage to the heart [11]. Interestingly, studies show that *a heart attack may be predicted a couple of hours before its onset by detecting changes in the Electrocardiogram (ECG) of the patient* [10]. This, indeed, is the starting point and the motivation for our innovation.

Traditional EKG/ECG tests are done in a special medical institution with the proper equipment and with professional medical personnel who will read the results and look for patterns and problems with the electrical activity of the patient"s heart. Recently, with the advent of new advanced portable and wireless technologies, the medical institution need not be the only place where ECG tests are conducted [15]. The latest ECG sensors are wireless and easy to wear on the human body, they do not cause any discomfort and can be worn at all times and wherever the patient goes.

We have discussed challenges to develop an mHealth ECG monitoring solution [7] by analyzing the mobile application. Here we give details on realization of a cloud-based system that alerts an onset of a heartattack.

Even though there is much research on the subject of early prediction of heart attacks, to our knowledge there are no commercialized solutions to this problem. In Sect. 2 we propose an innovative solution for early warning and quick medical attention in case of a heart attack, by introducing a new product (sensor and smartphone application), and a new service (giving medical advice as a service) to patients with heart disease. Section 3 gives related work and Sect. 4 discusses the value, market potential and innovation impacts, and also compares our approach to the others. Finally, conclusions are given in Sect. 5.

2 ECGalert Solution

In this section we present a functional description and system architecture of the overall system that consists of a wearable ECG sensor, dew server (smartphone) and cloud-based server. The user attaches a sensor on the body and communicates to the smartphone and cloud web application.

2.1 Functional Description

The system presented in [7] consists of attaching a small portable wireless ECG biosensor on the patient's body and installing an application on the patient's smartphone that receives ECG signals from the sensor, passes them to the data center, which in turn, processes them and determines if there's any abnormal heart activity. The system is also connected to a web application on the cloud

(cloud computing) that communicates with doctors who are in charge of providing a 24 h remote medical care as a service. In case of a detected onset of a heart attack, doctors are notified by the system and may call the patient with instructions and medical advice. Additionally, 24 h medical care service centers are notified by the system to send an ambulance on the location given by the patient's smartphone. This timely medical attention may save a patient's life.

Even though there is much research on the subject of early prediction of heart attacks, still there is no commercialized solution to this problem. Wie propose an innovative solution for early warning and quick medical attention in case of a heart attack, by introducing a new product (sensor and smartphone application), and a new service (giving medical advice as a service) to patients with heart disease.

2.2   System Architecture

A wearable ECG sensor that is attached to the patient body transmits signals to the patient's smartphone (or mobile device) and then it transmits the signal to the cloud-based data center.

The overall architecture and organization of the processing modules is presented in Fig. 1.

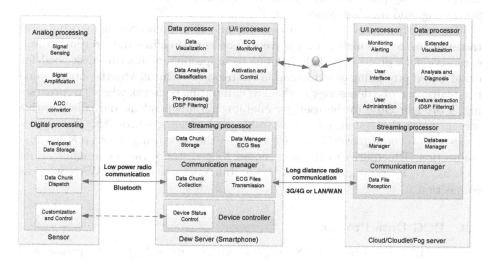

Fig. 1. System architecture and module organization

The sensor functions are based on two parts: the analog and digital processing. The analog part senses micro electrical signals on the human skin, amplifies them, and realizes analog to digital conversion. The digital part consists of temporal data sample storage in small buffers, their organization in small data chunks and then their dispatching via a Bluetooth channel. In addition the sensor realizes several customization and control functions, such as Bluetooth pairing and connection, and customization of several parameters including sampling frequency, time synchronization, etc.

The nearby smartphone receives the measured and dispatched data chunks via Bluetooth channel. Then, it stores them in temporal buffers and creates files for permanent storage on the local internal storage media. In case of internet connection, the files are transmitted to the higher level servers for further processing. In addition, the smartphone, realizes data processing by applying an efficient DSP filter and data analysis with QRS detection and heart rate determination. Besides this, it uses a UI interface to communicate with the user and realizes data visualization and ECG monitoring.

The cloud realizes data collection and processing features via a more complex data analysis for diagnosis and monitoring purposes. In addition, the web application needs user administration, and user interface to communicate with doctors and end-users.

The identified system modules are presented in Fig. 1. The smartphone is based on organization of the following five modules: device controller, communication manager, streaming processor, data processor and U/I processor. The cloud does not realize device control since all data is pushed by the smartphone and the only communication that might happen is alerting via standard voice telecommunication system. The other four modules (communication manager, streaming processor, U/I processor and Data processor) include extensive processing, and database storage.

The smartphone actually realizes data collection, preprocessing, storing, and transmitting to the higher level servers. The cloud realizes data reception and extensive data processing. This belongs to a specific architecture described for cloud-based processing of streaming IoT sensors [4]. This is why the smartphone is a realization of a dew server in the dew computing scenario, where the processing is brought close to the user, by enabling an independent and collaborative server. Since the system uses a wearable and wireless sensor, the dew server needs also to be a moveable device close to the user using a low power radio connection (Bluetooth) to the sensor and long distance radio communication, such as 3G/4G communication to the mobile operator or LAN/WAN wireless connection.

2.3 ECG Data Processing

A typical ECG consists of a QRS complex (with identified Q, R and S points) and P and T waves (Fig. 2). The data path and processing of ECG data is organized as presented in Fig. 3. The preprocessing and feature extraction is realized in two paths. The first path uses a DSP bandpass filter (5–20 Hz) to eliminate the noise and reduce feature space, and the other just to eliminate the noise by a DSP bandpass filter (0.5 Hz–30 Hz). As DSP filters we use an improved pipelined wavelet implementation for filtering ECG signals [13].

The first feature extraction phase is the QRS detection algorithm to detect the R peaks on the signal with reduced feature space. The R peak within the QRS complex is detected by an adaptive differential technique to compare the

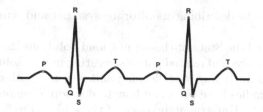

Fig. 2. A typical PQRST waveform of an ECG.

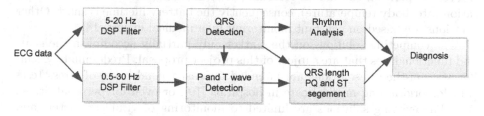

Fig. 3. Organization of the software

slope and adapt its magnitude to the signal energy, using a partial pattern matching algorithm [6] The second phase of the feature extraction follows a detected R peak with detection of the other characteristic waves, such as P and T are determined by appropriate analysis on the signal with eliminated noise.

Once the characteristic peaks are determined, the results from this feature extraction phase are forwarded to morphological processing to determine the QRS length, PQ and ST segments and other morphological features. In parallel, the rhythm is analyzed for the occurrence of the R peaks to determine regularity of the heart rate and rhythm.

Both the morphological and heart rate analysis are used to establish the diagnosis and support the monitoring and alerting system. This phase will diagnose arrhythmia and finishes with determination of a set of heart malfunction diagnoses and eventually an onset of heart attack. A heart attack can be detected by ST segment elevation in the morphological analysis of the ECG [1].

This system will not result with detecting of all possible diagnoses, due to the limitation of the used wearable sensor to detect only one ECG channel that exposes most characteristic features. To establish a more elaborated diagnosis, one needs more channels and analyze 3D heart positioning by more precise detecting the source of the identified problems. Note that by using two wearable sensors and principles of the physics, one can reconstruct 12 lead ECG [20].

3   Related Work

For our proposed solution, there are numerous clinical trials and scientific studies published in the form of proof-of-concept [2]. In this section we overview latest

ECG sensors, patents describing monitoring systems and similar commercial products.

We have analyzed the State-of-the-art of Cloud Solutions Based on ECG Sensors [5]. including analysis of related patents, existing mobile solutions and cloud-based architectures, including, cloudlet, fog and dew computing approaches.

So far, many studies have proposed how to develop a remote ECG monitoring system, in theory. Unfortunately, none of these ideas are fully implemented in the way that we propose. For instance, most of the solutions on the market [14,16,19], perform continuous recording of vital signs (ECG, heart rate, respiration rate, body temperature), consequently the battery lifetime is short. Other solutions are based on constant monitoring by a medical expert [14].

The competing solutions on the market only partially accomplish the goals and functionalities that are subject of this project proposal. Predominately, similar state-of-the-art solutions are intended for easier conducting of stress tests [14], for prioritizing medical care in hospitals [19], or even for personal fitness [16]. The existing solutions are limited to monitoring of vital parameters and signaling on simple preprogrammed thresholds. This basic data processing is insufficient for detecting the onset of a heart attack.

For instance, the sensor Shimmer3 ECG [19] provides a continuous recording of vital signs, but does not process the data and has no system for early warning. Similarly, NeuroSky [14] provides a recording of vital signs but is more intended for conducting stress tests in hospitals. QardioCore [16] is a multi-sensor platform that besides ECG, measures body temperature, the level of activity, and other vital signs, and thus presents a more complete picture of the patient's health. Nonetheless, the data is only presented, and not processed at all. ZIO XT Patch [8] is a sensor in the form of a replaceable patch where the battery lasts for up to 14 days and sends data to a clinical software iRhythm, where medical professionals set a diagnosis and therapy. While this is a good starting point, no warning in case of emergencies, however, is offered with ZIO XT Patch. Savvy [18] is a commercially available solution and is used for realization of the ECGalert project.

On the same note, many scientific studies deal with wireless, mobile, and remote ECG biosensors. One such comprehensive study [2] examined 120 different ECG biosensors from several aspects. The authors conclude that this area is a hot research topic, and that innovative, applicative solutions may seize a unique market chance.

The end-user benefit is the reduced mortality rate due to early alerting of potential heart attack, and prolonged patient life. The benefit of the medical experts that will actively participate is in the possibility to react faster and be more successful in the treatment of the patients, and in financial incentives with the monthly fee subscription for providing a medical care on the basis of early alerting. This is a kind of a 'win-win' situation where both the care providers and clients will benefit of the proposed system.

The development of wireless sensors that can communicate with personal electronic devices is still at an early phase of development. In Macedonia there are no suppliers of such an equipment yet. It is, therefore, an excellent market position/opportunity to use sophisticated technology (sensors in combination with personal smart devices) to early alert of potential medical risks, as well as to provide a medical care as a service (web application on cloud). This, coupled with the chance to emerge first on the market significantly increases the probability of success of the project.

Although much research exists on this topic, none of it resulted with a definite commercialized solution. Partial, incomplete, or un-implemnted solutions include the design of a wireless ECG system for continuous recording and communication with clinical alarm station [3], the procedure for self-test to detect a heart attack using mobile phone and wearable sensors [10], and an overview of mobile phone sensing systems [9].

4   Discussion

This section discusses the product value, market potential, foreseen impacts, and compares our proposed solution with the others available commercialized products.

4.1   Value and Market Potential

The value for customers is in developing a system that will alert early of a potential medical emergency (heart attack). Interestingly, it has been scientifically proven that heart attacks can be detected at least two hours prior to their onset, by patterns in the heart's electrocardiogram [11], and that a quick medical attention significantly reduces the dangers of tissue degeneration and possible death [10]. Recently, small, wireless and wearable ECG biosensors have emerged on the market that can keep track of the patient's heart output and other vital parameters. In this context, our system makes use of newly available technologies and advances their application. We propose a solution for early alerting of heart abnormal function, doctor's access to the patient's history and recent ECGs, and a quick medical treatment.

The idea is to sell such sensors, which will be connected with a smartphone application and a system for early alerting of abnormal heart function. The business model is based on selling the smartphone application together with the sensor, whereas the 24 h medical care, based on the application's early alerting of abnormal heart function, will be charged on a monthly subscription basis. The main customers are patients with cardiovascular diseases, under the risk of heart attack.

The main partners are clinics and ambulances for cardio-vascular diseases. This system is based on an active participation of the medical professionals, not only in their recommending of the product to patients, but also in giving medical expertise at regular intervals, or intervening in the case of an alert of abnormal heart function - services for which the patients will be charged a monthly fee.

From informal discussions with patients, we have concluded that our proposed system is of interest to a wider circle of clients, especially keeping in mind the fact that in recent years the dynamic lifestyle and stress have shifted the age limit for incidences of heart attack from people of 65 years of age to people of 40 years of age or older.

4.2   Foreseen Impacts

The proposed innovation will contribute in the area of applying ICT solutions for medical care as a service, based on remote sensors. Here we address two important areas: the use of wearable sensors in combination with smart personal devices for health monitoring, and timely action in case of an emergency.

The designed system does not affect or harm the environment. Smartphones are electrical devices used daily by everyone in society, and the sensors do not pose any environmental threat. The development of smartphone application and web application in the cloud also do not harm the environment.

Implementation of this system will have a significant impact in lowering the mortality rate of patients with cardiovascular diseases, and in prolonging their life. Therefore, it will have positive social and societal implications. Estimates show that at least 80% of all heart disease and stroke could have been prevented.

It is known that heart attacks can be detected at least two hours prior to their onset [11], but at the moment ECG devices are fixed, stationary equipment confined in medical facilities. Latest state-of-the-art advances in technology enable the production of portable, mobile and wireless sensors, which are worn by patients as a sticker patch.

4.3   Comparison to the Other Solutions

Our idea is to translate these results in an efficient IT solution. Making use of the fact that the ECG sensors and smartphones are more sophisticated in recent years, we intend to use the personal network established by the smartphone to accept these ECGs and transfer them to the data center. A specialized web service will process the data and alert on abnormal heart function. The doctor will have the opportunity to see the patient's medical history and recent ECGs, establish a proper diagnosis and react accordingly.

Differences to other competing products are:

1. Instead of continuous recording and sending ECG data, the sensor works in time intervals (20, 30 or 60 min) and thus considerably saves battery life;
2. Instead of a complex smartphone application, the smartphone receives sensor data via a personal network and sends data to data center via Internet;
3. Instead of continuous monitoring, the doctor on duty monitors on regular time intervals and on alerts send by the alerting system;
4. Instead of using a stationary equipment in the medical institution, diagnosis of cardiovascular diseases can be done remotely, using the sensors and the system for early alerting;

5. Instead of a specialized device with sensor and communicator to the data center, our solution uses sensor accessed by the personal network and mobile smartphone to access WiFi and the data center.

5 Conclusion

Our approach goes beyond this by a product innovation in that a constant medical attention is not necessary. ECG data is sent periodically via a smartphone to a data center, prolonging in this manner, the sensor's battery life. In addition, we introduce an application, based on specially developed algorithms, which analyzes the QRS signals from the ECG and determines if there is abnormality. Therefore, there is no need of a constant monitoring by a medical expert, but only in the case of emergency.

In this paper, we presented the system architecture and organization of the software modules, along with details on data processing of the signals. We have realized a prototype solution and are currently in a process of testing and improving the solution, that will be elaborated in a future work.

Precisely this idea of a response only in case of an emergency is a service innovation. In such a case, the doctor on duty would be able to remotely get a visual of the ECG diagram, call the patient and if necessary call an ambulance. To enable this remote health care system with a doctor on duty who will intervene in the event of an emergency, patients will be charged a low monthly fee not exceeding that of other standard communal utilities, such as fixed-phone line, or Internet subscription, or selected TV channel subscription by a cable operator.

References

1. Anderson, J.L., Adams, C.D., Antman, E.M., Bridges, C.R., Califf, R.M., Casey, D.E., Chavey, W.E., Fesmire, F.M., Hochman, J.S., Levin, T.N., et al.: ACC/AHA 2007 guidelines for the management of patients with unstable angina/non-st-elevation myocardial infarction. J. Am. Coll. Cardiol. **50**(7), e1–e157 (2007)
2. Baig, M.M., Gholamhosseini, H., Connolly, M.J.: A comprehensive survey of wearable and wireless ECG monitoring systems for older adults. Med. Biol. Eng. Comput. **51**(5), 485–495 (2013)
3. Fensli, R., Gunnarson, E., Hejlesen, O.: A wireless ECG system for continuous event recording and communication to a clinical alarm station. In: 26th Annual International Conference of the IEEE Engineering in Medicine and Biology Society, IEMBS 2004, vol. 1, pp. 2208–2211. IEEE (2004)
4. Gusev, M.: A dew computing solution for IoT streaming devices. In: IEEE Conference Proceedings MIPRO, 40th International Convention on ICT, pp. 415–420. IEEE (2017)
5. Gusev, M., Guseva, A.: State-of-the-art of cloud solutions based on ECG sensors. In: Proceedings of IEEE EUROCON 2017, pp. 501–506. IEEE (2017)
6. Gusev, M., Ristovski, A., Guseva, A.: Pattern recognition of a digital ECG. In: Kulakov, A., Stojanov, G. (eds.) ICT Innovations 2016, AISC (Advances in Intelligent and Soft Computing). Springer (2017, in press)

7. Gusev, M., Stojmenski, A., Chorbev, I.: Challenges for development of an ECG m-Health solution. J. Emerg. Res. Solutions ICT **1**(2), 25–38 (2016)
8. iRhythm: ZIO XT patch: continuous cardiac monitoring option (2017). http://irhythmtech.com/zio-services.php
9. Khan, W.Z., Xiang, Y., Aalsalem, M.Y., Arshad, Q.: Mobile phone sensing systems: a survey. IEEE Commun. Surv. Tutorials **15**(1), 402–427 (2013)
10. Leijdekkers, P., Gay, V.: A self-test to detect a heart attack using a mobile phone and wearable sensors. In: 21st IEEE International Symposium on Computer-Based Medical Systems, CBMS 2008, pp. 93–98. IEEE (2008)
11. Lewine, H.: New test may speed detection of heart attacks. Harvard Health Publications (2012). http://www.health.harvard.edu/blog/new-test-may-speed-detection-of-heart-attacks-201208155166
12. Løgstrup, S., O'Kelly, S.: European Cardiovascular Disease Statistics 2012 Edition, 2012th edn. European Heart Network, Brussels (2012)
13. Milchevski, A., Gusev, M.: Improved improved pipelined wavelet implementation for filtering ECG signals. Pattern Recogn. Lett. **95**, 85–90 (2017)
14. NeuroSky: Neurosky ECG biosensor (2017). http://neurosky.com/biosensors/ecg-sensor/
15. Pantelopoulos, A., Bourbakis, N.G.: A survey on wearable sensor-based systems for health monitoring and prognosis. IEEE Trans. Syst. Man Cybern. Part C Appl. Rev. **40**(1), 1–12 (2010)
16. Qardio: Qardiocore wearable EKG/ECG monitor (2017). https://www.getqardio.com/qardiocore-wearable-ecg-ekg-monitor-iphone/
17. Razdan, A.: Rethinking heart health (2016). https://experiencelife.com/article/rethinking-heart-health/
18. Saving: Savvy ECG biosensor (2017). http://savvy.si/
19. Shimmer: Shimmer3 ECG unit (2017). http://www.shimmersensing.com/shop/shimmer3-ecg-unit
20. Trobec, R., Tomašić, I.: Synthesis of the 12-lead electrocardiogram from differential leads. IEEE Trans. Inf Technol. Biomed. **15**(4), 615–621 (2011)
21. World Health Organization, others: Cardiovascular diseases: Data and statistics (2014)

An Event-Based Messaging Architecture for Vehicular Internet of Things (IoT) Platforms

Meera Aravind[1], Gustav Wiklander[2], Jakob Palmheden[3], and Radu Dobrin[1(✉)]

[1] Mälardalen University, Västerås, Sweden
radu.dobrin@mdh.se
[2] Uppsala University, Uppsala, Sweden
[3] Scania AB, Södertälje, Sweden

Abstract. Internet of Things (IoT) has revolutionized transportation systems by connecting vehicles consequently enabling their tracking, as well as monitoring of driver activities. Such an IoT platform requires a significant amount of data to be send from the on-board vehicle to the off-board servers, contributing to high network usage. The data can be send at regular intervals or in an event-based manner whenever relevant events occur. In interval-based approach, the data is send even if it is not relevant for reporting leading to a wastage of network resources, e.g., when the data does not change considerably compared to the previously sent value. In this paper, we investigate the possibility of using an event-based architecture to send data from the on-board system to the off-board system. The results show that our event-based architecture improves the accuracy of data available at the off-board system, by a careful selection of events. Moreover, we found that our event based architecture significantly decreases the frequency of sending messages, particularly during highway driving, leading to reduced average data transfer rates. Our results enable a customer to perform trade-offs between accuracy and data transfer rates.

1 Introduction

The advent of technology and wide availability of Internet has resulted in an increased number of interconnected devices. The network of physical devices, like sensors, actuators or other embedded devices, interconnected using Internet so that they can communicate with each other is referred to as Internet of Things (IoT). The advances in IoT has further motivated the connection of critical infrastructures such as transportation systems to the Internet. The IoT platform for most vehicles typically consists of (1) an on-board system, (2) an off-board system and (3) and mobile devices that communicate with both systems. The on-board system consists of the communication unit, sensors and a set of ECU's that are interconnected using a CAN network [8]. The communication unit (e.g. at Scania) communicates using 2G and 3G mobile networks, in addition to Bluetooth or Wifi for short range communications. The off-board system consists of the applications deployed on the servers (e.g., cloud) that processes the data send by the communication unit over the Internet. Mobile devices are used

© Springer International Publishing AG 2017
D. Trajanov and V. Bakeva (Eds.): ICT Innovations 2017, CCIS 778, pp. 37–46, 2017.
DOI: 10.1007/978-3-319-67597-8_4

to interact with the vehicle (on-board and off-board systems) and the off-board and on-board systems together are called as *the extended vehicle*.

The extended vehicle concept represents the coherent entirety of the road vehicle, off-board systems, external interfaces and the data communication between the vehicle and off-board systems [7]. The shortcomings of the telecom network can be covered-up by the off-board unit making the data available to data consumers even in case of network coverage problems. The sensors in the on-board system measure driver behaviors and the state of vehicle. The communication unit transfers the different data collected continuously during the operation of the vehicle to the off-board system. The current state-of-practice at Scania is to send the data at pre-defined time intervals. In the rest of the paper, this is referred to as *interval-based messaging*. One challenge of using telecom networks to send data to the off-board system is the scarce availability of network resources, e.g. bandwidth and coverage. Moreover, sending data over the telecom network may incur high data usage charges, particularly when the vehicle is "roaming". Consequently, sending data periodically may imply that redundant data is send, thus increasing the network usage (e.g., when the data remains the same over a long time interval).

Vehicular Internet of Things is an emerging field of research that has the potential to provide a wide range of benefits, e.g., improve fuel efficiency by understanding the driver behaviors. Gerla et al., [5] discusses the evolution of urban fleet of vehicles from a collection of sensor platforms to a network of connected vehicles that can exchange their information in order to improve performance and safety with minimum impact on environment. Several algorithms and techniques have been designed to estimate the position of a vehicle using the input from various sensors. GPS is the most commonly used vehicle positioning system. Li and Xu [10] proposes a novel fusion positioning strategy for estimating position by integrating multiple low-cost sensors such as GPS, MEMS-based (Micro-Electro-Mechanical Systems) Inertial Measurement Unit and a sliding mode observer. Chio et al., [3] presents a dead-reckoning (DR) positioning approach based on inertial-measurement-unit (IMU) and improving the accuracy of the position using Character Recognition algorithm by extracting road names of street signs. A *dead reckoning algorithm* estimates the current position of a moving object using previously known position and calculates the progress made using other known or estimated values such as speed, direction of travel etc. Several researches were conducted to integrate map details along with the sensor data to improve the accuracy of position estimation. Davidson et al., [4] proposes a particle filter based map matching algorithm and compares it with the position data obtained from GPS. A similar approach of estimating position using low cost sensor equipment and digitally stored map is discussed in [6].

All of these algorithms are for estimating position from the vehicle, with the help of different sensor values that can be read from the vehicle. However, we require an algorithm that can estimate the position from the off-board system as well which will not have access to the latest sensor values. Leonhardi et al., [9] solves a similar problem and proposes different event-based predictor corrector approaches for reducing the number of update messages for transmitting location information of mobile objects.

Sending data in an event-based manner, i.e. only when the data is relevant for reporting instead of fixed intervals, is expected to reduce the network usage. If the amount of data transferred is lowered, it will also substantially reduce the associated costs. It can also potentially improve the accuracy of the data since the data is transmitted whenever there is a change. *Position data* of a vehicle, for example, is obtained from GPS and send to off-board system at regular intervals. It consists of vehicle id, latitude, longitude, speed, heading, altitude and a timestamp. This data is used, e.g., for live tracking and behavior analysis of vehicles in different road conditions. Therefore, in this research we focus on designing an event-based messaging architecture for connected vehicles and to evaluate it with respect to already existing interval-based messaging system for sending position data, in terms of accuracy and data consumption.

In this paper, we propose an event-based messaging architecture for connected vehicles, that requires data transmissions only when it is relevant for reporting (for example, deviates from estimated value or change considerably compared to previously sent value), resulting in a reduction of network usage. If the amount of data transferred is lowered, it will also substantially reduce the cost of sending data. We use a model-based dead reckoning approach for estimation of the position of the vehicle, assuming that the vehicle is always on the road. The evaluation shows that our proposed approach achieves a higher data accuracy while reducing the overall data consumption compared to the currently used interval-based messaging approach.

In Sect. 2, we present our proposed architecture and algorithm. Sections 3 and 4 describe the evaluation set-up and results before concluding in Sect. 5.

2   An Event-Based Architecture for Position Data Management

In this section, we present in detail different proposed architectural components with respect to the position data, as illustrated in Fig. 1:

1. Estimator: uses a suitable dead reckoning algorithm to estimate the position of the vehicle using the previous position (previously estimated position or known position from GPS), velocity, heading and a digital map. The algorithm used in the estimator is explained in detail in Sect. 2.1. The same algorithm is executed both in on-board and off-board device. The off-board estimator uses the latest available value from the database.
2. Error Detector: compares the estimated value of position from the estimator with the actual value obtained from the GPS. If the deviation of estimated position is larger than the user defined tolerance level, then an event is triggered to send the actual value to the off-board system. The tolerance level can be the max distance between the actual and the estimated positions, or the max heading difference between the actual and estimated value, if both are on the same road.

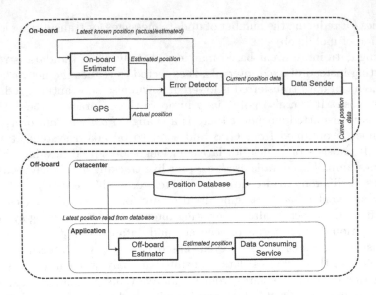

Fig. 1. The proposed architecture of the event-based system for position data.

3. Data Sender: sends the data to the datacenter in case an event occurs and the actual position needs to be sent to the off-board system.
4. Position Database: stores all the position data received from various vehicles.
5. Data Consuming Service: There are different services which use live data from the truck. One example is the fleet management portal where the customers can track live position of the vehicle. These services use the data provided by the off-board estimator.

2.1 Proposed Position Estimation Algorithm

The off-board and the on-board estimators employ the same algorithm to calculate the estimated position data. Our proposed algorithm, called POS_ESTIMATE (Algorithm 1), estimates the position using the last transferred GPS position, speed and heading, assuming the vehicle is on the road. The estimation begins using the initial position obtained from the vehicle when the vehicle starts, and is repeated every second. These steps repeats using the newly estimated position in case no update happened from the on-board to off-board system. If the actual position was sent from on-board system, then both on-board and off-board systems uses this newly updated position for estimation. A detailed explanation of the implementation of algorithm can be found in the thesis report [2].

2.2 Delaying the Off-Board Vehicle Representation

At times, the estimation can go wrong, depending on the route chosen by the driver, and the users may observe jumping updates of the vehicle positions on

---

Algorithm 1. POS_ESTIMATE: The position estimator algorithm.

1: Compute the distance travelled in 1 second time using last known speed.
2: Identify the direction of travel using last known heading.
3: Identify the road on which the last known position is.
4: Trace the calculated distance in the direction of travel on the same road.
5: Identify the coordinates of the newly estimated position.
6: Check the estimated point against the set requirements and determine if the actual position need to be send to the off-board system.

---

the map when the actual position is sent from the on-board system. This phenomenon occurs when the newly estimated point is on a different road than the actual vehicle (for example, while estimating at the intersections), or if the distance between the actual and estimated position is large. A delay can be used to reduce the number of cases displaying incorrect routes to the user. The user application executes the estimation algorithm after a certain delay relative to the on-board system. However, the on-board system runs the estimation algorithm without any delay. Hence, whenever there is a deviation from the actual route of truck, the latest position is send to the off-board system. The client can use this newly received value which will be ahead of the displayed value on the client. Using this newly sent value from the on-board system, the user can follow the correct route without any estimation until the displayed position of vehicle reaches the newly received point. The data will be sent from the vehicle to the off-board system only if the difference of current time and the time at which last position data was send is greater than or equal to the delay. An additional benefit of using the delay is that the number of points sent to the off-board system can be decreased, hence reducing the data transfer.

3   Evaluation Setup

In this section, we describe the simulator that we implemented to compare the proposed event-based architecture with the interval-based architecture that is currently being used at Scania. We use the map APIs by "Here", which is a company owned by Volkswagen, BMW and Daimler [1]. The input to the simulator consists of the GPS data of a truck, which also contains its speed and heading, collected at an interval of 1 second for a duration of one hour. This data is considered as the live GPS data during simulation.

The proposed architecture and estimation algorithm is evaluated on the basis of 5 metrics. The results obtained from simulations for both interval-based algorithm and event-based algorithm are discussed here. The 1 min interval values of the interval-based architecture is the basis for comparison for all 5 metrics. The metrics for evaluation are:

1. **Average off-board error distance:** It is the average of the great circle distance between the real position of the vehicle at that time and the position displayed on the off-board system. In case of the time-based algorithm

that is currently being used in Scania, the average error distance is calculated between the actual position of the vehicle and the last known position displayed on the off-board system. The distance between two points is measured using Haversine formula [11].

2. **Number of incorrectly estimated routes:** The number of times the off-board displays a position which is not the prospected route of the vehicle due, e.g., to a wrong route selection by the estimator. It is measured by counting the segments covered by the estimated points which are not traversed by the actual vehicle.

3. **Average message size (bytes)/minute:** Average size of the messages, in bytes, transferred during 1 min.

4. **Average message transfer interval:** It is the average time difference between the messages containing position data that will be send from the on-board to the off-board system.

5. **Total number of messages:** It is the total number of messages sent from the on-board system to the off-board system.

The experiments were conducted by varying the maximum allowed error distance and delay. Maximum allowed error distance is the user defined acceptable distance between the actual and estimated points in metres. The delay determines the time (seconds) for which the display on the client-side is behind the actual position of the truck. The values used for maximum allowed error distance were 50 m, 75 m, 100 m, 125 m, 250 m, 500 m and 1000 m. The values used for delay were 0 s, 5 s, 10 s, 15 s, 20 s, 30 s, 45 s and 60 s. We also considered the effect of network latency, that was obtained by extracting the existing position data from Scania's database for the last two years and compared their message creation time and database update time.

4   Evaluation Results

All 5 metrics discussed in Sect. 3 were evaluated for different combinations of maximum allowed error distance and delay with and without latency. A detailed explanation of all the results is given in the technical report [2].

Figure 2 shows different values of average off-board error distance for different combinations of delay and maximum allowed error distance, both set by the user. As observed from the graphs, for each of the maximum allowed error distances, average off-board error distances increases with increase in delay. However, for higher values of delay like 45 and 60 s, the maximum allowed error distance have negligible effect on the average off-board error distance. The average off-board error distance values are similar for higher values of delay. The black line represents the average off-board error distance value for the interval-based algorithm. The average error distance is much below the black line for most of the combinations except for high values of delay like 30, 45 and 60 s.

In Fig. 3 we compare the incorrectly off-board estimated routes generated with no delay, with the ones generated by the simulation runs after adding different values of delay. The number of incorrect estimated routes decreased

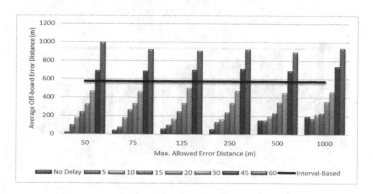

Fig. 2. Average off-board error distance variation with changing maximum allowed error distance and delay in event-based algorithm with latency.

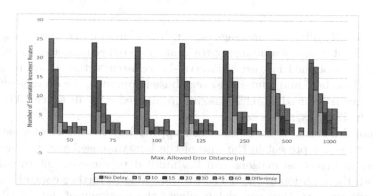

Fig. 3. The difference in number of incorrect routes with changing maximum allowed error distance and delay in event-based algorithm with latency. (Color figure online)

when we added delay compared to the simulation runs with no delay. However, the effect of latency increased the number of incorrect routes. The differences in the number of incorrect routes for different delays and maximum allowed error distances compared to the case with no latency is shown in Fig. 3 in red colour. No incorrect routes will be generated in case of interval-based approach.

In Fig. 4 the black line represents the average message size in bytes/minute for the interval-based architecture. The average message size in bytes/min in case of event-based system exceeded the interval-based value only in 5 cases out of the total 56 cases. With the increase in maximum allowed error distance, the average message size in bytes/minute decreases. Moreover, by adding delay, number of messages sent decreased resulting in decrease of average message size/minute.

The evaluation results show that the adoption of an event-based architecture can improve the accuracy of data available at the off-board system. Suitable values of maximum allowed error distance and delay have to be set depending on the user defined accuracy and data reduction goals. The total number of

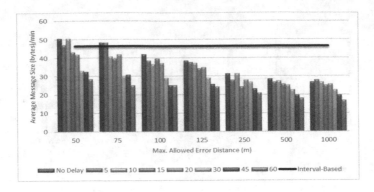

Fig. 4. Average message size in bytes/minute variation with changing maximum allowed error distance and delay in event-based algorithm with latency.

messages directly affect the message size in bytes/minute and the average message transfer interval. By sending approximately 46 bytes/min, average off-board error distance obtained in case of interval-based architecture is approximately 522 m. On the other hand, under the event-based approach, the average off-board error distance is reduced by 92% (maximum allowed error distance 100 m) even though only 42 bytes/min is sent.

A minor drawback of the event-based approach is the incorrectly estimated routes which is not present in the case of the interval-based architecture. However, these incorrect estimations can be handled to some extent by adding a delay. With the addition of the delay, we were able to further control the number of messages sent. However, delay affects the accuracy of data available at off-board system. Using event-based architecture we were able to obtain an average off-board error distance of around 505 m by sending 29 bytes/min. In case of interval-based architecture, around 46 bytes had to be sent to obtain a similar average-off-board error distance. Hence, similar accuracy was achieved in event-based architecture by using only 63% of the data.

Figure 5 shows the relative average off-board error distance plotted against relative data consumption for each of the simulation runs with latency. The average off-board error distance and average message size/minute of interval-based architecture with 1 min interval represents 100% of x and y axis respectively. The values obtained from different simulation runs for the event-based architecture are represented by different dots. Dots within the red box are the ones having lesser average off-board error distance, i.e., better accuracy, and lesser data consumption compared to the interval-based architecture. Most of the values with delay lesser than 45 s falls within the box. Therefore, delays of up to 30 s are better, on an average, for event-based architectures.

We investigated the average message size/minute under event-based and interval-based approaches for the same average off-board error distance. We varied the maximum allowed error distance and delays of the simulator in order to achieve an average off-board error distance similar to that of the interval-based

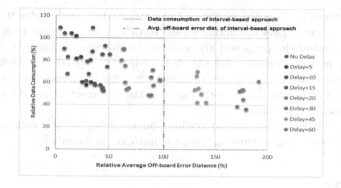

Fig. 5. Relative average off-board error distance vs relative data consumption

architecture for intervals from 1 to 5 min. This was obtained for the cases in which the maximum allowed error distance was fixed as 500 m and for delay values 30, 60, 80, 120 and 160 s. The average message size/minute required to achieve similar accuracy in interval-based and event-based approach is compared in Fig. 6. From the figure, it is clear that the event-based approach can provide the same level of accuracy with lesser data consumption. The average message transfer interval for each of the above discussed cases is also shown in the figure.

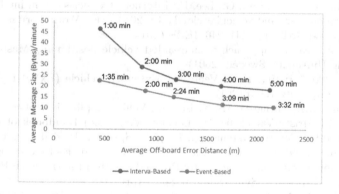

Fig. 6. Comparison of the average message size/minute in interval-based and event-based approaches

5   Conclusions

In this paper we proposed an event-based architecture for Internet of Vehicles' messaging platform. The proposed event-based architecture consists of an on-board and off-board estimator executing the same estimation algorithm. The estimated value is verified against the actual sensor value in the on-board system. An event is triggered whenever the estimated value deviates more than a threshold, and eventually the actual sensor data is send to the off-board system. We then evaluated the proposed architecture in the specific context of position data, which mainly consists of location coordinates, speed and heading. The proposed event-based architecture together with the dead reckoning estimation algorithm was found to significantly improve the accuracy of the data available for the off-board applications.

Future work will focus on improving the accuracy of the estimation, and implementing the architecture on the real platform at Scania.

References

1. Here - maps for developers (2017). https://here.com
2. Aravind, M.: Event-based messaging architecture for vehicular internet of things (IoT) platforms. Master's thesis, Mälardalen University, Sweden (2017)
3. Chiou, Y.S., Tsai, F., Yeh, S.C.: A dead-reckoning positioning scheme using inertial sensing for location estimation. In: 2014 Tenth International Conference on Intelligent Information Hiding and Multimedia Signal Processing, pp. 377–380, August 2014
4. Davidson, P., Collin, J., Raquet, J., Takala, J.: Application of particle filters for vehicle positioning using road maps. In: 23rd International Technical Meeting of the Satellite Division of The Institute of Navigation, Portland, OR, pp. 1653–1661 (2010)
5. Gerla, M., Lee, E.-K., Pau, G., Lee, U.: Internet of vehicles: From intelligent grid to autonomous cars and vehicular clouds. In: 2014 IEEE World Forum on Internet of Things (WF-IoT), pp. 241–246. IEEE (2014)
6. Hall, P.: A bayesian approach to map-aided vehicle positioning. Master's thesis, Linköping University, Sweden (2001)
7. ISO/DIS 20077–1:2017: Road Vehicles - Extended vehicle (ExVe) methodology. ISO, Geneva, Switzerland (2017)
8. Johansson, K.H., Törngren, M., Nielsen, L.: Vehicle applications of controller area network. In: Hristu-Varsakelis, D., Levine, W.S. (eds.) Handbook of Networked and Embedded Control Systems, pp. 741–765. Springer, Heidelberg (2005)
9. Leonhardi, A., Nicu, C., Rothermel, K.: A map-based dead-reckoning protocol for updating location information. In: Proceedings 16th International Parallel and Distributed Processing Symposium, p. 8pp, April 2002
10. Li, X., Xu, Q.: A reliable fusion positioning strategy for land vehicles in GPS-denied environments based on low-cost sensors. IEEE Trans. Ind. Electron. PP(99), 1 (2016)
11. Veness, C.: Calculate distance and bearing between two latitude/longitude points using haversine formula in javascript. Movable Type Scripts (2011)

Internet of Things Based Solutions for Road Safety and Traffic Management in Intelligent Transportation Systems

Arnav Thakur[1], Reza Malekian[1(✉)], and Dijana Capeska Bogatinoska[2]

[1] Department of Electrical, Electronic and Computer Engineering, University of Pretoria,
Pretoria 0002, South Africa
u11100037@tuks.co.za, reza.malekian@ieee.org
[2] University of Information Science and Technology St. Paul the Apostle,
Ohrid, Republic of Macedonia
dijana.c.bogatinoska@uist.edu.mk

Abstract. Road safety, traffic congestion and efficiency of the transport sector are major global concerns. Improving this is the primary objective of intelligent transport systems (ITS). Having Internet of things (IoT) based solutions for ITS would enable motorists to obtain prior contextual guidance to reduce congestion and avoid potential hazards. IoT based solutions enabling collection of data from client nodes in a wireless sensor network in the transport environment implementing ITS goals is studied. The parameters to be monitored, type of sensors and communication related design parameters are identified to develop an effective IoT based solution. Road safety techniques studied include distance sensing, improper driving detection and accident prevention, weather related events and negligent driving detection and accident avoidance. Vehicle to vehicle communication and vehicle to infrastructure based channels are studied. Wireless communication technologies suitable for the channels are studied. Additional benefits and services that can be added to a system with the IoT approach are also studied. The effectiveness of such a system is studied with the use of validation framework. Multiple case studies of current and future IoT based ITS along with the challenges in the application is discussed.

Keywords: Internet of Things (IoT) · Intelligent Transport Systems (ITS) · Peer to Peer (P2P) · Roadside Unit (RSU) · Vehicle to Vehicle (V2V) communication · Vehicle to Infrastructure (V2I) communication · Wireless Sensor Network (WSN)

1 Introduction

Vehicular accidents are a major global phenomenon and cause human fatalities and damage to property. This directly has an impact on human life and has financial repercussions. It was found that up to 60% of the collisions can be avoided if the motorist receives advanced warnings 1.5 s before the occurrence of the event [1]. The road safety mechanisms currently implemented in the transport industry include sensors mounted on the body of vehicles for lane keeping, maintaining safe driving distance and blind

© Springer International Publishing AG 2017
D. Trajanov and V. Bakeva (Eds.): ICT Innovations 2017, CCIS 778, pp. 47–56, 2017.
DOI: 10.1007/978-3-319-67597-8_5

spot assistance [2]. A system is required that can detect and warn motorists of potential accident events, will save human life and prevent loss of property.

Traffic congestion reduces efficiency in the transport sector. Traffic management and location based alerts are used to minimize congestion and are currently reliant on cellular networks, web based platforms and mobile applications such as Waze and mapping services [3, 4]. A dedicated system to convey traffic related location based alerts, warn about high risk accident zones, communicate precautions and provide location of emergency services and public infrastructure would effectively solve the problem and improve overall driving experience [4]. The above problems of improving road safety and improving efficient management of traffic are goals of Intelligent Transport Systems (ITS) and are studied in this paper.

Internet of things (IoT) based approaches are evolving with growing sensor and wireless communication technologies. The sensory data is processed in real time by remote handlers and the clients are given feedback. Application of such an approach for vehicular clients in the transport industry is helping to develop increasingly effective ITS solutions and is studied in this paper [2, 3].

Sensors are used in ITS to detect potential accident causing events such as sudden braking, blind spot obstacle detection, to assist during lane change events, non-line of sight factors, and uneven road surface [2]. Weather related sensors also form part of such a system and feedback from it can prevent potential accidents. Moreover traffic monitoring sensors will help in real time management of congestion. IoT based processing is performed and alerts are conveyed to clients by onboard units (OBU) placed in vehicles, road side device/units (RSU) and a centralized server forming a WSN. Vehicle to vehicle (V2V) and vehicle to infrastructure (V2I) based communications are used between the nodes of the WSN [1].

The remainder of the paper is organized as follow: Sect. 2, gives a brief overview of road safety aspect of IoT based ITS and describes the sensors involved, services and data processing methods. Traffic control and management aspect in IoT based ITS is discussed in Sect. 3. The communication system is studied in Sect. 4. With growth in electric vehicle penetration in vehicular fleets, its impact on the power grid along with IoT Enabled Smart Grid Optimized for Electric Vehicles is studied in Sect. 5. In Sect. 6, the results from the studied literature are discussed and finally in Sect. 7, a conclusion is drawn.

2   Road Safety System

The road safety system components in an IoT based ITS is discussed in this section. This includes sensor systems to detect distance to obstacles, improper driving practice based accident prevention, negligent driving practice detection based accident avoidance and weather related warning components.

Sensors are used to detect potential accident causing events and the processed readings are used to determine preemptive steps required to be taken to avoid an accident. Accidents are caused due to lack of maintenance of safe driving distance, blind spots, misinterpretation of road signs due to lack of awareness and negligent driving due to fatigue [4]. Sensing such events is thus required and is studied in the following.

2.1 Distance Sensing

Distance sensors are used to maintain safe driving distance between vehicles in the same lane and in neighboring lanes. Distance sensors are mounted on the front and rear end of vehicles to keep safe distances between vehicles in the same lane and in parking zones. Distance sensors can also be mounted on the left and right side body panels of vehicles to keep safe driving distance between vehicles in neighboring lanes and avoid side collisions especially at road intersections. This can be beneficial to avoid accidents caused by obstacles in the blind spots, especially during a lane change event [4].

Distance is sensed by using electromagnetic fields, light or sound based techniques. Light based distance sensing techniques use light dependent transducers. This includes photo transistors, lasers and infrared LEDs. Due to their drawbacks, light based techniques cannot be used to detect distance to obstacles in dynamic environments for vehicles. Sound based techniques use sound waves of high frequency such as ultrasonic which is directed in the direction of obstacle detection. High frequency sound wave transmitters emit the sound wave and receivers detect the received wave reflected by the obstacle. Since the sound based technique is not affected by the lighting conditions and texture, color and reflectivity of obstacles, it is the most reliable obstacle distance measuring technique for application in the dynamic environment of vehicles [6].

2.2 Improper Driving Practice Detection and Accident Prevention

Sudden braking, improper lane events and uneven road surfaces when not handled appropriately by the motorist and it's peer are also a major cause of accidents [2]. Threshold based classification and neural network based algorithms use sensor values to determine sudden braking, lane change events and presence of speed humps and potholes [11]. For this the accelerometer sensor is used.

2.3 Weather Related Alerts

Poor weather conditions are also known to increase risk of accidents [14]. This is because of visibility being impacted during thunderstorms, snowstorms, sandstorms and during windy and foggy weather. Precipitation detection sensors are used to detect rain and snow. Visibility sensors detect foggy weather and sandstorm related hazards. The sensors are integrated into roadside units (RSU), which detect the events and use machine learning algorithms to compute accident risk probability [8]. The risk factor probability is used to generate precautionary message and send weather related driving alerts to the motorists in the coverage area of the RSU.

2.4   Negligent Driving Practice Detection and Avoidance

Road sign boards can be ignored by motorists due to negligence and in poor lighting conditions. Sign boards are also misinterpreted by motorist due to lack of awareness and ignorance [6]. This increases accident risk. Image processing algorithms are used to detect and interpret road signs. The interpretation of the sign board symbols are conveyed to the motorists though audio alerts [13].

3   Traffic Control and Management System

Traffic management and congestion control is also solved in ITS containing connected vehicles. Current traffic analysis algorithms use data collected from surveillance cameras and embedded loop detectors to determine traffic density [8, 9]. The traffic density estimation techniques used to monitor traffic conditions is summarized in the following diagram (Fig. 1).

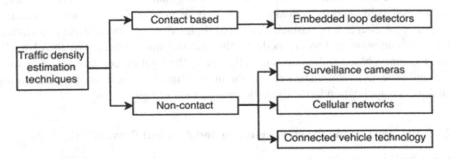

Fig. 1.   Traffic density estimation techniques for traffic management.

The use of data collected from vehicles with V2I connectivity gives more accurate estimation of traffic condition and is an effective method that can be used to develop reliable solutions for traffic control and management.

Vehicle platooning based concept is another method for traffic control and management. The advantages of such a system include increased performance of vehicles due to improved aerodynamic effectiveness and improved fuel consumption of vehicles. The system will lead to increase in the capacity of the road infrastructure by reducing driving distances between vehicles and stabilize traffic flow. The system can be integrated into new vehicle technologies for self-driving autonomous automobiles.

4   Communication System

In this section, vehicle to vehicle (V2V) communication, vehicle to infrastructure (V2I) communication channel, the wireless commutation technologies and communication of alerts are studied.

4.1 Vehicle to Vehicle (V2V) Communication

V2V communication is used to communicate with the following peer motorist and warn about a potential accident event [10]. The leading vehicle functions as the server to the following vehicle which is the client. Peer to peer data sharing is used to convey contextual messages to the client through the V2V channel. Moreover warning messages relevant for multiple vehicles in the same traffic direction are also relayed though the V2V channel [10]. This is beneficial to avoid multiple vehicle collisions and accidents caused by non-line of sight factors.

4.2 Vehicle to Infrastructure (V2I) Communication

On the other hand vehicle to infrastructure (V2I) communication is used to effectively manage traffic and improve overall driving experience of motorists. The roadside units (RSU) controlled by the centralized server with data processing capabilities manages client requests and collects data from road side sensors and sensors placed in vehicles. The RSU provides location based alerts, weather related precautions, recommended speed for the section of the road, enforcement of speed limits and implementing variable road signs [11–14]. Figure 2 shows the WSN describing the V2I communication channel.

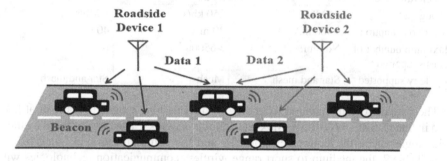

Fig. 2. WSN containing the V2I communication channel in a star network topology (From [4]).

4.3 Wireless Communication Technologies

While selecting a wireless communication technology for V2V and V2I communication channel, the factors to be considered are the range, throughput, association times, power consumption, number of connections supported and the network topology supported by the protocol. The communication technologies in the radio frequency (RF) bands reserved in the industrial, scientific and medical (ISM) radio band include Wi-Fi (IEEE 802.11), Bluetooth (IEEE 802.15) and ZigBee (IEEE 802.15.4). The wireless communication technologies not in the ISM band available for V2V and V2I communications include cellular networks, worldwide interoperability for microwave access(WiMax-IEEE 802.16), dedicated short range communications (DSRC-IEEE 802.11p) wireless access in vehicular environment (WAVE- IEEE 1609). Tables 1 and 2 summarize the wireless communication technologies that can be used for V2V and V2I communication channels.

Table 1. Comparison of long range wireless communication technologies that can be used for the V2V and V2I communication channels with ranges of at least 1 km [4, 6, 22]

Parameter	Cellular networks	WiMax	DSRC	Wave
Carrier frequency	Multiple spectrum bands	2.5/3.5 GHz	5.855–5.92 5 GHz	5.855– 5.925 GHz
Range	<15 km	<50 km	1 km	1 km
Data rates	<2 to 100 Mb/s	40 Mb/s	27 Mb/s	27 Mb/s
Topology	GSM/mobile	Point-multipoint	Point-point	Point-multipoint

Table 2. Comparison of wireless communication technologies that can be used for the V2V and V2I communication channels and support ranges up to 1 km [4, 22]

	Wi-Fi	ZigBee	Bluetooth
Carrier frequency	2.4/5 GHz	2.4 GHz	2.4 GHz
Range	100 m–1 km	100 m	10 m
Association time	4 s	30 ms	600 ms
Throughput	54 Mb/s	250 kb/s	1 Mb/s
Power consumption	>400 mA	30 mA	40 mA
Maximum number of nodes supported	No limit	>65000	8
Topology supported	Star and mesh	Mesh	Star and mesh

The wireless communication technologies offering long range coverage of at least 1 km include DSRC, WAVE, WiMax and cellular networks. The range offered by these technologies meets the requirements of V2V and V2I communication channels [6].

In Table 2, the medium to short range wireless communication technologies with coverage of up to 1 km is studied. The technologies include Wi-Fi, ZigBee and Bluetooth. From Table 2, it can be observed that the Bluetooth has the shortest range of just 10 m. The short range, large association time and low number of connections supported are the disadvantages of Bluetooth. This makes it unsuitable for V2V and V2I communication [5, 7]. The high data rate is an advantage of Bluetooth and makes it suitable for usage in intra-vehicle communication. Wi-Fi and ZigBee support large ranges. They both have low association times and is an advantage for communication between moving entities. ZigBee has a disadvantage of significantly lower throughput when compared to Wi-Fi. This is a disadvantage for ZIgBee for usage in the V2I communication channel as multiple alerts and sensor warnings are exchanged [6, 7]. Thus ZigBee is more suitable for V2V commutation and WI-Fi is more suitable for V2I communication [4].

The factor associated with vehicles moving at high velocities also needs to be considered. Remaining time available for communication after successfully establishing of connection between moving communicating devices needs to be studied. In Fig. 3, the remaining time available for communication between entities after establishment of a connection is compared.

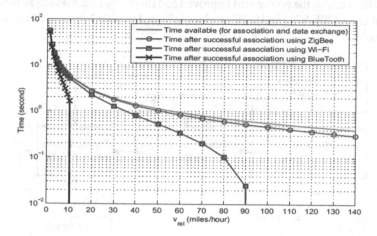

Fig. 3. Comparison between wireless commutation technologies for maximum available time for communication between moving entities (from [2]).

4.4 Data Processing and Alert System

Driver assist system with V2V, V2I and sensor systems are integrated into onboard device units (OBU). A graphical user application optimized to reduce driving related distractions serves as the human machine interface for the client motorist [17–19]. Audio based warning systems are used. A standardized communication protocol is used to exchange information between client OBUs and server. Cloud computing based approaches are applied for users to grant ITS based services securely [14]. The lack of a global standard on method for exchange of information in ITS is a concern.

5 IoT Enabled Smart Grid Optimized for Electric Vehicles

Adoption of electric powered vehicles is increasing with rising awareness of greenhouse gas emissions and urban air pollution originating from conventional gasoline powered vehicles.

  With the deployment of smart grid, the opportunity of using the energy stored in the battery of idle electric vehicles to supply power to the grid during peak hours is being explored to increase efficiency and reliability of the power distribution system. Energy generation from renewable energy sources as wind and solar is volatile and excess energy generated is wasted if not utilized or stored. Shortfall in supply is also possible from such sources. Surplus power can be stored in batteries of electric vehicles and can be utilized during shortfall in power generation. This is achieved by integrating the electric vehicles into the power grid through vehicle to grid (V2G) services. Vehicles for personal usage have long rest hours which are as high as 90% [21]. This makes deployment of V2G services practical to implement as charging and discharging of batteries can be done locally in the rest hours of the vehicles. Thus energy wastage is reduced, efficiency of the power grid improved and this helps in achieving holistic environmental goals [20]. V2G and G2V services in an IoT enabled smart grid are summarized in Fig. 4.

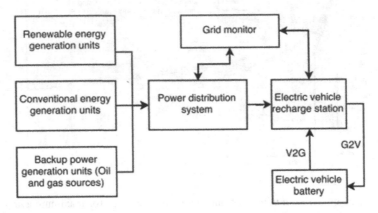

Fig. 4. V2G and G2V integration into smart grid.

6   Discussion

Road safety is ensured by detecting accident causing events due to improper driving distance, negligent driving and weather related factors. Algorithms process the sensory data to generate warnings and the motorist is notified on the user interface of the vehicle onboard unit. Warning from peer motorists is also used along with location based alerts received from roadside units.

The use of connected vehicle based traffic management is more effective in estimating traffic condition then traditional methods due to reliability of the data. In simulations it was found with even low penetration of the connected vehicle based data collection technology of as low as 20%, the minimum traffic density estimation has accuracy of 85% and traditional methods have an estimation capability of 80%. The benefit of a more reliable and larger data set will lead to higher accuracy in estimating traffic density as the algorithm can be trained more effectively [17, 23]. With improved traffic estimation capability in IoT based ITS approaches, congestion detection and management techniques can be improved.

Wireless commutation technologies operating in the ISM RF bands can successfully be used for communication between moving and stationary communicating devices. The wireless communication technology suitable for V2I and V2I commutation for operation at speeds up to 150 km/hr are Wi-Fi and ZigBee. Since Wi-Fi has larger throughput then ZigBee, it is more suitable for V2I communication as exchanged information contains sensory information and multiple alerts.

Establishment of a global standard for IoT based ITS would serve as a framework to build robust solutions [17, 19]. Integrating IoT based road safety mechanisms catering to non-line of sight factors in autonomous self-driving vehicle is a potential field for application [17]. Traffic control and management algorithms on IoT based ITS can also be adapted to efficiently handle congestion to reduce vehicular pollution [17]. The use of V2I and V2V communication along with IoT based processing can be successfully used in smarter road safety and traffic management strategies [15, 16].

7 Conclusion

Road accidents caused by improper driving practices, negligent driving, lack of maintaining of safe driving distance and weather related factors are detected in IoT based ITS. The ability to warn peer motorists about potential accidents reduces probability of accidents by providing heads up warning and preemptive steps can be taken. The use of machine learning algorithms further enhances the performance of the road safety aspect. The use of connected vehicle based data collection for traffic management enables accurate traffic estimations. Thus IoT based approaches successfully solve the road safety and traffic management objectives of ITS and will revolutions the transport sector.

References

1. Rawat, D.B., Bista, B.B., Yan, G., Olariu, S.: Vehicle-to-Vehicle connectivity and communication framework for vehicular ad-hoc networks. In: 2014 Eighth International Conference on Complex, Intelligent and Software Intensive Systems, Birmingham, pp. 44–49 (2014)
2. Djajadi, A., Putra, R.J.: Inter-cars safety communication system based on Android smartphone. In: 2014 IEEE Conference on Open Systems (ICOS), Subang, pp. 12–17 (2014)
3. Zhu, Z., Loo, J., Chen, Y., Chai, K.K., Zhang, T.: Recent advances in connected vehicles via information-centric networking. In: IET International Conference on Intelligent and Connected Vehicles (ICV 2016), Chongqing, pp. 1–8 (2016)
4. Hirose, K., Ishibashi, K., Yamao, Y., Hirayama, Sawada, M.: Low-power V2M communication system with fast network association capability. In: 2015 IEEE 2nd World Forum on Internet of Things (WF-IoT), Milan, pp. 204–209 (2015)
5. Jeong, S., Baek, Y., Son, S.H.: A hybrid V2X system for safety-critical applications in VANET. In: 2016 IEEE 4th International Conference on Cyber-Physical Systems, Networks, and Applications (CPSNA), Nagoya, pp. 13–18 (2016)
6. Padmini, K., Dorle, S., Keskar, A., Chakole, M.: Micro-controlled based vehicle safety system using dedicated short range communications (DSRC). In: 2009 Second International Conference on Emerging Trends in Engineering & Technology (2009)
7. Deng, Y., Li, L., Wang, H., Liu, Z., Wang, S., Shen, J.: Visible light communication based collision avoidance scheme for vehicle lane-changing in intelligent transport systems. In: 2015 14th International Conference on Optical Communications and Networks (ICOCN), Nanjing, pp. 1–3 (2015)
8. Khan, S.M., Dey, K.C., Chowdhury, M.: Real-time traffic state estimation with connected vehicles. IEEE Trans. Intell. Transp. Syst. 1–13 (2017)

9. Stanciu, E.A., Moise, I.M., Nemtoi, L.M.: Optimization of urban road traffic in intelligent transport systems. In: 2012 International Conference on Applied and Theoretical Electricity (ICATE), Craiova, pp. 1–4 (2012)
10. Prinsloo, J.: Accurate vehicle location system using RFID, an internet of things approach. Sensors MDPI 16(6), 1–24 (2016). Article no. 825
11. Zhu, F., Ukkusuri, S.V.: An optimal estimation approach for the calibration of the car-following behavior of connected vehicles in a mixed traffic environment. IEEE Trans. Intell. Transp. Syst. 18(2), 282–291 (2017)
12. Fouchal, H., Bourdy, E., Wilhelm, G., Ayaida, M.: A framework for validation of cooperative intelligent transport systems. In: 2016 IEEE Global Communications Conference (GLOBECOM), Washington, DC, pp. 1–6 (2016)
13. Fouchal, H., Wilhelm, G., Bourdy, E., Wilhelm, G., Ayaida, M.: A testing framework for intelligent transport systems. In: 2016 IEEE Symposium on Computers and Communication (ISCC), Messina, pp. 180–184 (2016)
14. Malekian, R., Moloisane, N.R., Nair, L., Maharaj, B.T., Chude-Okonkwo, U.A.K.: Design and implementation of a wireless OBD II fleet management system. IEEE Sensors J. (ISI/SCIE, Impact factor: 2.512) 17(4), 1154–1164 (2017)
15. Kolosz, B., Grant-Muller, S.: Sustainability assessment approaches for intelligent transport systems: the state of the art. IET Intel. Transport Syst. 10(5), 287–297 (2016)
16. Kabashkin, I.: Reliability of bidirectional V2X communications in the intelligent transport systems. In: 2015 Advances in Wireless and Optical Communications (RTUWO), Riga, pp. 159–163 (2015)
17. Ning, Y., Zhong-Qin, W., Ru-Chuan, W., Abdullah, A.H.: Design of accurate vehicle location system using RFID. Elektron. ir Elektrotechnika 19(8) (2013)
18. Ye, N., Wang, Z., Zhang, Y., Wang, R.: A method of vehicle route prediction based on social network analysis. J. Sens. 15, 1–10 (2015)
19. Malekian, R., Kavishe, A., Maharaj, B.T., Gupta, P., Singh, G., Waschefort, H.: Smart vehicle navigation system using hidden markov model and RFID technology. Wirel. Pers. Commun. 90(4), 1717–1742 (2016)
20. Waraich, R.: Plug-in hybrid electric vehicles and smart grid: investigations based on a microsimulation. Transp. Res. Part C: Emerg. Technol. 28, 74–86 (2013)
21. Yoon, S., Park, K., Hwang, E.: Connected electric vehicles for flexible vehicle-to-grid (V2G) services. In: 2017 International Conference on Information Networking (ICOIN), Da Nang, pp. 411–413 (2017)
22. Hannan, M., Habib, S., Javadi, M., Samad, S., Muad, A., Hussain, A.: Inter-vehicle wireless communications technologies, issues and challenges. Inf. Technol. J. 12(4), 558–568 (2013)
23. Ye, N., Zhang, Y., Wang, R., Malekian, R.: Vehicle trajectory prediction based on hidden markov model. KSII Trans. Internet Inf. Syst. 10(7), 3150–3170 (2016)

Local Diffusion Versus Random Relocation
in Random Walks

Viktor Stojkoski[1(✉)], Tamara Dimitrova[1], Petar Jovanovski[1], Ana Sokolovska[1], and Ljupco Kocarev[1,2]

[1] Macedonian Academy of Sciences and Arts, 1000 Skopje, Macedonia
{vstojkoski,tdimitrova,pjovanovski,sokolovska,lkocarev}@manu.edu.mk
[2] Faculty of Computer Science and Engineering,
Ss. Cyril and Methodius University, 1000 Skopje, Macedonia

Abstract. We study a class of random walks on graphs with two mechanisms: local diffusion and random relocation. Such mechanisms are common in search algorithms, an example being the PageRank. The rate with which two mechanisms are mixed is called damping factor. It determines the first-passage time, defined as the time required for a walker starting from a source node to find a given target node. We provide bounds on the stationary distribution of random walks. Global mean first-passage time as a function of the damping factor is computed in closed form for a simple example. The results provide new insights on the search engines.

Keywords: Random walks on graphs · PageRank · First-passage time

1 Introduction

Random walks on graphs [1–3] have gained enormous interest in various disciplines ranging from physics to computer science and from biology to economics. A standard random walk chooses for every vertex uniform probability distribution among its outgoing edges. Different stochastic processes have been introduced similar to the standard random walks on graphs, for example, self-avoiding walks (for which, the walker chooses its next move excluding nodes which were already visited) and maximal entropy random walks (for which, the walker chooses its next move assuming uniform probability distribution among all paths in a given graph). Here we consider a special class of random walks on graphs with two competing mechanisms: local diffusion and random relocation. At every time step the walker chooses the next move with probability α by selecting uniformly at random among its neighboring nodes (local diffusion) and with probability $1-\alpha$, the walker restarts (or teleports) to another node chosen (uniformly) at random (random allocation). A random walk in this class can be understood as a mixture of the Markov chain representing the graph connectivity structure with a fully connected graph [4], or as a random walk that occurs in a two-layer network with the same set of nodes in both layers: the physical layer and

© Springer International Publishing AG 2017
D. Trajanov and V. Bakeva (Eds.): ICT Innovations 2017, CCIS 778, pp. 57–66, 2017.
DOI: 10.1007/978-3-319-67597-8_6

the teleportation layer [5]. The stationary distribution of these random walks is called PageRank, while the parameter α is called damping factor.

PageRank has been used for an enormous number of applications within data mining [6] and machine learning [7], as well as the broader science community including physics [8], biology [9], chemistry [10], neuroscience [11], sociology [12], and many others [13]. In computer science, machine learning, and data mining (information retrieval), a particularly popular search method is based on the PageRank random walk model in which a combination of local and long-range search is implemented: a walker visits local neighbors but occasionally by performing (possible) long jumps visits different regions of the graph. A fundamental quantity related to random walks is first-passage time (FPT), representing a key indicator of how fast information diffuses in a given graph. The mean first-passage time (MFPT) is defined as the average of FPTs over all source nodes in the graph, which is a useful tool to analyze the behavior of random walks, including target search [14,15]. MFPT has been intensively studied for traditional standard (unbiased) random walks [16–20] as well as for some biased random walks [21–24].

Global mean first-passage time (GMFPT), defined as the average of MFPTs over all nodes in a graph, is a single quantity that characterizes search processes on graphs. This paper studies how global mean first-passage time depends on the damping parameter α. A mathematical analysis of PageRank when α changes was recently suggested by Boldi and his coworkers [25]. For search engines, the choice of α is eminently empirical, and in most cases the original suggestion $\alpha = 0.85$ by Brin and Page is still used. The authors of [25] provided closed-form formula for PageRank derivatives of any order and showed how to obtain an approximation of the derivatives. Here, we derive bounds for stationary distributions of random walks with two diffusion mechanisms: local diffusion and relocation-assisted diffusion. Two methods are reviewed for calculating the mean first-passage time. By studying in detail an analytically solvable example, we show that by tuning the parameter α minimum searching time can be reached. The results of the paper provide new insights on the Page Rank algorithm. Albeit mainly theoretical in nature, they also provide efficient ways to study the search behavior in graphs. In future, we plan to develop approximate techniques for computing global mean first-passage time for large graphs.

2 Inequalities Involving Stationary Distributions

Let $G = (V, E)$, be a simple strongly connected directed graph with n nodes, with V and E respectively denoting the sets of nodes and edges; let $A = [a_{ij}]$ be the $n \times n$ adjacency matrix of the graph G, that is, $a_{ij} = 1$ if and only if $i \rightarrow j \in E$. We write S_i^{out} for the set of out-neighbors of i, $S_i^{out} = \{j : i \rightarrow j\}$ and let $d_i^{out} = |S_i^{out}|$ be the out-degree of the node i. Consider a random walk on G with two diffusion mechanisms: local diffusion and relocation-assisted diffusion, described with a transition matrix $Q = [q_{ij}]$. A walker starts at a node X_0; if at the tth step the walker is at a node $X_t = i$, she/he moves/jumps to other node j with probability

$$p_{ij} = \alpha \frac{a_{ij}}{d_i^{out}} + (1 - \alpha)q_{ij}. \tag{1}$$

Clearly, $\{X_t\}$, $(t \geq 0)$ is a discrete time Markov chain with state space V and transition matrix $P = [p_{ij}]$ which can be written in more compact form as

$$P = \alpha R + (1 - \alpha)Q. \tag{2}$$

The matrix R describes local diffusion: the walker moves to her/his neighbor j only if there is a link $i \rightarrow j$. On the other hand, the matrix Q reflects relocation (long-range jumps, teleportation) to other nodes (neighbors are not excluded). The parameter $\alpha \in [0, 1]$ tunes the relative strength of the two competing mechanisms, local diffusion and random relocation. Both matrices R and Q are assumed to be left stochastic, so that P is also left stochastic. We further assume that both the local Markov chain (specified only with the transition matrix R) and the Markov chain $\{X_t\}$, $(t \geq 0)$ (with transition matrix P) are regular (finite, irreducible and aperiodic). It is well known that such Markov chains have a unique stationary distribution, which is also the limiting distribution of the Markov chain. Let $\pi^T = (\pi_1, \pi_2, \ldots, \pi_n)$ be the stationary probability vector of the Markov chain $\{X_t\}$, $(t \geq 0)$ and let $\rho^T = (\rho_1, \rho_2, \ldots, \rho_n)$ be the stationary probability vector of the local Markov chain (with transition matrix R).

Several possibilities for the matrix Q have been considered. First, $Q = (1/n)\mathbf{1}\mathbf{1}^T$ where $\mathbf{1}$ is a column vector whose entries are all 1. In this case the walker is equally likely to "jump" or "teleport" to any node on the graph. Second, $Q = \mathbf{v}\mathbf{1}^T$, where $\mathbf{v}$ is a vector with non-negative coordinates, satisfying $\mathbf{1}^T\mathbf{v} = 1$, called seed or personalization vector. In the third possibility, the matrix Q is seen as a transition matrix of a random walk performed on another graph with the same vertex set (representing a layer in the multilayer/multiplex structure). More precisely, let $A^1 = [a_{ij}^1]$ be $n \times n$ adjacency matrix of a graph $G^1 = (V, E^1)$; then $Q = [a_{ij}^1/d_i^{1,out}]$ where $d^{1,out}$ is the out-degree of the node i in the layer (graph) G^1.

We show that when $Q = \mathbf{v}\mathbf{1}^T$, the following inequalities hold:

$$\frac{1 - \alpha}{1 + \alpha}\|\mathbf{v} - \rho\|_1 \leq \|\pi - \rho\|_1 \leq \|\mathbf{v} - \rho\|_1,$$

where

$$\|\mathbf{v} - \rho\|_1 = \sum_j |v_j - \rho_j|.$$

Let $\mathbf{v}$ be a distribution, which we will call the preference distribution (by distribution we mean a vector with non-negative entries and L_1-norm equal to 1). We rewrite Eq. (2) as

$$P = \alpha R + (1 - \alpha)\mathbf{v}\mathbf{1}^T. \tag{3}$$

PageRank is defined (up to a scalar) by the eigenvector equation $P\pi = \pi$. The damping factor $\alpha \in [0, 1]$ determines how often the Markov chain follows the graph rather than moving at a random node according to the preference vector $\mathbf{v}$. The preference vector is used to bias PageRank with respect to a selected set of trusted pages, or might depend on the user's preferences, in which case one speaks of personalised PageRank. The PageRank vector can be expressed as $\pi = (I - \alpha R)^{-1}(1-\alpha)\mathbf{v}$ for $\alpha < 1$. Let ρ be the eigenvector of R corresponding to its largest eigenvalue 1. Note that $(I-\alpha R)\rho = I\rho - \alpha R\rho = I\rho - \alpha\rho = (1-\alpha)\rho$. Therefore, $\rho = (I - \alpha R)^{-1}(1-\alpha)\rho$. It follows that

$$\pi - \rho = (1-\alpha)(I - \alpha R)^{-1}(\mathbf{v} - \rho),$$
$$(I - \alpha R)(\pi - \rho) = (1-\alpha)(\mathbf{v} - \rho).$$

It is well-known [26] that

$$\|I - \alpha R\|_1 = 1 + \alpha \text{ and } \|(I - \alpha R)^{-1}\|_1 = \frac{1}{1-\alpha}.$$

Now,

$$(1-\alpha)\|\mathbf{v} - \rho\|_1 = \|(I - \alpha R)(\pi - \rho)\|_1 \leq (1+\alpha)\|\pi - \rho\|_1,$$
$$\frac{1-\alpha}{1+\alpha}\|\mathbf{v} - \rho\|_1 \leq \|\pi - \rho\|_1.$$

In a similar fashion

$$\|\pi - \rho\|_1 = (1-\alpha)\|(I - \alpha R)^{-1}(\mathbf{v} - \rho)\|_1 \leq \|\mathbf{v} - \rho\|_1.$$

Combining the last two inequalities

$$\frac{1-\alpha}{1+\alpha}\|\mathbf{v} - \rho\|_1 \leq \|\pi - \rho\|_1 \leq \|\mathbf{v} - \rho\|_1.$$

It is easy to see that if $v_j = \rho_j$ for all j then $\pi = \rho$. Moreover, if $|v_j - \rho_j| \leq \epsilon$ and ϵ is sufficiently small, π and ρ will be close. Figure 1 illustrates our findings for two different graphs: (a) Directed Erdos-Renyi, and (b) Directed Barabasi-Albert.

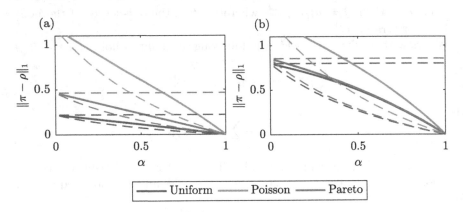

Fig. 1. (a) Inequalities involving stationary distributions for a sample directed Erdos-Renyi Graph and three different types of $\mathbf{v}$ - Uniform, Poisson and Pareto. Filled lines indicate real values for $\|\pi - \rho\|_1$, whereas dashed lines are lower and upper bounds. (b) Same as (a), only for a directed Barabasi-Albert graph (a–b). Both graphs have 100 nodes and average degree 10.

3   Global Mean First-Passage Time

Let T_{ij} be the first-passage time random variable from state i to state j , i.e. $T_{ij} = \min\{t \geq 1$ such that $X_t = j$ given that $X_0 = i\}$, so that T_{ii} is the "first return to state i". The irreducibility of the Markov chain ensures that the T_{ij} are all proper random variables and under the finite state space restriction, all the moments of T_{ij} are finite. Let $M = [m_{ij}]$ be the matrix of mean first-passage times, where m_{ij} is defined as expected value of the first-passage time random variable from state i to state j, i.e. $m_{ij} = E[T_{ij}|X_0 = i]$ for all $i, j \in V$. We also consider global mean first-passage time (GMFPT) $gm = \frac{1}{n(n-1)} \sum_j \sum_{i \neq j} m_{ij}$ defined as the average of MFPTs over all nodes in the network.

We define the search time as the time needed by a walker starting at node i to reach a node j, that is, search time is a random variable that coincides with the first-passage time random variable T_{ij}. There are two approaches to study the first-passage time and its moments (in particular, the mean first-passage time): by directly analyzing the transition matrix P or by assuming that the node j is absorbing trap and analyzing an $(n-1) \times (n-1)$ sub-matrix obtained by suppressing the j–th row and the j–th column of the matrix P; we denote this matrix with P_j. Computation of the matrix M can be done in two ways: by computing the so called generalized inverses (g-inverses) of the matrix $I_n - P$ (where I_n is $n \times n$ identity matrix), or by computing the $(n-1) \times (n-1)$ fundamental matrix $N_j = [n_{ik}]$, where $i, k \neq j$, defined as $N_j = I_{n-1} + P_j + P_j^2 + \ldots$, which in our notation, leads to $m_{ij} = \sum_{k \neq j} n_{ik}$, for all $i \neq j$. Note that $m_{jj} = 1/\pi_j$.

The matrices P and P_j (for all j) play crucial role in determining the matrix M. Moreover, the sum of every row of P_j is strictly less than 1, the largest eigenvalue of P_j is less than 1, so that $P_j^t \to 0$ as $t \to \infty$. However, the associated right eigenvector equals the unique quasi-stationary distribution for the discrete-time Markov chains with one absorbing state and finite set of transient states obtained from the graph by making the node j an absorbing node (replacing all its out-links with one self-loop). Thus, the eigenvector associated with the largest eigenvalue of the transition matrix P determines the stationary distribution π while the eigenvector associated with the largest eigenvalue of the matrix P_j controls the limiting behavior of the chain as time tends to infinity conditional on absorption not yet having taken place. For this reason, the distribution is called quasi-stationary: although the walker will eventually reach the absorbing state, the quasi-stationary distribution describes behavior before the walker reaches the absorbing state.

3.1   Direct Calculations of the Mean First-Passage Times

It is well known, see for example [27], that generalized inverses of the Markovian kernel $I - P$ have useful properties in determining the mean first-passage times. In particular, M satisfies the matrix equation

$$(I - P)M = \mathbf{1}\mathbf{1}^T - PM_d, \qquad (4)$$

where $M_d = [m_{ii}]$, is a diagonal matrix with elements the diagonal elements of M such that $m_{ii} = 1/\pi_i$. The solution of equations of the form of Eq. (4) can be effected using g-inverses of $I - P$. Any g-inverse Γ of $I - P$ has the form

$$\Gamma = [I - P + \mathbf{t}\mathbf{u}^T]^{-1} + \mathbf{1}\mathbf{f}^T + \mathbf{g}\pi^T,$$

where $\mathbf{u}^T \mathbf{1} \neq 0$, $\pi^T \mathbf{1} \neq 0$ and $\mathbf{f}, \mathbf{u}, \mathbf{t}$ and $\mathbf{g}$ are arbitrary vectors.

3.2   Absorbing Markov Chains

We assume that the absorbing node is j and write $\{\tilde{X}_t\}$, $t \geq 0$ for a homogeneous discrete-time Markov chain for which the state space is $V = \{1, 2, \ldots n\}$ consisting of an absorbing state $j \in V$ and a finite set of transient states $S = V \setminus \{j\}$. We let $P_j = [p_{kl}]$, where $k, l \in S$, be $(n-1) \times (n-1)$ matrix of one-step transition probabilities of the (sub)Markov chain on S. Note that p_{ij}, for all $i \in S$ are the probabilities of absorption into state j. Since all states in S are assumed to be transient, at least one of these probabilities must be positive and absorbing is certain. As is well known, the eigenvalue $\lambda \equiv \lambda(P_j)$ with maximal real part (the Perron – Frobenius eigenvalue of P_j) is real and nonnegative. Since all states in S are transient we have $0 < \lambda < 1$. Although $P_j^t \to 0$ as $t \to \infty$ (since $0 < \lambda < 1$), if S is aperiodic, the $t-$step transition matrix P_j^t satisfies

$$\lim_{t \to \infty} \lambda^{-t} P_j^t = A,$$

where A is $(n-1) \times (n-1)$ matrix which depends only on left and right eigenvectors associated with λ. Moreover, the right (column) eigenvector $\mathbf{u}$, where $\mathbf{u}^T = (u_i, i \in S)$, is the unique quasi-stationary distribution for the discrete-time Markov chains with one absorbing state and a finite set S of transient states. A probability distribution $\mathbf{u}$ is said to be a quasi-stationary distribution for $\{\tilde{X}_t\}$, $t \geq 0$ if the distribution of $\tilde{X}_t$, conditional on absorption not yet having taken place, is constant over t when $\mathbf{u}$ is the initial distribution.

The $(n-1) \times (n-1)$ matrix $N_j = [n_{ik}]$, where $i, k \neq j$, defined as

$$N_j = I_{n-1} + P_j + P_j^2 + \ldots = (I_{n-1} - P_j)^{-1},$$

is called the fundamental matrix for the absorbing chain. n_{ik} is the expected number of periods that the chain spends in the kth non-absorbing state given that the chain began in the ith non-absorbing state. Therefore, $m_{ij} = \sum_{k \neq j} n_{ik}$, for all $i \neq j$, is the expected number of periods before absorption (into the absorbing state j) given that the chain began in the ith non-absorbing state.

3.3   Example

Let $P = [p_{ij}]$ be the transition matrix of a Markov chain with state space $\{1, 2, 3\}$, where

$$P = \alpha \begin{bmatrix} 0 & a_1 & 1 - a_1 \\ 1 - b_1 & 0 & b_1 \\ c_1 & 1 - c_1 & 0 \end{bmatrix} + (1 - \alpha) \begin{bmatrix} 0 & v_1 & 1 - v_1 \\ 1 - v_2 & 0 & v_2 \\ v_3 & 1 - v_3 & 0 \end{bmatrix},$$

which can be rewritten in a compact form as

$$P = \begin{bmatrix} 0 & a & 1-a \\ 1-b & 0 & b \\ c & 1-c & 0 \end{bmatrix},$$

where $a = \alpha a_1 + (1-\alpha)v_1$, $b = \alpha b_1 + (1-\alpha)v_2$, and $c = \alpha c_1 + (1-\alpha)v_3$. Let $\Delta_1 = 1 - b(1-c)$, $\Delta_2 = 1 - c(1-a)$, $\Delta_3 = 1 - a(1-b)$, and $\Delta = \Delta_1 + \Delta_2 + \Delta_3$. It is easily shown that the stationary probability vector is $(\pi_1, \pi_2, \pi_3) = \frac{1}{\Delta}(\Delta_1, \Delta_2, \Delta_3)$. The mean first-passage time matrix can be computed, see the Appendix, as:

$$M = \begin{bmatrix} \frac{\Delta}{\Delta_1} & \frac{2-a}{\Delta_2} & \frac{1+a}{\Delta_3} \\ \frac{1+b}{\Delta_1} & \frac{\Delta}{\Delta_2} & \frac{2-b}{\Delta_3} \\ \frac{2-c}{\Delta_1} & \frac{1+c}{\Delta_2} & \frac{\Delta}{\Delta_3} \end{bmatrix}.$$

Therefore, for the global mean first-passage time, we have

$$gm(\alpha) = \frac{1}{6} \left[\frac{3-c+b}{\Delta_1} + \frac{3-a+c}{\Delta_2} + \frac{3-b+a}{\Delta_3} \right].$$

For the special case when $a_1 = 0$, $b_1 = c_1 = 1$, $v_1 = 1$ and $v_2 = v_3 = 0$, after some algebra, it easy to show that

$$gm(\alpha) = \frac{1}{6} \left[\frac{3}{1-\alpha+\alpha^2} + \frac{2+2\alpha}{1-\alpha^2} + \frac{4-2\alpha}{2\alpha-\alpha^2} \right].$$

The minimum of $gm(\alpha)$ can be easily calculated to be $gm(0.5) = 2$, see Fig. 2. A more detailed analysis of the example shows that the dependency of the global mean first-passage time gm on α can be complicated and the minimal value can be reached (depending on the parameters) for any value of $\alpha \in [0, 1]$, see also Fig. 3. Concretely, Fig. 3 (a), (b), and (c) depict the global mean first-passage time gm as a function of $\alpha \in [0.2, 0.8]$ and one of the parameters a_1, b_1, c_1 (while keeping the other two fixed) and assuming $v_1 = 1$ and $v_2 = v_3 = 0.1$. On the same figure we plot the components of the stationary distribution vector, for this example, the components of the vector $(\pi_1, \pi_2, \pi_3) = \frac{1}{\Delta}(\Delta_1, \Delta_2, \Delta_3)$. Figures (d), (e) and (f) show the PageRank vector π as a function of a_1 and α, while the other parameters are set in the same way as for the Fig. 3(a). Note that contour patterns of the global mean first-passage time gm on Fig. 3(a) are similar to those for π_1 on Fig. 3(d). The question of how the global mean first-passage time is correlated with the stationary distribution vector in this example will be addressed in a future study.

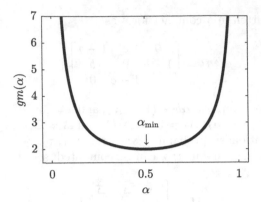

Fig. 2. Global mean first-passage time gm as a function of α. In this example, $a_1 = 0$, $b_1 = c_1 = 1$, $v_1 = 1$ and $v_2 = v_3 = 0$.

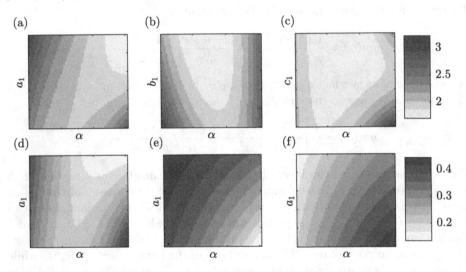

Fig. 3. (a) Global mean first-passage time gm as a function of $\alpha \in [0.2, 0.8]$ and $a_1 \in [0, 1]$, where we fix $b_1 = c_1 = 1$. (b) Same as (a), only now we vary $b_1 \in [0, 1]$ and fix $a_1 = 0$ and $c_1 = 1$. (c) Same as (a), only now we vary $c_1 \in [0, 1]$ and fix $a_1 = 0$ and $b_1 = 1$. (d) Heat map for the element π_1 of the PageRank vector π, where the parameters are set in the same way as (a). (e) Same as (d), only for π_2. (f) Same as (d), only for π_3. (a–e) We fix $v_1 = 1$ and $v_2 = v_3 = 0.1$.

Acknowledgments. We thank DFG for support through the project "Random search processes, Lévy flights, and random walks on complex networks".

Appendix

We now assume that 1 is absorbing state/node; then the transition matrix P transforms to

$$P = \begin{bmatrix} 1 & 0 & 0 \\ 1-b & 0 & b \\ c & 1-c & 0 \end{bmatrix},$$

and P_1, the matrix of one-step transition probabilities of the (sub)Markov chain on the set $\{2,3\}$, is given by

$$P_1 = \begin{bmatrix} 0 & b \\ 1-c & 0 \end{bmatrix}.$$

The Perron – Frobenius eigenvalue of P_1 is $\lambda = \sqrt{b(1-c)} < 1$ and its normalized eigenvector $\mathbf{u}$ is

$$\mathbf{u}^T = \left(\frac{\lambda - b}{1-c-b}, \frac{1-c-\lambda}{1-c-b} \right).$$

Since the set of transient states $S = \{2,3\}$ is irreducible, $\mathbf{u}$ the (unique) quasi-stationary distribution for the discrete-time Markov chains with one absorbing state and a finite set S of transient states.

The fundamental matrix for the absorbing chain is

$$I + P_1 + P_1^2 + \ldots = I \left[1 + b(1-c) + b^2(1-c)^2 + \ldots \right] +$$
$$+ P_1 \left[1 + b(1-c) + b^2(1-c)^2 + \ldots \right]$$
$$\equiv \frac{1}{1 - b(1-c)} \begin{bmatrix} 1 & b \\ 1-c & 1 \end{bmatrix} \equiv \begin{bmatrix} n_{22} & n_{23} \\ n_{32} & n_{33} \end{bmatrix}.$$

recovering m_{21} and m_{31} from the matrix M, that is

$$m_{21} = \sum_{k \in S} n_{2k} = \frac{1+b}{1-b(1-c)}, \qquad m_{31} = \sum_{k \in S} n_{3k} = \frac{2-c}{1-b(1-c)}.$$

The other elements of the matrix M can be computed in a similar way.

References

1. Lovasz, L.: Random walks on graphs: a survey. In: Combinatorics, Paul Erdos in Eighty, vol. 2 (1993)
2. Noh, J.D., Rieger, H.: Random walks on complex networks. Phys. Rev. Lett. **92**(11), 118701 (2004)
3. Burioni, R., Cassi, D.: Random walks on graphs: ideas, techniques and results. J. Phys. A Math. Gen. **38**(8), R45 (2005)
4. Berkhout, J., Heidergott, B.F.: Ranking nodes in general networks: a markov multi-chain approach. Discret. Event Dyn. Syst. 1–31 (2017). Special Issue on Performance Analysis and Optimization of Discrete Event Systems
5. Pedroche, F., García, E., Romance, M., Criado, R.: Sharp estimates for the personalized multiplex pagerank. J. Comput. Appl. Math. (2017, in press)

6. Pan, J.Y., Yang, H.J., Faloutsos, C., Duygulu, P.: Automatic multimedia cross-modal correlation discovery. In: Proceedings of the Tenth ACM SIGKDD International Conference on Knowledge Discovery and Data Mining, pp. 653–658. ACM (2004)

7. Zhou, D., Bousquet, O., Lal, T.N., Weston, J., Schölkopf, B.: Learning with local and global consistency. In: Advances in Neural Information Processing Systems, pp. 321–328 (2004)

8. Garnerone, S., Zanardi, P., Lidar, D.A.: Adiabatic quantum algorithm for search engine ranking. Phys. Rev. Lett. **108**(23), 230506 (2012)

9. Singh, R., Xu, J., Berger, B.: Global alignment of multiple protein interaction networks with application to functional orthology detection. Proc. Natl Acad. Sci. **105**(35), 12763–12768 (2008)

10. Mooney, B.L., Corrales, L.R., Clark, A.E.: Molecularnetworks: an integrated graph theoretic and data mining tool to explore solvent organization in molecular simulation. J. Comput. Chem. **33**(8), 853–860 (2012)

11. Zuo, X.N., Ehmke, R., Mennes, M., Imperati, D., Castellanos, F.X., Sporns, O., Milham, M.P.: Network centrality in the human functional connectome. Cereb. Cortex **22**(8), 1862–1875 (2011)

12. Leskovec, J., Lang, K.J., Dasgupta, A., Mahoney, M.W.: Community structure in large networks: natural cluster sizes and the absence of large well-defined clusters. Internet Math. **6**(1), 29–123 (2009)

13. Gleich, D.F.: Pagerank beyond the web. SIAM Rev. **57**(3), 321–363 (2015)

14. Jasch, F., Blumen, A.: Target problem on small-world networks. Phys. Rev. E **63**(4), 041108 (2001)

15. Bénichou, O., Loverdo, C., Moreau, M., Voituriez, R.: Intermittent search strategies. Rev. Mod. Phys. **83**(1), 81 (2011)

16. Tejedor, V., Bénichou, O., Voituriez, R.: Global mean first-passage times of random walks on complex networks. Phys. Rev. E **80**(6), 065104 (2009)

17. Agliari, E., Burioni, R., Manzotti, A.: Effective target arrangement in a deterministic scale-free graph. Phys. Rev. E **82**(1), 011118 (2010)

18. Meyer, B., Agliari, E., Bénichou, O., Voituriez, R.: Exact calculations of first-passage quantities on recursive networks. Phys. Rev. E **85**(2), 026113 (2012)

19. Lin, Y., Julaiti, A., Zhang, Z.: Mean first-passage time for random walks in general graphs with a deep trap. J. Chem. Phys. **137**(12), 124104 (2012)

20. Hwang, S., Lee, D.S., Kahng, B.: First passage time for random walks in heterogeneous networks. Phys. Rev. Lett. **109**(8), 088701 (2012)

21. Wu, B., Zhang, Z.: Controlling the efficiency of trapping in treelike fractals. J. Chem. Phys. **139**(2), 024106 (2013)

22. Yang, Y., Zhang, Z.: Random walks in unweighted and weighted modular scale-free networks with a perfect trap. J. Chem. Phys. **139**(23), 234106 (2013)

23. Fronczak, A., Fronczak, P.: Biased random walks in complex networks: the role of local navigation rules. Phys. Rev. E **80**(1), 016107 (2009)

24. Bonaventura, M., Nicosia, V., Latora, V.: Characteristic times of biased random walks on complex networks. Phys. Rev. E **89**(1), 012803 (2014)

25. Boldi, P., Santini, M., Vigna, S.: Pagerank: functional dependencies. ACM Trans. Inf. Syst. (TOIS) **27**(4), 19 (2009)

26. Kamvar, S., Haveliwala, T.: The condition number of the pagerank problem. Technical report, Stanford InfoLab (2003)

27. Ben-Israel, A., Greville, T.N.: Generalized Inverses: Theory and Applications, vol. 15. Springer Science & Business Media, New York (2003)

Network-Dependent Server Performance Analysis of HTTP Adaptive Streaming

Sasho Gramatikov[✉]

Faculty of Computer Science and Engineering, Skopje, Republic of Macedonia
sasho.gramatikov@finki.ukim.mk

Abstract. The HTTP adaptive streaming (HAS) is a popular mechanism for delivery of live and on-demand video contents encoded with different qualities and divided into segments with equal length. The mechanism adapts the requested segment qualities to the quality of the link, providing uninterrupted service even in congested network conditions. In this work, we analyze the HAS for delivery of Video on Demand (VoD) contents from server performance point of view for different segment lengths and different network conditions. For that purpose, we created an environment for real-case measurements of the server performance and measured performance parameters like CPU utilization, generated in-bound and out-bound traffic and number of established TCP connections. From the analysis of the obtained data, we conclude that streaming of shorter video segments generates more appropriate and predictable traffic pattern, but requires more CPU power and TCP connections. Therefore, the shorter contents are suitable for streaming in networks with very low packet losses. Longer video segments, on the other hand, tend to require more resources only at the beginning of the streaming session, which they release before the end of the session, and hence, alleviate the network equipment. The main advantage of using long segments is that they can achieve uninterrupted streaming experience even in harsh network environments such as congested wireless networks.

Keywords: Adaptive streaming · Video-on-Demand · Performance

1 Introduction

The HTTP adaptive streaming (HAS) emerges as a response to the large popularity of the video streaming services and the necessity to provide uninterrupted video experience in networks with different bandwidth capacities for devices with different capabilities. It is a hybrid mechanism that takes the advantages of the traditional and the progressive streaming.

In the HAS, each video is encoded with several different qualities, and each quality is divided into segments with the same length, but different size. The segments typically have duration between 2 and 10 s. Apart from division of the videos, a manifest file containing relevant information about the available

© Springer International Publishing AG 2017
D. Trajanov and V. Bakeva (Eds.): ICT Innovations 2017, CCIS 778, pp. 67–78, 2017.
DOI: 10.1007/978-3-319-67597-8_7

qualities of the video and the URL of the segments is created. The clients request the manifest file, and based on its content, their video player proceeds requesting segments with size appropriate to the network conditions. Hence, the devices with high resolution displays and high-speed Internet connection require the best quality segments, while the mobile devices with limited bandwidth require the lowest quality segments. The main feature of the HAS that makes it superior to the other streaming mechanisms is that the devices can switch to lower qualities whenever the player detects that the network conditions deteriorate and switch back to the maximal quality when the network recovers. Although reducing the quality of the video reduces the Quality of Experience (QoE), it keeps the clients satisfied because it still provides uninterrupted service. Just like the progressive streaming, the HAS uses the HTTP as application protocol to deliver the video segments, i.e., the segments are equally treated as web pages. Therefore, the videos can be hosted on ordinary web servers for hosting web contents, instead of use of proprietary streaming servers typical for the traditional streaming.

Another huge advantage of the adaptive streaming comes from the fact that the HTTP uses TCP as a transport protocol, which guarantees that the data between the server an the client will arrive in order and with no errors. Moreover, since the TCP uses the well known port number 80, the video contents can easily pass through the firewalls of even the most restrictive administrators.

Unlike the progressive streaming, which tends to maximally utilize the available throughput enabling the clients to download significantly higher portion of the video at the very beginning of the transfer, the HAS tends to entirely download only one segment at a time with the maximal available speed. Therefore, whenever the clients desire to stop the video session, no downloaded data is wasted as it is the case with the progressive streaming.

The only disadvantage of the HAS streaming is that it requires extra storage space for hosting many qualities of the some video. However, taking into account the cost of storage per byte nowadays, this disadvantage is not a great concern for the video service providers.

There are many commercial HAS implementations such as Apple's HTTP Live Streaming (HLS), Microsoft's HTTP Smooth Streaming (HSS) and Adobe's HTTP Dynamic Streaming (HDS) used in the popular video streaming platforms. In order to overcome the different segment and manifest file formats, the HAS was standardized by MPEG into the open standard MPEG Dynamic Adaptive Streaming over HTTP (MPEG-DASH) [10]. The standard defines the format of the manifest file called Media Presentation Description (MPD). The DASH standard is currently widely accepted by a large community of leading streaming companies which formed the DASH Industry Forum (DASH-IF) [1]. The DASH-IF provides open-source software tools for implementation of the DASH, which is the reason why we use DASH as a representative of the HAS mechanisms in our work.

The main interest of our work is to find out how the streaming server performs when streaming videos with different segment lengths under different network conditions. There is a vast community that treats the HAS streaming with main focus on performance of various algorithms used by the players to choose the appropriate video qualities. The goal of these algorithms is, on one hand, to provide uninterrupted service even in unfavourable network conditions, and on the other hand, to reduce the number of quality switches. Thus, in [4,6,8,11] the authors analyze the QoE for DASH streaming of a single video session for various bandwidth restrictions, packet delay and packet loss probability. In [5,9,12], the authors propose various algorithms for choosing the optimal quality levels in order to achieve the aforementioned goals, comparing not only a single session, but also competitive requests that share the same link by varying the network conditions and other algorithm-specific parameters. However, to our knowledge, there are no articles in the literature that treat the measurement of server performance for simultaneous service of large number of clients. Therefore, the goal of this paper is to measure and compare the resources required by the server for VoD streaming using HAS, emphasizing the effect of the network conditions and the video segment duration. For that purpose, we created test environment consisting of a streaming server and clients that simultaneously generate requests for videos from the server. Then, we measure the resources required by the streaming server such as CPU utilization, out-bound, in-bound bandwidth and number of established TCP connections in different emulated network conditions. The results of these measurements give an overall picture of the efficiency of the HAS, which can serve the VoD service providers in profiling their streaming servers under different network conditions.

The rest of the paper is organized as follows: in Sect. 2 we describe the network environment, the software and the content adaptation required for conducting the measurements. Afterwards, in Sect. 3, we analyze the obtained results and give the conclusions from our work in Sect. 4.

2   Measurement Environment

In order to create real-case scenarios for measurements of adaptive streaming, we designed a simple network consisting of a web server that hosts videos with different segment duration and clients that generate requests for the videos. The server is a virtual machine with 4 CPU cores and 16 GB or RAM. Since the adaptive streaming does not require dedicated video streaming server, we use the Apache web server that runs on Ubuntu 16.04 operating system.

We use 3 client machines that are connected to the server on the faculty campus network via 4 hops. The overall capacity of the links between the clients and the server is limited to the speed of Fast Ethernet used to connect the clients to the network. The small number of clients generate frequent requests in a short time interval to simulate the streaming process of large number of clients that request a single video. There are 42 clients that generate requests for videos with a rate of 1 request per second.

In order to measure the server performance in different network conditions, we use the built-in Linux Kernel module *netem* [7] that runs on the streaming server. It emulates different network conditions by adding configurable delay or packet loss probability to every packet that passes through the network interface.

We measure and record the system performance during the streaming process by using a dedicated program that runs on the server. The program uses the SIGAR library [3] which provides a cross-platform, cross-language programming interface to low-level information on computer hardware and operating system activity. The program periodically gathers the performance data and writes it in a text file that is used for latter analysis.

The clients use the Mozilla Firefox web browser to request videos and its native video player to play the received data. The main reason for choosing Mozilla Firefox is that it is among the few browsers that allow establishing a configurable number of simultaneous TCP connection to the same host. This feature is essential for conducting the experiment where only a few clients establish large number of simultaneous connections to the same server.

Since the HAS streaming is not explicitly supported by the native players of the web browsers, we use an open-source JavaScript library for DASH streaming, developed and maintained by the DASH Industry Forum [1]. The library processes the requested manifest file and uses its contents to request video segments which are further played by the native video player.

The server hosts one sample 720×680 video with duration of 60 s. The video is encoded with the H264 codec in MP4 format with playback rate of 1.1 Mbps and total size of 8.2 MB. We used the open-source MP4Box tool [2] to divide the video in segments with duration of 2, 5, 10, 30 and 60 s, and to create a MPD manifest file for each duration of the segments. We consider only a single quality video, i.e., there are no rate switches when the network conditions change.

3   Results and Analysis

For each simulation scenario, we conduct three independent measurements and use their mean value in the analysis. In order to reduce the periodic spikes that appear in most of the analyzed parameters, we apply a smoothing function that calculates a moving average of the proximate measurement data.

In our first scenario, we measure the out-bound traffic originating from the streaming server for delivery of videos with different segment length. The results from the measurements are shown in Fig. 1. The figure shows that as the segment length increases, so does the peak value of the traffic requested by the server. This peak is reached earlier in the time for longer segments. On the contrary, for segments with length of 2 s, the amount of traffic increases slowly and reaches the peak in the interval when all the clients in the system play the video. The peak value in this moment equals the sum of the playback rates of all clients in the system. The adaptive streaming with small segment length has the properties of traditional streaming, which downloads the video data with rate equal to its playback rate.

In the case when the segment length equals the length of the video, the peak is reached shortly after the first initiated requests as a result of the tendency of the TCP to maximally utilize the available bandwidth between the two end-points. Thus, every client buffers the entire video at the beginning of the session

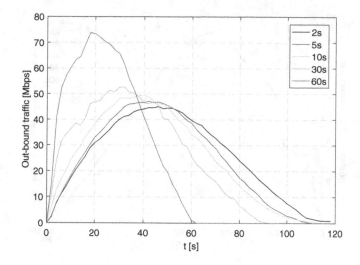

Fig. 1. Out-bound traffic for different segment duration

and generates no traffic until the end of the session. Therefore, the curve for out-bound traffic reaches value zero shortly after the last request for video. In the moments when all the clients are simultaneously playing the video, there is no load on the server. The system has similar behaviour when the clients request videos with segment length of 30 s, but the out-bound traffic reaches the peak at the moment when part of the clients that reached half of the video request the second half, and the clients that initiate the session in that moment request the first half.

From the results of Fig. 1, we can conclude that having short segments does not overload the network, requiring maximum bandwidth that equals the total playback rate of the videos. In this situation the traffic demands for a certain number of simultaneous session can be easily predicted, unlike the long segments which can lead to initial delays due to the tendency to use the available bandwidth. Another advantage of short segments is that if the clients decide to end the session at the middle of the streaming, there is no waste of transferred data since the video segments are downloaded as they are watched. In the case of long segments, the link is pushed to its capacity limits. However, the clients initially obtain large amounts of the video, which gives them enough time to load the rest of the video in case that the link saturates and, hence, the download rate becomes lower than the playback rate. Other disadvantage of long segments is that there is waste of the downloaded data in case that the client decides to end the session.

Another aspect of the traffic in the network is the number of established TCP connections for delivery of the video segments with different lengths shown in Fig. 2. Although the short segments proved to generate lowest amount of peak out-bound traffic, they dominate with the number of established concurrent

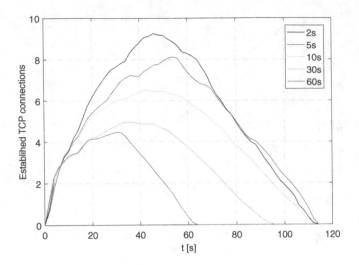

Fig. 2. Number of established TCP connection for different segment duration

TCP connections. As the playback of the video proceeds, the player establishes a new connection for every subsequent segment. Upon reception of the segment, the player releases the connection. Hence, when the segment duration is very short, the connection establishment frequency increases, which leads to high number of concurrent connections. As expected, the peak number is reached when all the clients are receiving a video, which is at the middle of the first request and the end of the last session. Although there are 42 simultaneous streaming sessions at the time of the peak value, there are only approximately 9 connections since at the time of the measurements only that number of clients are downloading segments while the rest of the active clients are playing the buffered segments downloaded previously. The same analogy applies to the low number of connections for long video segments. In this case there are fewer segments that have to be downloaded, which imply lower number of established connections. Therefore, in Fig. 2, the segments with duration of 60 s reach a peak number of approximately 4 connections. Each connection in this case is used to load entire segment and is never used by the client later during the session.

The CPU utilization is another performance parameter considered in our work. For every video, the web server initially receives request for the MPD manifest file, and based on its content, continues receiving requests for the subsequent video segments. The number of requests depends on the length of the video, and therefore, it is expected that the web server will utilize different processing power to handle the requests. The CPU utilization for serving videos with different segment lengths is shown in Fig. 3. The CPU is loaded the most when serving videos with segment length of 2 s. It reaches the peak value before the last request for video and slowly decreases to zero utilization at the end of the last session. The curves for CPU utilization for longer segments have similar shape in the initial part of

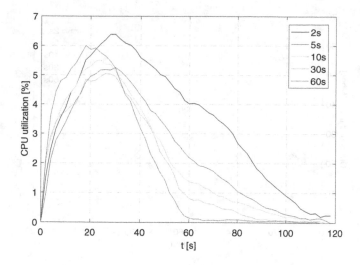

Fig. 3. CPU utilization for different segment duration in perfect network conditions

the measurements since all sessions start with request for the MPD manifest file. The maximum difference in their peak values is approximately 1.5%. What makes the curves different is the descending shape after the peak value is reached. The videos with segment length of 60 s rapidly release the CPU before the sessions end, while the shorter segments start releasing the CPU with higher intensity, and then, slowly decrease to value zero. It is interesting to note that the peak of the CPU utilization does not temporally coincide with the out-bound traffic and the number of TCP connections peaks. From the figure, we can conclude that longer segments are more appropriate for use than shorter segments from CPU utilization point of view because not only they reach lower peak values than the short segments, but they also keep the server's CPU busy shorter time.

In order to analyze the server performance in harsh network conditions, we conducted the same measurements, but we introduced emulated packet loss with probabilities of 2, 5, 10, 12 and 15%.

In Fig. 4, we show the out-bound traffic behaviour in unfavorable network conditions with packet loss probability of 10% for streaming videos with different segment lengths. The figure shows that the segments with lengths of 2 s are not appropriate for use in networks with high packet loss, which can be proved by the fact that there is still significant amount of out-bound traffic after the expected finish time of all sessions. This traffic implies that the clients experience stalls during the session. The reason for the stalls is that within the short duration of each segment, there is not enough time for the lost packets to be re-transmitted. Apart from the increased stall time, we can observe that the peak values of the traffic are reduced compared to the perfect network conditions as shown in Fig. 1, since the average time for delivery of one segment is increased, and hence, the average download rate per session is decreased.

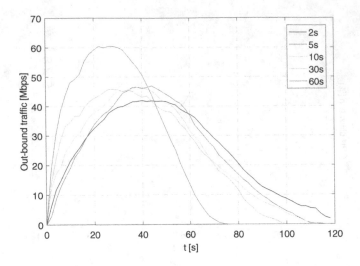

Fig. 4. Out-bound traffic during time for different segment duration with packet loss probability of 10%

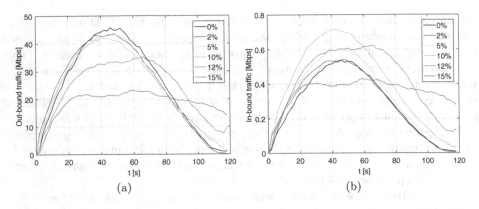

Fig. 5. (a) Out-bound and (b) in-bound traffic for different packet loss probabilities for segments with duration of 2 s

The dependence of the out-bound and in-bound traffic on the packet loss probability for segments with duration of 2 s is shown in Fig. 5(a) and (b), accordingly. In Fig. 5(a), we can see that the short segments can be delivered to the clients in time only in perfect network conditions since, as the packet probability increases, the peak out-bound traffic reduces. This fact implies delay in completing the whole segments, and hence, interruptions in the playback of the video. Another indicator of the increased number of session interruptions is the presence of traffic after the predicted end of the last session.

The packet loss probability has also a significant impact on the server inbound traffic as shown in Fig. 5(b). The reason for the increased amount of

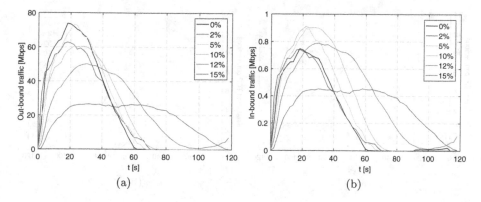

Fig. 6. (a) Out-bound and (b) in-bound traffic for different packet loss probabilities for segments with duration of 60 s

received traffic for higher loss probabilities is that the clients send more requests for re-transmission of the unacknowledged TCP segments.

Unlike the short segments, the longer segments are more resistant to packet loss, which can be seen from the traffic patterns for segments with duration of 60 s in Fig.6(a) and (b). We can witness that the streaming process can withstand packet loss probabilities as high as 10% without interruptions of the watching experience. The only drawback of the long segments in harsh network conditions is that the clients may suffer a short initial delay until there is enough buffered data to play the video, however, this issue is mostly tolerable as there are no further interruptions. The figure also shows that for packet loss probabilities above 12% the clients will suffer intolerable number of interruptions.

In Fig. 7(a) and (b), we show the impact of the packet loss probability on the number of established concurrent TCP connections for segment duration of 2 and 60 s. We can see that high packet loss probability implies higher number of TCP connections compared to the low packet loss probability. The reason for this behaviour is the fact that each client holds the connection longer time as a result of re-transmission of the unacknowledged segments. The connection is released upon a complete video segment delivery, and therefore, the more packets are lost, the longer it will take the server to deliver the entire segment.

Figures 8(a) and (b) show the impact of the packet loss probability on the server CPU utilization for serving the videos for segment duration of 2 and 60 s. The general conclusion from the figures is that, in cases of better network conditions, the CPU is kept longer time busy serving the requests that arrive with appropriate timing, unlike the cases with harsh conditions where some of the requests are delayed, and so is their processing by the CPU.

Increasing the packet loss probability implicates more interruptions of the streaming sessions in general, however, the results obtained from our measurements show that certain segment lengths are more tolerant to packet loss than others. Table 1 shows an overview of the dependence of the service

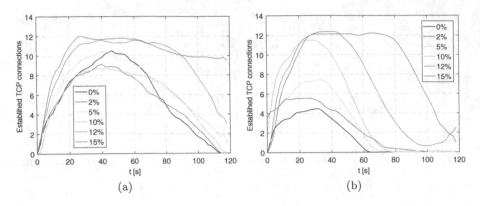

Fig. 7. Number of established TCP connection for different packet loss probabilities for segments with duration of (a) 2 and (b) 60 s

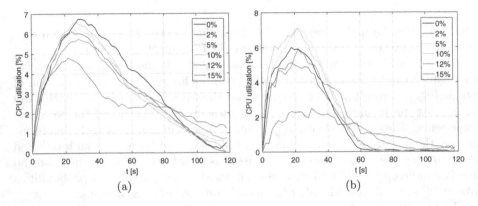

Fig. 8. Server CPU utilization for different packet loss probabilities for segments with duration of (a) 2 and (b) 60 s

continuity for different segment lengths on the packet loss probability. From the results we can conclude that short video segments are only suitable for delivery of videos in reliable networks. As the segment length increases, so does the resilience to the packet loss. According to Table 1, the optimal segment length for unreliable networks is 10 s. Longer video segments also show satisfactory results, but their main downside is that the clients my suffer long initial delays. From the results we conclude that the HAS streaming with segment length of 10 s can be used to provide uninterrupted streaming service in congested wireless networks.

Table 1. Overview of segment lengths eligibility for uninterrupted streaming for different packet loss probabilities

Seg. duration	Packet loss probability					
	0%	2%	5%	10%	12%	15%
2 s	yes	no	no	no	no	no
5 s	yes	yes	yes	yes	no	no
10 s	yes	yes	yes	yes	yes	no
30 s	yes	yes	yes	yes	no	no
60 s	yes	yes	yes	yes	no	no

4   Conclusion

In our work, we analyzed the influence of the video segment duration and the network conditions on the server performance for delivery of VoD contents using the HTTP adaptive streaming mechanism. For that purpose we created an environment for measurement of the server performance for serving clients in the campus network under different emulated network conditions.

From the analysis we concluded that hosting videos divided in shorter segments is more appropriate in better network conditions because they do not generate spikes in the required server traffic, i.e., the server traffic requirements are even and easily predictable. Their main disadvantage of short segments is that they require more CPU power and more resources for establishing TCP connections. Longer videos, on the other side, generate traffic patterns that can congest the network, but require less CPU power and fewer TCP connection. Their strong point is their resilience to unfavourable network conditions, which can be of great importance in congested wireless networks. The trade-off for providing predictable traffic patterns, fewer TCP connections, modest CPU utilization, and what is most important, uninterrupted watching experience is dividing the videos in segments with duration of 10 s. The results of our work can be a useful guideline for the VoD service providers in planing the server requirements for predefined demands and network conditions.

Acknowledgement. The author thanks the Faculty of Computer Science and Engineering at the Ss. Cyril and Methodius University in Skopje, under the EEAVS ("Energy Efficiency of Adaptive Video Streaming") project for financial support.

References

1. DASH Industry Forum (2016). http://www.dashif.org
2. MP4Box (2017). https://gpac.wp.imt.fr/mp4box/
3. SIGAR (2017). http://sigar.hyperic.com/
4. Biernacki, A., Tutschku, K.: Performance of HTTP video streaming under different network conditions. Multimed. Tools Appl. **72**(2), 1143–1166 (2014)

5. Garcia, S., Cabrera, J., Garcia, N.: Quality-control algorithm for adaptive streaming services over wireless channels. IEEE J. Sel. Topics Signal Process. **9**(1), 50–59 (2015)
6. Juluri, P., Tamarapalli, V., Medhi, D.: Measurement of quality of experience of video-on-demand services: A survey. IEEE Commun. Surv. Tutorials **18**(c), 682–686 (2015)
7. Jurgelionis, A., Laulajainen, J.P., Hirvonen, M., Wang, A.I.: An empirical study of NetEm network emulation functionalities. In: Proceedings - International Conference on Computer Communications and Networks, ICCCN (2011)
8. Kesavan, S., Jayakumar, J.: Dynamic adaptive streaming over HTTP (DASH)-comprehensive study and rate adaptation performance analysis. Int. J. Soft Comput. **10**(4), 261–273 (2015)
9. Li, Z., Zhu, X., Gahm, J., Pan, R., Hu, H., Begen, A.C., Oran, D.: Probe and adapt: Rate adaptation for HTTP video streaming at scale. IEEE J. Sel. Areas Commun. **32**(4), 719–733 (2014)
10. Vetro, A., Sodagar, I.: Industry and standards the MPEG-DASH standard for multimedia streaming over the internet. IEEE Multimed. **18**(4), 62–67 (2011)
11. Yitong, L., Yun, S., Yinian, M., Jing, L., Qi, L., Dacheng, Y.: A study on quality of experience for adaptive streaming service. In: 2013 IEEE International Conference on Communications Workshops (ICC), pp. 682–686 (2013)
12. Zhou, C., Lin, C.W., Guo, Z.: MDASH: A markov decision-based rate adaptation approach for dynamic HTTP streaming. IEEE Trans. Multimed. **18**(4), 738–751 (2016)

Universal Large Scale Sensor Network

Jakob Schaerer, Severin Zumbrunn, and Torsten Braun[✉]

Communication and Distributed Systems, Institute of Computer Science,
University of Bern, Neubrueckstrasse 10, 3012 Bern, Switzerland
{jakob.schaerer,severin.zumbrunn}@students.unibe.ch,
braun@inf.unibe.ch
http://cds.unibe.ch

Abstract. We developed a sensor node to support inter-node distances of up to 1 km, low-power consumption and low cost hardware. The design is modular and application specific sensors can be attached. It provides a flexible solution to a wide range of applications. An optimized MAC and routing layer (based on RPL and Trickle) supports interoperability among heterogeneous networks and low energy consumption over long distance transmissions of up to 1 km.

Keywords: Long-range wireless sensor networks · Internet of Things · Medium access control · Routing · Low power consumption

1 Introduction

Most Wireless Sensor Networks (WSNs) for the Internet of Things (IoT) use 2.4 GHz radio controllers today, which limits the distance between two nodes. WSNs that cover large areas [1] are mostly for predefined applications and consume significant energy. To fill the gap between those existing solutions, we created a solution called Universal Large Scale Sensor Network (ULSSN), which is able to transmit data over long distances, is energy efficient, has a modular design, and allows attaching different sensors. Its flexibility enables a variety of applications. The developed protocol stack is built of a PHY, MAC, routing, and application layer. This paper focuses on MAC and routing layer. In Sect. 2 we discuss the relevance of several existing approaches for the layers of our protocol stack. In Sect. 3 we briefly give an overview of our developed sensor node hardware. Sections 4 and 5.1 describe the implementations of our selected protocols and discuss the additionally included features. Finally, Sect. 6 presents some measurements of our implementation.

2 Related Work

In WSNs, the MAC layer needs to focus on low power consumption and collision avoidance. Several protocols use radio duty cycling (RDC) to reduce on-time of radio transceivers. WiseMAC [2] and B-MAC [3] use long preambles to announce

© Springer International Publishing AG 2017
D. Trajanov and V. Bakeva (Eds.): ICT Innovations 2017, CCIS 778, pp. 79–88, 2017.
DOI: 10.1007/978-3-319-67597-8_8

the packet. ContikiMAC [4] and X-MAC [5] send packets as announcements. Carrier Sense Multiple Access/Collision Avoidance (CSMA/CA) is the usual mechanism for collision avoidance in wireless networks. Collision avoidance is done by probing the channel before a transmission. If the probe detects a free channel, the packet will be transmitted. Otherwise the node waits for some back-off-time and repeats this step.

To limit overhead for storing routing information, many WSN protocols simply create a tree, where every node knows its parent node and passes all sensor data to it. One of the most important tree-based protocols is the "Routing Protocol for Low-Power and Lossy Networks" (RPL) [6]. RPL operates with constraints on energy and memory. RPL has been implemented in operating systems like Contiki or TinyOS [7]. Every RPL node may work as both router and endpoint. RPL has four modes of operation (MOP). MOP 0 does not maintain downward routes. Nodes only store information about the single parent node. MOP 1 (non-storing mode) features downward routes such that every node sends it's children addresses to the parent. In contrast to MOP 0 and 1, MOP 2 and 3 feature point-to-point (p2p) connections. MOP 0 uses Data Solicitation Information (DIS) and Data Information Object (DIO) packets for route creation and maintenance. The DIO packet is sent by each node in some interval t and holds data about the node position in the tree graph, also called Destination Oriented Directed Acyclic Graph (DODAG). An Objective Function (OF) then determines which node should be selected as a parent based on the lowest transmission costs.

Trickle [8] is used whenever an application requires to send as little information as possible. This is done by minimizing redundancy, using exponentially increased intervals and fast reactions to invalid information. Trickle fits perfectly to the DIO transmission in RPL, because it allows rapid adaptation of routing paths and saves energy by minimizing radio transmissions.

3   Sensor Hardware

We combined hardware and software development to have a perfectly adapted system. We developed a printed circuit board with a TICC1110 (low-power,

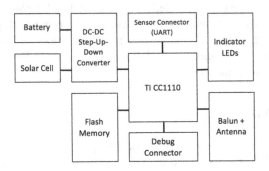

Fig. 1. Hardware overview

868 MHz, 10 dBm, system-on-chip) from Texas Instruments (TI), a debug connector, a connector for various types of sensors, which are connected via Universal Asynchronous Receive Transmit (UART), a 64 MB Flash memory, two indicator LEDs and a dc-dc step-up-down converter circuit (Fig. 1). The hardware was optimized for low power consumption based on TI design guidelines [9].

4 MAC Layer

4.1 Datarate-Range-Power Model

The signal range is determined by the output power of the device, the device attenuation, the antenna peak gain, the free space path loss, and the receiver sensitivity, which is higher at low data rates [9,10]. Therefore, a higher distance can be reached with a lower data rate. We have developed a model (Fig. 2) assuming free space propagation to determine the achievable range with a certain data rate and a specific output power level. The attenuation chain consists of twice the device attenuation V_{DEV} $[dB]$ (measured), twice the antenna gain V_{ANT} $[dB]$ (specification) and the free space path loss (FSPL, Eq. 1). r $[m]$ is the distance, f $[Hz]$ is the frequency and c $[\frac{m}{s}]$ is the speed of light. P_T $[dBm]$ is the output power set on the transmitting device. The receiver sensitivity P_{Rmin} $[dBm]$ is given by the radio module (see Sect. 6.6).

$$FSPL = 10 \log_{10} \left(\frac{4\pi r f}{c} \right)^2 \quad [dB] \quad (1)$$

Inequality 2 must hold to enable the receiver to demodulate the signal.

$$P_T - 2 \cdot V_{DEV} + 2 \cdot V_{ANT} - P_{Rmin} > FSLP \quad (2)$$

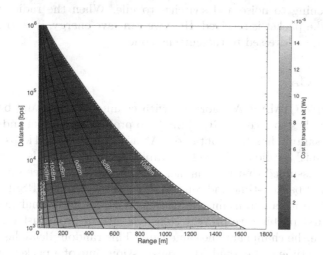

Fig. 2. Datarate-Range-Power model (Color figure online)

The maximum distance for given output power and data rate is bound by:

$$\log_{10}(r) < \frac{1}{20}(P_T - 2 \cdot V_{DEV} + 2 \cdot V_{ANT} - P_{Rmin})\frac{c}{4\pi f} \qquad (3)$$

Figure 2 shows the energy cost to transmit a single bit. Costs depend on data rate and current consumption (see Sect. 6.5) to transmit at a certain power level. Based on this model we decided to use a data rate of 2.4 kbps (red line). With this data rate we predict that the range between two nodes can reach up to 1.28 km in free space. This will be limited by trees, houses and other obstacles that hamper signal propagation.

4.2   Low Power Listening

On low data rates, preamble oriented protocols have the advantage that they can send a flexible amount of bits and, therefore, have a flexible interval t_i. The packet oriented announcement strategies lack this flexibility as they need to send full packets. Thus, we implemented radio duty cycling with a 24 bytes (80 ms for 2.4 kbps) long preamble to announce packets. According to [9] the receiver has to probe the channel for at least 757 μs resulting in a radio duty cycling of 1%. In our implementation the device probes the channel when it wakes up. This probing is done by listening to the channel and comparing the first valid Received Signal Strength Indication (RSSI) with a certain threshold. When the RSSI is below this threshold the channel is assumed to be inactive and the radio is turned off. When the channel power level is above the threshold the device continues to listen. We can use the preamble to qualify if the activity on the channel belongs to our protocol. This is done by the Preamble Quality Identifier (PQI) counter. It is incremented when two consecutive bits were not equal and decremented by 8 otherwise. If there is $PQI \leq 0$ after 8 bits, the node assumes that it is listening to noise and switches to idle. When the radio stays active the frame of the packet is received. Here we can save energy, if we only listen to packets that are addressed to the current node.

4.3   CSMA/CA

We use a simple CSMA/CA algorithm with channel probing and backoff-time. The radio switches to receive (RCV) mode to probe the channel and waits until a valid RSSI sample has been obtained. At a data rate of 2.4 kbps, this probe needs 757 μs until a valid RSSI level is established. If the RSSI is above a certain threshold it is assumed that the channel is occupied. The device waits for 80 ms (preamble time) backoff-time and repeats the process until the RSSI is below the threshold before it can transmit the packet. With an additional random delay prior to the first probing we prevent collisions after a broadcast request where all nodes probe the channel at the same time. This random delay lasts up to 5% of the preamble time and extends the transmission time of a packet by up to 1%.

4.4 Power Adaptation

The output power of the transmitter defines how far the signal will reach. The output power should be set to a level such that the signal can still be demodulated on the receiving node. The optimal output power for every link is different. The implementation of the power adaptation affects all non-broadcast and non-acknowledgment packets. The band between $-90\,$dBm and $-60\,$dBm defines the desired power level on the receiving node. When a received packet is not within this band a request to adapt the power is sent to the source node of the packet. Initially, all packets are sent using a well-defined standard power level. Each node maintains a list of its neighbor nodes and the according power levels. If a node receives an adaptation request, it adapts the power level of the requesting node in its own neighbor list.

5 Routing Layer

5.1 RPL

We use RPL as the basis of our lightweight routing protocol. Common RPL implementations use more than $50\,$kB memory [11]. Thus, we had to focus on program size since our micro-controller features only $32\,$kB of flash memory. Memory usage can be reduced by removing unnecessary functions and information in packets. For our application we require only a limited set of functionality. We decreased the DIO and DIS packet size by 84%, which results in total energy savings of $1.8\,$mJ per packet. Figure 3 depicts the modified RPL packet format as needed for our application-tailored RPL version. We implemented the minimum functionality of RPL by choosing RPL Mode 0, since we do not require peer-to-peer (p2p) connections. However, we need to send downward configuration messages from the base station. This is done by additional functionality described in Sect. 5.3. In our case, the parent selection is done by comparing the rank in the tree of different nodes and does not include any other metric to keep simplicity. We limit the number of different DODAGs to 1 and, thus, can completely remove the functionality for distinguishing DODAGs. This has a direct impact on the parent selection method, because a node that currently does not know about the DODAG will simply send a DIO message of maximum (infinite) rank. Another way of saving energy is extending the transmission interval for routing messages. However, when lowering the timeliness of routes, erroneous

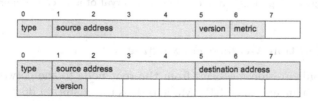

Fig. 3. Modified DIO and DIS packet definitions

transmissions can increase due to possible obsolete routes in nodes. Therefore, only extending intervals is not the appropriate solution but together with a mechanism that also addresses the issue of obsolete information it is possible to save a lot of energy. This is done with the Trickle Algorithm [8].

5.2   Trickle Algorithm

The Trickle Algorithm is a rather simple process:

1. Value I is randomly selected from an interval $[I_{min}, I_{max}]$, $I_{min} \in \mathbb{N}$ and $I_{max} \in \{I_{min} * 2^n, n \in \mathbb{N}^+\}$. The consistency counter c and the time counter a are set to zero and a redundancy constant k is defined. The value k describes the number of redundant routing packets tolerated in our network.
2. When an interval begins, c and a are set to zero, a transmission variable t is chosen randomly from our interval $[I/2, I)$ and a then starts counting up.
3. For each redundant packet (a packet that does not change the configuration) we increment c.
4. At time $a = t$, if $c < k$ we initiate a DIO transmission. Otherwise, the transmission is suppressed due to too much redundancy in our network.
5. When $t > I$, we double the interval $[0, I]$ to $[0, 2 * I]$ until $I > I_{max}$ then, we simply set the interval to $[0, I_{max}]$.
6. At any time, if we receive an inconsistent packet, we reset the Trickle Timer by setting the interval size back to $[0, I_{min}]$ and continue with step 2.

With these steps we can minimize the exchanged information. Let us assume a network of four nodes. All are interconnected and one is our base station, which has rank zero whereas all others have rank 1. Let $k = 1$. When the base station initiates a DIO message, all other nodes will receive it and increment c by 1. Next, every node will double it's interval and only one node (the one with the lowest value of t), will initiate another DIO message etc. (Fig. 4). In a perfect environment, we end up in sending only one routing message per hop.

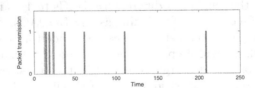

Fig. 4. Exponentially incremented interval of a Trickle Timer

5.3   Configuration Message Forwarding

To support configuration messages from the root node to the network, we need additional mechanisms, since RPL Mode 0 does not maintain downward routes. Since every configuration packet is sent as a broadcast message, it will be resent multiple times by every node. This causes a lot of overhead and will finally result in deadlock as the transmission buffers of the routing protocol overflow with copies of the same packet. To distinguish between upward and downward traffic we have added an ID to each configuration packet to identify the source of the packet. With that measure we are reducing the transmission of redundant information because every time a packet is received from a lower rank the packet should be forwarded downwards as it originated from root. Nodes on the same level will not respond to packets from each other as those are lossy broadcast packets. This mechanism not only provides a simple way for sending messages downwards the tree, but it also saves a lot of energy due to it's low overhead and fast algorithm for checking the timeliness of a packet.

6   Measurements

6.1   Low Power Listening

To measure the current drawn by our device we used a micro benchmark [4]. We measured the voltage over a 10 ohm shunt resistor with a Teledyne LeCroy HDO6104 oscilloscope to determine the current consumption of the node. Figure 5(a) shows the current consumption of a listening probe. In this measurement three parts can be identified. First, the controller starts up from sleep mode. In this part timers are updated and direct memory access channels are initialized. This wake-up of the operating system needs $\sim$590 μs. After wake-up, the radio device is calibrated. This is done at every 4th start of the radio and needs $\sim$754 μs. In the last step the radio is listening to the channel for $\sim$ 930 μs. Thus, the average probing time is 590 μs + 754/4 μs + 930 μs = 1.708 ms. Using this measurement we could determine that waking up from sleep and probing the channel needs 10.5 mA on average and lasts for 1.7 ms. The current consumption in sleep mode is 198 μA. With a listening interval t_i = 80 ms the resulting average current consumption is 420 μA as shown in Eq. 4.

$$\frac{1.7\,\text{ms} \cdot 10.5\,\text{mA} + (80\,\text{ms} - 1.7\,\text{ms}) \cdot 198\,\mu\text{A}}{80\,\text{ms}} = 420\,\mu\text{A} \tag{4}$$

6.2   Range Verification

To verify that we can reach a distance longer than 1 km between two nodes we sent packets over this distance. Therefore, we have chosen an area with only a few obstacles and the possibility of a line of sight (LOS) transmission over 1.03 km. We could transmit packets with a packet error rate of $\leq$4%.

6.3   Power Adaptation

To test the power adaptation algorithm we created a test setup consisting of three nodes. Nodes A and B are transmitting data and node C is listening to the channel and logging the power level, c.f. Fig. 5(b). To test the power adaptation, node A sends a packet to node B. If the received power level is above the desired power band, node B sends a request to A to reduce its output power. Node A adapts the output power for node B, but the request packet was above the desired power band too. Then, A requests B to lower its output power. This leads to a negotiation until both levels are within the power band.

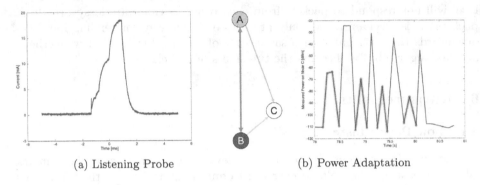

(a) Listening Probe   (b) Power Adaptation

Fig. 5. Listening probe and power adaptation

6.4   Response Times

We implemented a ping application [12] on the application layer to measure response times. Since we did not want to query single nodes, we only implemented broadcast support with a time-to-live (ttl) field. We can detect the nodes on different levels of the tree, e.g., with a ttl value of 2 we would only receive responses from nodes on rank 1 or 2, but no messages from higher ranks. Notation $1 : n$ means that 1 base station is connected to n nodes. The base station initiates the ping. From the measured average response times (Table 1) we see that the response time increases by 583 ms in average for each additional node. This is because by adding a node $(n + 1)$, the last packet is sent after a total time of $n * t_{window} + t_{window}$, where n is the number of nodes and t_{window} is the time slot for one ping packet, defined by parameters of the MAC layer.

Table 1. Average response times for a fully meshed network with 2, 3, 4 and 5 nodes

1:1	1065.9 ms
1:2	1701.1 ms
1:3	2470.2 ms
1:4	2752.5 ms

6.5 Transmission Costs

The energy to transmit a bit is calculated using the time to send the bit ($420\,\mu s$ at 2.4 kbps), the voltage (3 V) and the current consumption of a node. We used the same setup as in Sect. 6.1 to measure the current at different output power levels, c.f., Fig. 6(a).

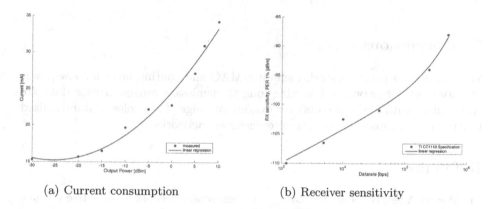

(a) Current consumption   (b) Receiver sensitivity

Fig. 6. Current consumption and receiver sensitivity

$$h_\theta = \theta_0 + \theta_1 x + \theta_2 x^2 \tag{5}$$

We used the hypothesis function (5) to find an approximation to these points. For the hypothesis h_θ function we found the parameters $\theta_0 = 2.45 \cdot 10^{-2}$, $\theta_1 = 0.73 \cdot 10^{-3}$ and $\theta_2 = 1.4 \cdot 10^{-5}$. The hypothesis function then h_θ has a mean squared error as shown in Eq. (6), where m is the number of samples, $x^{(i)}$ is the output power of sample i and $y^{(i)}$ is the current consumption of sample i.

$$J(\theta) = \frac{1}{m} \sum_{i=1}^{m} (h_\theta(x^{(i)}) - y^{(i)})^2 = 0.95 \tag{6}$$

Let $I(x) = h_\theta$ for the value of θ found above, then $I(x)$ in Ampere is the current consumption as a function of the output power. With function (7), where x is the output power in [dBm], s the data rate in [bps] and u the supply voltage in [V], we can predict the cost to transmit a single bit.

$$f(x, s, u) = I(x)\frac{u}{s} \tag{7}$$

6.6 Receiver Sensitivity

To estimate a continuous receiver sensitivity function we use linear regression (Eq. 8) on the receiver sensitivity specification of TI CC1110 [9,10] (Fig. 6(b)).

$$h'_\theta(x) = \theta_0 + \theta_1 x_1 + \theta_2 log(x) \tag{8}$$

For the hypothesis h'_θ function we found the parameters $\theta_0 = -1.267 \cdot 10^2$, $\theta_1 = 1.303 \cdot 10^{-5}$ and $\theta_2 = 2.423$. Then, the hypothesis function h'_θ has a mean squared error as shown in Eq. (9), where m is the number of samples, $x^{(i)}$ is the data rate of sample i and $y^{(i)}$ is the sensitivity of sample i.

$$J(\theta) = \frac{1}{m} \sum_{i=1}^{m} (h'_\theta(x^{(i)}) - y^{(i)})^2 = 0.7 \tag{9}$$

7   Conclusions

We described a protocol stack including MAC and routing layer for low-power and lossy sensor networks featuring long transmission ranges. Sensor data are transmitted with minimal costs over maximum range. The choice of standardized routing layer protocols supports heterogeneous networks.

References

1. Werner-Allen, G., et al.: Deploying a wireless sensor network on an active volcano. IEEE Internet Comput. **10**, 18–25 (2006)
2. El-Hoiydi, A., Decotignie, J.D.: WiseMAC: an ultra low power MAC protocol for the downlink of infrastructure wireless sensor networks. In: ISCC 2004 (2004)
3. Polastre, J., Hill, J., Culler, D.: Versatile low power media access for wireless sensor networks. In: ACM SenSys 2004, pp. 95–107 (2004)
4. Dunkels, A.: The ContikiMAC Radio Duty Cycling Protocol (2011)
5. Buettner, M., Yee, G.V., Anderson, E., Han, R.: X-MAC: a short preamble MAC protocol for duty-cycled wireless sensor networks. In: ACM SenSys 2006, pp. 307–320 (2006)
6. Winter, T., et al.: RPL: IPv6 routing power for low-power and lossy networks. RFC 6550, March 2012
7. Ko, J., Eriksson, J., et al.: ContikiRPL and TinyRPL: happy together. In: Extending the Internet to Low power and Lossy Networks, IP+SN 2011, April 2011
8. Levis, P., et al.: The trickle algorithm. RFC 6206, March 2011
9. Texas Instruments "CC1110Fx/CC1111Fx": Low-Power SoC (System-on-Chip) with MCU, Memory, Sub-1 GHz RF Transceiver, and USB Controller, January 2006
10. Hellan, S.: CC11xx Sensitivity versus Frequency Offset and Crystal Accuracy, August 2009
11. Clausen, T., et al.: A critical evaluation of the IPv6 routing protocol for low power and lossy networks (RPL). In: IEEE WiMob 2011, October 2011
12. Braden, R.: Requirements for internet hosts - communication layers. RFC 1122, October 1989

FPGA Implementation of a Dense Optical Flow Algorithm Using Altera OpenCL SDK

Umut Ulutas[1]($\boxtimes$) (iD), Mustafa Tosun[1], Vecdi Emre Levent[1],
Duygu Büyükaydın[2], and Toygar Akgün[2], and H. Fatih Ugurdag[1]

[1] Ozyegin University, Istanbul, Turkey
{umut.ulutas,mustafa.tosun.5245,emre.levent}@ozu.edu.tr,
fatih.ugurdag@ozyegin.edu.tr
[2] ASELSAN, UGES, Ankara, Turkey
{dcelebi,takgun}@aselsan.com.tr

Abstract. FPGA acceleration of compute-intensive algorithms is usually not regarded feasible because of the long Verilog or VHDL RTL design efforts they require. Data-parallel algorithms have an alternative platform for acceleration, namely, GPU. Two languages are widely used for GPU programming, CUDA and OpenCL. OpenCL is the choice of many coders due to its portability to most multi-core CPUs and most GPUs. OpenCL SDK for FPGAs and High-Level Synthesis (HLS) in general make FPGA acceleration truly feasible. In data-parallel applications, OpenCL based synthesis is preferred over traditional HLS as it can be seamlessly targeted to both GPUs and FPGAs. This paper shares our experiences in targeting a demanding optical flow algorithm to a high-end FPGA as well as a high-end GPU using OpenCL. We offer throughput and power consumption results on both platforms.

Keywords: Altera SDK for OpenCL · FPGA · High-Level Synthesis · Dense optical flow

1 Introduction

This work is unique in the sense that it benchmarks a dense optical flow algorithm on both FPGA and GPU using OpenCL language and related tools. This article may be useful to people interested in accelerating optical flow in particular or accelerating compute-intensive data-parallel algorithms using OpenCL on FPGA and/or GPU. Numerous works have been published in the literature that deal with acceleration of optical flow on FPGA versus GPU (e.g., [1,2]) and that deal with OpenCL based acceleration on FPGA versus GPU (e.g., [3,4]). However, this is the only work so far that covers acceleration of optical flow on FPGA versus GPU using OpenCL.

Optical flow is a motion analysis algorithm that takes two 2D images and finds the motion vectors of each pixel for the first image according to their movement by comparing it to the second image. Optical flow has been widely

© Springer International Publishing AG 2017
D. Trajanov and V. Bakeva (Eds.): ICT Innovations 2017, CCIS 778, pp. 89–101, 2017.
DOI: 10.1007/978-3-319-67597-8_9

studied in the literature. Two methods that are widely used for optical flow are Lucas-Kanade method [5] and Horn-Schunk method [6], which also happen to be the works that popularized the subject in the literature.

Dense optical flow is a type of optical flow that is designed to process all pixels in an image to compute the motion vectors, unlike sparse optical flow, which looks for interesting regions to estimate the motion. Since dense optical flow works on every pixel and gives a motion vector for each of them, it requires a higher amount of memory and computational power. It is suitable to compute dense optical flow in platforms that allow parallel computation in higher degrees such as GPUs [7–9] and FPGAs [10,11].

Open Computing Language (OpenCL) is a language that is mainly used for multi-core programs. OpenCL can be used on many different platforms such as FPGA, CPU, GPU, and DSPs. This opens up many possibilities. It is possible to run the same program on different platforms, greatly increasing its availability and portability. Also, it makes it easier to carry out performance tests and benchmarking across platforms for a given project. On FPGAs, OpenCL code can synthesized and converted to HDL. That means the user can work on a reconfigurable platform using C-like languages without directly writing RTL code. Since it is more desirable, FPGA companies developed toolchains/SDK based on OpenCL and research was conducted in the literature on optimizing OpenCL for FPGAs [10,11].

The paper proceeds by first presenting the particular optical flow algorithm we have implemented. In Sect. 3, we present some implementation details. Section 4 contains the results of our FPGA/GPU comparisons, and Sect. 5 concludes.

2   Dense Optical Flow Algorithm

The optical flow estimation algorithm of "Anisotropic Huber-L^1" has been selected as it ensures good quality and is easily parallelizable. According to Middlebury optical flow benchmark results [12], it has an average rating of 62.6. It is also one of the algorithms that can provide very fast results, making it very suitable for video processing.

The starting point of calculations is to find a solution to the optical flow problem based on L^1 data term and isotropic TV regularization, which is disparity preserving and spatially continuous formulation. For two input images I_0 and I_1 defined on a regular rectangular domain, $\Omega \in$ the function can be formulized as Eq. 1:

$$min\left\{\int_\Omega \sum_{d=1}^2 |\nabla u_d| + \lambda |I_1(\boldsymbol{x} + \boldsymbol{u}(\boldsymbol{x})) - I_0(\boldsymbol{x})|d\boldsymbol{x}\right\} \tag{1}$$

where $\boldsymbol{x} = (x_1, x_2)$ is the vector that contains coordinates of corresponding pixels in the image and $\boldsymbol{u}(\boldsymbol{x}) = (u_1(\boldsymbol{x}), u_2(\boldsymbol{x}))$ is the flow vectors that include displacement values. Free parameter λ is used to balance the relative weight of the data and regularization term.

L^1 data term is the distance between motion warped I_1 image by using flow vectors $I_1(x + u(x))$ and I_0 image. Also, regularization term is the sum of absolute gradients of 2D flow/displacement vectors. To further simplify the minimization calculation, new auxiliary variables are introduced. Vector v and connecting term to guarantee that v is in near proximity of v. Then, anisotropic Huber regularization is applied to the formula and some other mathematical operations to simplify it further, which is explained in detail in the anisotropic Huber-L^1 optical flow paper. This results in:

$$\min_{u,v} \sup_{|p_d| \leq 1} \left\{ \int_\Omega \sum_{d=1}^{2} [(D^{\frac{1}{2}} \nabla u_d) \cdot p_d - \epsilon \frac{|p_d|^2}{2} + \frac{1}{2\theta}(u_d - v_d)^2] + \lambda|\rho(v(x))|dx \right\} \tag{2}$$

Anisotropic Huber in this Eq. (2) is the regularization term that points to the final shape, because it is image-driven and discontinuity preserving, also uses Huber cost. Huber cost for small differences is quadratic, while for large differences it is linear. A new equation is produced by solving this optimization problem iteratively using alternating minimization procedure. For fixed v solution yields to:

$$u_d^{n+1} = v_d^n + \theta div(D^{\frac{1}{2}} p_d^{-n+1}) \tag{3}$$

$$p_d^{-n+1} = \frac{p_d^{-n} + \tau(D^{\frac{1}{2}} \nabla u_d^{-n+1} - \epsilon p_d^{-n})}{max\{1, |p_d^{-n} + \tau(D^{\frac{1}{2}} \nabla u_d^{n+1} - \epsilon p_d^{-n})|\}} \tag{4}$$

Then, the solution for u results:

$$\min_{v} \left\{ \int_\Omega \frac{1}{2\theta} \sum_{d=1}^{2}(u_d - v_d)^2 + \lambda|\rho(v(x))|dx \right\} \tag{5}$$

A final thresholding part is applied with three cases results in a direct solution:

$$v^{-n+1} = u^{-n+1} + \begin{cases} \lambda\theta\nabla I_1, if \ \rho(u^{-n+1}) < -\lambda\theta|\nabla I_1|^2 \\ -\lambda\theta\nabla I_1, if \ \rho(u^{-n+1}) > -\lambda\theta|\nabla I_1|^2 \\ -\rho(u^{-n+1})\frac{\nabla I_1}{|\nabla I_1|^2}, else \end{cases} \tag{6}$$

Equations 3, 4, and 6 are the formulae to be used to calculate the flow, by replacing the corresponding operations with discrete values that are calculated on 2D image grids.

3   Implementation Details

We have used Altera OpenCL SDK to implement this application, which is an OpenCL synthesis tool for FPGAs. Altera OpenCL SDK's programming model includes three parts: host application, OpenCL kernels, and a custom platform that provides the board design as shown in Fig. 1. The language is C/C++ for the host application and a specific C-like language for the kernels. They are compiled by Altera Offline Compiler (AOC). Altera OpenCL puts a hold to compilation

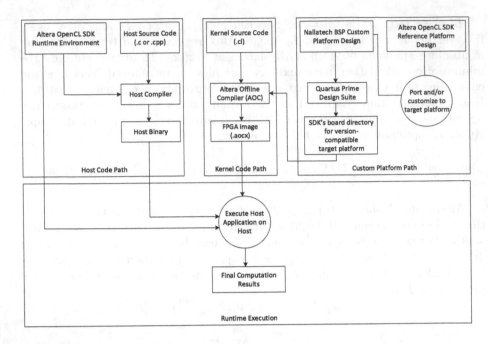

Fig. 1. Schematic diagram of the Altera OpenCL SDK programming model

until the last phase of the design, decreasing the productivity bottlenecks of VHDL tools. The design can easily be ported to other FPGAs that supports Altera OpenCL. The compiler also optimizes the design flow itself. Ability to work through a C-like language for the hardware makes it easier to design specialized hardware architectures without relying on managing complicated RTL codes [13].

3.1   Overview

OpenCL kernels contain the core algorithm. Then, AOC converts those kernel codes to hardware design. Kernels are used in the host application. The OpenCL host application uses standard OpenCL runtime APIs to manage device configuration, data buffers, kernel launches, and synchronization. Host application also contains functions such as file I/O, or portions of the source code that do not run on an accelerator device [14]. The host application is linked with the board-specific Memory-Mapped Design (MMD). This MMD is the communication layer between the upper Altera OpenCL SDK software and accelerator board. The MMD provided by Nallatech OpenCL BSP is used as custom platform in this application [15], and Nallatech 385A Arria10 FPGA Accelerator Card is used as the FPGA board in this work. Nallatech 385A Arria10 FPGA has 1506 system logic elements, ensuing a large space for calculations. Its peak floating point performance is 1366 GFLOPS [16].

AOC also provides an emulation option to debug and test the designs beforehand. It takes the information of the targeted FPGA and creates an emulation as if the design is working on the targeted FPGA. In VHDL, debugging is done via verifying specific bits in simulations, however, AOC provides GNU debugger support in a software environment. While emulating the hardware design on the host computer, the designer is able to debug the prototype by using regular debugging methods with GNU debugger, such as creating software breakpoints and tracing variables. Another perk of this emulation is that, on traditional VHDL systems, the design and test of host code begins after kernels are targeted on to FPGA. This creates two different design and testing processes to be handled. While, in AOC emulation, both kernels and the C/C++ host code that operates/uses kernels are designed and tested together, making it possible to analyze and set the timing and power during the design phase of kernels.

The technical details of the FPGA implementation are shown in Tables 1 and 2. In Table 1, the results show when the design is compiled normally, while in Table 2, the results show when the design is compiled using another method for optimization. In here, the optimization method used is "fp-relaxed" (float point relaxed), a method which further optimizes the floating point declarations in kernels, resulting in slightly fewer total RAM and DSP block usage in the FPGA.

Table 1. Resource utilization of Arria10 FPGA without any optimization directive

Logic utilization (in ALMs)	213,940/427,200 (50%)
Total registers	446,217/1,708,800 (26%)
Total RAM blocks	1,947/2,713 (72%)
Total DSP blocks	763/1,518 (50%)
Total PLLs	60/112 (54%)

Table 2. Resource utilization of Arria10 FPGA with fp-relaxed optimization directive

Logic utilization (in ALMs)	212,076/427,200 (50%)
Total registers	446,217/1,708,800 (26%)
Total RAM blocks	1,939/2,713 (71%)
Total DSP blocks	714/1,518 (47%)
Total PLLs	60/112 (54%)

A total of 10 OpenCL kernels are used in the OpenCL application, 3 of which can be categorized as main kernels and 7 as auxiliary kernels. In the host code, to find the optical flow of the video, the video is first read using OpenCV. Two frames are read in succession and the optical flow is calculated using OpenCL on these frames. Once the optical flow is calculated, the values are stored in a file. In addition, these values are displayed on screen as a new video where the optical flow can be visually seen on screen as it is calculated. Those are the kernels that are used in the process are those, first 3 of which (Warping, Compute-P and Loop Flow) are the main kernels:

- Warping
- Compute-p
- Loop Flow
- Gaussian (3 variations)
- Upscale
- Copy (2 variations)
- Zerosize

The calculations are done on 3 Gaussian pyramid levels for each frame as shown in Fig. 2. Warping is called 10 times for each level, compute-P and loop flow are called 50 times for each warp. Auxiliary kernels are also used many number of times each frame when necessary. To explain the application, it is better to go over kernels one by one, explaining each of them while also explaining the overall process.

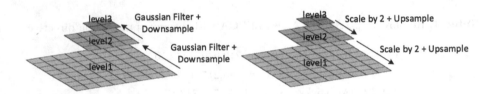

Fig. 2. Image and flow vector pyramids

3.2   Warping

The warping part is the part in which we update the 2D image based on the motion vectors calculated so far. Here, the kernel takes two images and their motion vectors (u and v vectors) as input. The image is updated using the motion vectors u and v. That is, each pixel in the frame is matched with the corresponding motion values from the vectors u and v, and new values are calculated using those values. The warping kernel also calculates difference between the second frame and the warping applied first frame and creates a vector of it.

In the warping process, it is not possible to directly move to pixel to its motion warped location and read the values from there. Because the values in motion vectors are usually floating point numbers rather than integers. Instead, each pixel in the first frame is moved according to the vector u and v, and the value is taken from the values of pixels in a 2×2 window at the target point. The values are calculated using bilinear interpolation from the 2×2 pixels in the window. Warping kernel calculates this interpolation manually.

As input, kernel takes two frames, and u and v motion vectors. Picture type is set to unsigned char*, and the type of the vector is set to float*. The kernel also takes width, height, and stride parameters to do corner detection, which their type is set to int. The kernel outputs x, y, t, and gradSqr vectors, whose types are float* similar to other vectors.

When using OpenCL kernels, global and local workgroup sizes are specified. Local workgroups are small units within the global workgroup, working in parallel within themselves. The size of the global workgroup has been chosen to be the same as the height and width in the warping kernel, and the size of the local workgroup is 64×4. Hence, OpenCL processes 64×4 pixels at the same time in a frame. The buffer sizes of all other parameters (vectors and squares) that the kernel receives as input, except for the width, height, and step parameters, are set to be the same as the global workgroup size.

In the process; the pixel is mapped according to the values in the u and v motion vectors and their values are assigned to a newly created fx and fy floating points. These values are increased by one to produce fx1 and fy1 values. Those values are converted to integer and four points are created using those values (fx, fy), (fx, fy1), (fx1, fy), (fx1, fy1). The values at these four points enter the linear interpolation and the value of the current pixel's warped value at global index is calculated. This warping process is executed 10 times each pyramid level is shown in Fig. 3.

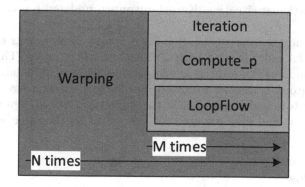

Fig. 3. Algorithm design flow

3.3 Compute-p

P vectors are calculated using the motion vectors u and v, and the vector w. Later, p vectors are also used in the Loop Flow kernel to calculate these u, v, and w vectors. Compute-p kernel calculates and updates the p vectors. Then the Loop Flow kernel comes in and uses the previously computed p vectors and updates the vectors u, v, and w. These two kernels do calculations one after another, passing their output data/vectors to each other in 50 iterations is shown in Fig. 3. It is aimed to produce the most accurate result in this way.

One p vector holds the x and y change values separately for the vectors u, v, and w. Therefore, 6 p vectors are defined. 3 of those vectors are for horizontal displacements while the other 3 are vertical. The calculation process of p vectors can be explained by showing the calculation of one of the six. For point (i, j), the vertical p vector that uses the derivative of u motion vector is calculated as follows:

$$p[i][j] = p[i][j] + (u[i][j + 1] - u[i][j]) * f1$$

After calculating all p vectors, the kernel normalizes the values between 1 and 0. While computing the p vectors, threads need to read the same pixel values when processing adjacent pixels, the kernel can take advantage of the local memory buffer. For this reason, this kernel is prepared using local memory cache. For vector type, float4 data type was used in this kernel. It is a data type in OpenCL that provides 4 floating point numbers.

The kernel takes 6 p vectors, u and v motion vectors and w auxiliary vector as input. All of these are specified as float4, as indicated. It also takes the auxiliary parameters height, width, and stride values as input. These auxiliary parameters are of type int. The stride parameter is used to calculate the coordinate values in the vector using the global indexes. The height and width parameters are also used in corner detection, that is, to determine whether the algorithm's pixel window is on the corner of the frame, where different treatments are applied to solve this particular situation. Kernel outputs 6 updated p vectors with new calculations.

For this kernel, the size of the global workgroup is chosen as the height of the image, and its width is one-fourth of the width of the image. This is because float4 data type is used, so there are 4 calculations at each point. The size of the local workgroup is also set to 32 × 4. Similarly, the buffer sizes of all other parameters (vectors and squares) that the kernel receives as input, except for the width, height, and step parameters, are set to be the same as the global workgroup size.

3.4   Loop Flow

Loop Flow kernel runs in the same iteration with Compute-p kernel, and they feed each other. As previously said, while Compute-p kernel calculates p vectors using u, v, and w vectors, Loop Flow kernel calculates u, v, and w vectors using p vectors. Also, in Loop Flow kernel x, y, and t vectors, which are the output vectors of Warping kernel, are used in the calculations of u, v, and w vectors. After calculating u, v, and w vectors, the kernel applies thresholding.

Inputs and outputs of the kernel are similar to Compute-P kernel with some reversed roles and extra inputs. The kernel receives 6 p vectors, u and v motion vectors and w auxiliary vector as input. All of these are specified as float4. Auxiliary parameters width and stride are also used as input. As stated, it takes vectors x, y, and t computed in warping kernel as input and uses them to update u, v, and w vectors. Their type is also specified as float4. The kernel also outputs updated u, v, and w vectors.

3.5 Auxiliary Kernels

Copy. It takes a data source as an input, copies this data source, and outputs it. The types are specified as float4. There are two variations of this kernel, one of them produces one copy and other produces two copies of the data source.

Zerosize. This kernel takes a data source in the form of float4 as input, then makes all the bits of the source equal to zero, then outputs it.

Upscale. It takes a data source in float4 type and a scale value in integer type as input. First, it upsamples the data source by 2, which fills the empty gaps with duplicating pixels, then scales the data source by the scale value. It then outputs the result.

Gaussian. This kernel takes two 2D image vectors and one filter as input. It applies the filter to the first image source and writes the filtered results to the second image. It then gives the second image as output.

4 Results

The specifications of the environment used for the tests are:

- **CPU:** Intel Xeon 1650 v3 3.5 GHz
- **GPU:** Nvidia GeForce GTX 980 Ti with 6 GB GDDR5 Memory, 336.5 GB/sec memory bandwidth and 1000 MHz clock speed
- **FPGA:** Nallatech 385A Arria10 FPGA Accelerator Card with 8 GB DDR3 on-card memory, up to 1.5 TFlops and 2133 MHz clock speed

All results were obtained on the same host computer. During measurements, execution is the only system function that spends serious amount of system resources. Measurements were done in three categories depending on the device, where the main calculations are made: CPU, GPU, and FPGA. GPU and FPGA ran the same OpenCL kernels with some modification in the host code to make it suitable for the desired device. The OpenCL application is implemented using double precision floating point numbers. CPU measurements were done in MAT-LAB, and it is also double precision because of MATLAB being double precision inherently.

In all the tests, respectively 50 and 2 iterations per warp were carried out; those are the iterations consisting of computing p vectors and calculating loop flow as explained above. Here are the values of the constants that are used in all three tests: $\lambda = 40$, $\beta = 0.01$, $\tau = 1/\sqrt{8}$ and $\epsilon = 0$. All measurements are done on the same videos' three variations in different resolutions.

Frame execution times of CPU, GPU, and FPGA are shown in Table 3. These results show that the GPU offers a much better performance compared to the FPGA. Note that the FPGA used (15 billion transistors in 20 nm) was even more high-end than the GPU (8 billion transistors in 28 nm). The GPU frame execution times we have obtained in this work are quite competitive; for instance, you may compare them to the frame execution times reported in [7] by some of the authors of this paper.

We were able to obtain higher performance on the FPGA than GPU in [17] when the design was directly done in Verilog instead of OpenCL. Therefore, this work shows that with an OpenCL based design flow, speedup potentials of FPGAs may not be fully utilized although impressive gains are possible in design time.

Table 3. CPU, FPGA, GPU frame execution times in sec. for 50 iterations per warp

Resolution	CPU	FPGA	GPU
576×384	107	2.734	0.128
640×480	156	3.594	0.168
768×576	471	4.854	0.254

The results in Table 4 were obtained after the GPU was slowed down to the same frame execution times as the FPGA Table 5. These results show that even with an OpenCL based flow FPGAs offer significantly lower power consumption compared to GPU implementations. Note that these power consumption figures were obtained after subtracting the power consumption of the idle system from the actual power consumption. We tried to lower GPU power consumption by slowing down its clock instead of adding sleep cycles. This did not seem possible on Nvidia GPUs, whereas it may be possible on, for example, AMD GPUs.

Table 4. FPGA and GPU power consumptions with equal frame execution times for 50 iterations per warp

Resolution	Frame time	FPGA power	GPU power
576×384	2734 ms	15 W	58 W
640×480	3294 ms	19 W	58 W
768×576	4854 ms	22 W	60 W

The results in Table 5 fare even better for the FPGA. That is, when the compute requirements of the algorithm is lowered (2 versus 50 iterations), then the FPGA starts producing more acceptable frame rates and yet become even more power efficient compared to the GPU.

Table 5. FPGA and GPU power consumptions with equal frame execution times for 2 iterations per warp

Resolution	Frame time	FPGA power	GPU power
576 × 384	201 ms	10 W	56 W
640 × 480	269 ms	11 W	57 W
768 × 576	362 ms	11 W	59 W

Output of the FPGA implementation is shown in Fig. 4. The left image is the input frame, while the right image is a visualization of the motion vectors produced from an input frame and its previous frame. Stationary parts are shown in white, while moving pixels of the frame are color-coded indicating different motion directions and speeds.

Fig. 4. FPGA output

5   Conclusions

In this work, anisotropic Huber-L optical flow estimation algorithm was implemented on Nallatech 385A Aria10 FPGA board using Altera OpenCL. Details of the algorithm was explained, and implementation process on the FPGA was discussed. Performance and power consumption results were presented for GPU and FPGA.

On the basis of performance results, when FPGA design is compared with the CPU implementation, FPGA design provides 30–100× better performance in execution times. On the other hand, when FPGA Altera OpenCL design is compared to GPU OpenCL design, a decline in the performance is observed. Although the result depends on the input videos' resolution, GPU implementation can process a single frame up to 20× faster on the average than the FPGA implementation. The warping parameters and iteration numbers can be further reduced to achieve faster frame execution times, but this also reduces the quality of the optical flow estimation, resulting in a trade-off between execution time and quality. Note that better performance on FPGA can be obtained with domain-specific synthesis tools even when OpenCL is used [18].

We have found that the power consumption is a lot less in FPGA implementation, resulting around 4x better power consumption compared to the GPU implementation depending on the input resolution. In the end, it may be more suitable to use the FPGA implementation based on the needs of a designer, since although FPGA implementation results in higher frame execution times, it has much better power consumption. It can suit projects where low power consumption is required or designs that the extra performance that is provided by the GPU is not needed.

Acknowledgments. This work has been at the crossroads of multiple projects, namely, European Union Artemis JU Project called ALMARVI (GA 621439) [19] and TÜBİTAK (The Scientific and Technological Research Council of Turkey) projects ARDEB-114E343, KAMAG-114G029, and TEYDEB-9140015. More specifically, U. Ulutas is supported by KAMAG; M. Tosun and V.E. Levent were supported by ARDEB; D. Büyükaydın and T. Akgün were supported by TEYDEB; and H.F. Ugurdag has assumed leadership roles in the above Artemis, KAMAG, and ARDEB projects.

References

1. Pauwels, K., Tomasi, M., Diaz, J., Ros, E., Van Hulle, M.M.: A comparison of FPGA and GPU for real-time phase-based optical flow, stereo, and local image features. IEEE Trans. Comput. **61**(7), 999–1012 (2012)
2. Bodily, J., Nelson, B., Wei, Z., Lee, D.-J., Chase, J.: A comparison study on implementing optical flow and digital communications on FPGAs and GPUs. ACM Trans. Reconfigurable Technol. Syst. **3**(2), Article 6 (2010)
3. Muslim, F.B., Ma, L., Roozmeh, M., Lavagno, L.: Efficient FPGA implementation of OpenCL high-performance computing applications via high-level synthesis. IEEE Access **5**, 2747–2762 (2017)
4. Hoffman, H.N.: Program acceleration in a heterogeneous computing environment using OpenCL, FPGA, and CPU. University of Rhode Island DigitalCommons@URI, Open Access Master's thesis (2017)
5. Lucas, B.D., Kanade, T.: An iterative image registration technique with an application to stereo vision. In: Proceedings of International Joint Conference on Artificial intelligence (IJCAI), San Francisco, CA, USA, vol. 2, pp. 674–679 (1981)
6. Horn, B.K.P., Schunck, B.G.: Determining optical flow. Technical report, Massachusetts Institute of Technology, Cambridge, MA, USA (1980)
7. Büyükaydın, D., Akgün, T.: GPU implementation of an anisotropic Huber-L1 dense optical flow algorithm using OpenCL. In: Proceedings of International Conference on Embedded Computer Systems: Architectures, Modeling, and Simulation (SAMOS), Samos, Greece, pp. 326–331 (2015)
8. Shiralkar, M.: A self organization-based optical flow estimator with GPU implementation. All dissertations, p. 630 (2010)

9. Wedel, A., Pock, T., Zach, C., Bischof, H., Cremers, D.: An improved algorithm for TV-L^1 optical flow. In: Cremers, D., Rosenhahn, B., Yuille, A.L., Schmidt, F.R. (eds.) Statistical and Geometrical Approaches to Visual Motion Analysis. LNCS, vol. 5604, pp. 23–45. Springer, Heidelberg (2009). doi:10.1007/978-3-642-03061-1_2
10. Janik, I., Tang, Q., Khalid, M.: An overview of Altera SDK for OpenCL: a user perspective. In: Proceedings of IEEE Canadian Conference on Electrical and Computer Engineering (CCECE), Halifax, NS, Canada, pp. 559–564 (2015)
11. Zohouri, H.R., Maruyamay, N., Smith, A., Matsuda, M., Matsuoka, S.: Evaluating and optimizing OpenCL kernels for high performance computing with FPGAs. In: Proceedings of International Conference for High Performance Computing, Networking, Storage and Analysis, Salt Lake City, UT, USA, pp. 409–420 (2016)
12. Middlebury Optical Flow webpage. http://vision.middlebury.edu/flow/. Accessed June 2017
13. Hill, K., Craciun, S., George, A., Lam, H.: Comparative analysis of OpenCL vs. HDL with image-processing kernels on Stratix-V FPGA. In: Proceedings of IEEE International Conference on Application-Specific Systems, Architectures, and Processors (ASAP), Toronto, ON, Canada, pp. 189–193 (2015)
14. Altera SDK for OpenCL Programming Guide. https://www.altera.com/ja_JP/pdfs/literature/hb/opencl-sdk/aocl_programming_guide.pdf. Accessed June 2017
15. NALLATECH OpenCL A10 BSP Reference Guide. http://www.nallatech.com/store/pcie-accelerator-cards/nallatech-385a-arria10-1150-fpga. Accessed June 2017
16. Altera product selection guide. https://www.altera.com/content/dam/altera-www/global/en_US/pdfs/literature/pt/arria-10-product-table.pdf. Accessed June 2017
17. Güzel, A.E., Levent, V.E., Tosun, M., Özkan, M.A., Akgün, T., Büyükaydın, D., Erbas, C., Ugurdag, H.F.: Using high-level synthesis for rapid design of video processing pipes. In: Proceedings of East-West Design & Test Symposium (EWDTS), Yerevan, Armenia (2016)
18. Özkan, M.A., Reiche, O., Hannig, F., Teich, J.: FPGA-based accelerator design from a domain-specific language. In: Proceedings of International Conference on Field-Programmable Logic and Applications (FPL), Lausanne, Switzerland (2016)
19. Artemis JU Project, ALMARVI: Algorithms, Design Methods, and Many-Core Execution Platform for Low-Power Massive Data-Rate Video and Image Processing, GA 621439. http://www.almarvi.eu

Persistent Random Search on Complex Networks

Lasko Basnarkov[1,2]([⊠]), Miroslav Mirchev[1], and Ljupco Kocarev[1,2]

[1] Faculty of Computer Science and Engineering,
Ss. Cyril and Methodius University in Skopje,
Rudjer Boshkovikj 16, P.O. Box 393, 1000 Skopje, Republic of Macedonia
{lasko.basnarkov,miroslav.mirchev,ljupco.kocarev}@finki.ukim.mk
[2] Macedonian Academy of Sciences and Arts,
P.O. Box 428, 1000 Skopje, Republic of Macedonia

Abstract. Searching of target based on random movements in space is an interesting topic of research relevant in different fields. For searching in complex networks besides the classical random walk, various biasing procedures have been applied for reducing the searching time. We propose one such biasing algorithm that favors movements towards more distant nodes, while penalizing going backward. Using Monte Carlo numerical simulations we demonstrate that the proposed algorithm provides lower Mean First Passage Time for several types of generic and real complex networks.

1 Introduction

Random walk is a concept of motion or change of state in general, introduced a century ago [1] and since then it was applied in a wide range of science topics. Its relevance was proven fruitful for modeling of Brownian motion in physics [2], or motion of animals in biology [3]. Engineering fields have found numerous applications as well. For example, random walk can be conveniently applied for modeling of electrical networks [4] or for image segmentation [5]. It is at the root of the Page Rank algorithm [6] and can be used for better understanding of the dynamics of gross national product [7]. It should not come as surprise that at the moment of preparation of this paper typing the words "Random walk" in Google's Scholar searching engine returned nearly one million results.

The research path of random walk on lattices has progressed from symmetrical spaces, like one or more dimensional regular lattices, towards more irregular structures such as complex networks. One of the most important concepts related to random walks is the First Passage Time that roughly speaking denotes the number of steps or time needed for the random walker to reach a previously chosen target (point) starting from another initial point or state. Due to the probabilistic nature of the individual movements, one is usually more interested to the related mean value obtained by averaging across all possible realizations of the random choices – the Mean First Passage Time (MFPT). The first rigorous results for the MFPT were proved for one and more dimensional regular lattices [8]. Closed form solution for the MFPT were then obtained for the Markov

© Springer International Publishing AG 2017
D. Trajanov and V. Bakeva (Eds.): ICT Innovations 2017, CCIS 778, pp. 102–111, 2017.
DOI: 10.1007/978-3-319-67597-8_10

chains, where the transition between any pair of states is allowed and possibly with a nonidentical probability [9].

The classical random walk assumes that at each iteration all possible moves have equal probability and is thus the simplest and the most tractable approach. In order to find the optimal strategies that result in smallest MFPT various biasing alternatives have started to appear. One approach considers Lévy flights that are long jumps towards distant points, which in the classical case can be achieved only with more steps, rather than a single one [10]. However, it was recently shown that a so-called persistent random walk is more efficient in a random search of a target placed on a regular square lattice [11]. The persistence within this algorithm denotes favoring the repetition of the direction of the search instead of changing it (going back or aside). This kind of random walk is not only a useful mathematical concept but can be observed in nature as well, which means that its efficiency is exploited by the living organisms. When animals are surveying the environment for food, they make several steps forward, then scan locally, which is followed again with few steps and so on [12]. It is well known that even simple organisms like the bacteria *E. Coli* perform a kind of *run and tumble* motion [13]. The periods of local wandering (the tumble part) are interspersed with nearly strait line motion (the runs).

Complex networks theory has emerged as one of the most popular research areas in the last two decades. The first insights have shown that for modeling of many natural or man-made systems of large number of interacting units, one should use networks that among their other properties possess two key features: short paths (small world property) [14] and a power law degree distribution [15]. There have been numerous other problems about the topological features of the complex networks, but also various dynamical processes running on top on such structures, where the random walk appears one of the most exploited ones. Random walks on connected networks whether of classical or biased kind can be cast in the language of Markov chains. This convenience allows one to apply the powerful machinery of Markov chains to obtain certain analytical results [16,17].

In this work we present a possible generalization of the idea of persistent random walk on complex networks. Since the concept of direction in non-spatial complex networks can not be applied one can not easily make a theoretical analysis of the walk and determine the MFPT analytically. Therefore, we have performed extensive Monte Carlo simulations and verified that the introduction of *persistence* leads to shortening of the MFPT, like in the case of regular networks. The improvement is noted by a comparison with the generic random walk for which we have calculated the MFPT by using the analytical formulas from Markov chains theory. The paper is organized as follows. In Sect. 2 we define the problem, give a short overview of the MFPT formula for Markov chains and describe the algorithm. Then, in Sect. 3 we provide some numerical results with a discussion and we close the paper with the conclusion.

2 Model

2.1 Problem Statement

The random walk is a model of motion in a given space whether in the real world or in some abstract space of states. It consists of a sequence of steps with an identical or variable size made consecutively at discrete iterations. The directions of successive steps are chosen by chance, which is the reason for the attribute *random* of the resulting path. Due to the stochastic nature of the obtained paths one is usually interested in their statistical properties. A great number of problems arising from different fields using random walk can be stated as a calculation of the mean number of iterations needed to reach some state for the first time, when the walk begins from another arbitrarily chosen starting state. A related statistical quantity is the above mentioned MFPT. When one is interested in studying random walks on complex networks, one such path is identified with a sequence of nodes of a network under study. Usually consecutive steps are allowed only between neighboring nodes. In some applications jumps to non-neighboring nodes are also permitted, especially when the network is not connected like in the Page Rank algorithm [6]. The generic random walk is the simplest case that corresponds to the situation when the probability for choosing the next node in the sequence is identical for all neighbors and equals to the inverse of the degree of the current node $p = 1/d_i$. The MFPT $m_{i,j}$ on a network is the average number of steps needed to reach the target node j, starting from node i. It depends on the properties of the whole network and on the location of the pair of nodes within the network. Hence, the MFPT for the same target node depends on the starting node, and one can find an average across all nodes in the network taken as starters

$$g_j = \frac{1}{N} \sum_{i=1}^{N} m_{i,j}. \tag{1}$$

This quantity is called a Global MFPT (GMFPT) [18], which is a property of a single node in a network considered as a target. It is known that for networks having a good mixing [19], which means that any initial distribution of random walkers quickly converges to the stationary one, GMFPT depends on the stationary density w_j corresponding to the target node. Moreover, for a generic random walk on undirected networks the stationary density of each node $w_j = d_j/2L$ depends on the degree of that node d_j and the total number of links in the network L [16]. This observation can explain the intuitive reasoning that hubs are easier to arrive at, and thus have a smaller GMFPT.

Similarly, in order to make a comparison between different networks one can make an average over all GMFPT and obtain the network MFPT (NMFPT)

$$G = \frac{1}{N} \sum_{j=1}^{N} g_j. \tag{2}$$

The last quantity characterizes the searching propensity of the network, or easiness to find randomly chosen target starting generally from another randomly chosen starting node.

2.2   MFPT Formula for Random Walk on Complex Networks

Random walk on connected networks whether generic or biased, but with time invariant transition probabilities between neighboring nodes can be treated as random walk on Markov chain. Then one can apply the tools of the Markov chains and use the needed formulas. Here we will briefly introduce the key quantities necessary for calculation of the MFPT as given in [9]. We assume that the transition matrix between neighboring nodes $\mathbf{P}$ is known and it has constant elements. The element $p_{i,j}$ represents the probability to jump from node i to its neighbor j. For example for generic random walk the transition probabilities are $p_{i,j} = 1/d_i$ where d_i is the number of neighbors, or degree of node i. The random walk is called biased when transition probabilities are different from the inverse of the node degrees. When the network is connected, one can reach any node from any other one by going from one node to a neighbor and so on. For such networks there is a unique stationary probability distribution to find a random walker at each node w_i. The stationary distribution for all nodes is encoded in a row vector $\mathbf{w}$. Moreover, it is the same regardless of the starting distribution. From the stationary vector one constructs a matrix $\mathbf{W}$ of rows identical to $\mathbf{w}$. Next, for the network one can calculate the so called fundamental matrix

$$\mathbf{Z} = (\mathbf{I} - \mathbf{P} + \mathbf{W})^{-1}, \tag{3}$$

where I is an identity matrix with a corresponding size. The MFPT between any pair of start-target nodes is now given by

$$m_{i,j} = \frac{z_{j,j} - z_{i,j}}{w_j}, \tag{4}$$

where w_j is the stationary probability corresponding to node j, and $z_{i,j}$ are the matrix elements of the fundamental matrix $\mathbf{Z}$. The MFPTs from the last formula can be then used to calculate the GMFPT and NMFPT for generic or biased random walks. We will use these equations for calculations only in the generic case since our algorithm does not satisfy the Markov property, which means that the elements of the transition matrix $\mathbf{P}$ are not constants.

2.3   Persistent Random Walk on Complex Networks

This approach is inspired by the result that a so called persistent random walk was proven as very efficient in random search of a target. It was shown that in a problem of location of randomly placed target over regular square lattice it outperforms even the Lévy flights based random walk [11]. The persistence within this algorithm denotes favoring the repetition of the direction of the search instead of changing it. It can be also seen as a strategy preferring 'going forward' instead of 'side wandering'. In the context of random walk on complex networks this strategy could translate to one preferring moving towards nodes that are more distant to some node that was visited recently – in our case one step before. We remind that in complex networks vocabulary a convenient measure of distance is the number of consecutive links in the shortest path between two nodes. Another thing to have in mind is that this kind of movement needs a memory – at least the node that was visited in the previous step.

One possible way for applying the persistence idea is the following. Consider a situation when the walker has been at node i in the previous step and it is currently at node j. The favoring of repetition of direction in a regular lattice can be understood as a mean to go further away. In the complex networks' setting persistence can be mimicked by giving larger transition probability p_2 to the neighbors of j that are second neighbors to i, but not first. The nodes that are neighbors to i and j simultaneously can be seen as complex network peers to the side nodes in the regular scenario. These nodes will be associated with a transition probability $p_1 < p_2$. Returning should be disfavored but not forbidden to avoid trapping at leaf nodes, so transition probability towards i from j would be some p_0 satisfying $0 < p_0 < p_1$. Assuming that within the neighborhood of j there are N_1 nodes that are also neighbors of i, while N_2 are not, the probability normalization condition would be

$$p_0 + N_1 p_1 + N_2 p_2 = 1. \tag{5}$$

The last equation constraints the choice of the transition probabilities and thus one is left to experiment with only two free parameters. Introducing the following ratios $p_2/p_1 = \alpha$ and $p_0/p_1 = \beta$ one obtains the following parameterized normalization condition

$$p_1 \left(\beta + N_1 + \alpha N_2 \right) = 1. \tag{6}$$

Hence, depending on the parameters α and β, and the number of neighbors $1 + N_1 + N_2$, the transition probabilities can be uniquely determined.

3   Results

To estimate the potential of the presented algorithm in this section we provide the results of the numerical calculations of the NMFPT of the random walk over different generic and real complex networks. These results are compared with the respective NMFPT for the same networks for the generic random walk, which are calculated from the analytical formulas. Since the algorithm we study employs Monte Carlo simulations, in order to estimate its accuracy we have also performed numerical calculations of the NMFPT for the generic random walk. In the Monte Carlo simulations the NMFPT is calculated using a very large number of arbitrarily chosen pairs of nodes and it is averaged using many different network instances having some predefined properties. For the generic networks we average over 100 instances for each setting and 5000 node pairs from each instance. We note again that as the proposed biasing approach breaks the Markov property that the classical random walk has, one can not use the formulas for determination of MFPT based on transition matrix and is left to perform Monte Carlo simulations.

The search algorithm under study was tested on various complex networks: purely random, small world and scale free. To generate these networks we have used the algorithms from the NetworkX library in Python [20]. The random graphs are complex networks created with the Erdős-Rényi algorithm where every pair of nodes i and j is linked with some prescribed probability p that appears as a parameter of the graph, besides the number of nodes N [21]. As the probability p becomes larger the obtained graph is becoming more connected. The scale free networks were made on base on the Barabasi-Albert preferential attachment algorithm that sequentially builds the network by adding nodes one by one [15]. The seed of the mechanism is a fully connected network of m_0 nodes and every newly added node has m links to the growing network[1]. The probability to connect a new node to an existing node is proportional to the number of existing links of that existing node. The small world networks were build following the Watts and Strogatz approach [14]. It starts with a regular ring-like network with N nodes where each of them is connected with K neighbors, and then by randomly rewiring the links with some probability p. In all considered generic networks the number of nodes is $N = 1000$.

First, we examine the impact of the parameters α and β on the persistent random walk on a scale free network in Fig. 1. The number of links added for each new node is $m = 3$. The effects of varying β is examined using fixed $\alpha = 1$, and it can be seen that as β decreases the NMFPT is reduced. For the simulations with different values of α we set $\beta = 1$ and it can be noticed that the increase in α also leads to a reduction in the NMFTP. Hence, we can conclude that both small β and higher α help in reducing the NMFPT.

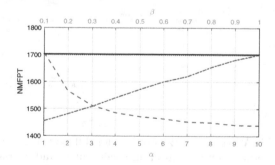

Fig. 1. NMFPT in a Barabasi-Albert network with $m = 3$ for different values of α and β, compared to the values obtained with the 'Classical' approach (dotted magenta line), and the 'Analytical' method (solid black line). (Color figure online)

---

[1] Parameters m_0 and m here are used as the authors Barabasi and Albert have denoted them and are different from the elements of the MFPT matrix m_{ij}.

In Fig. 2a we show how the NMFPT reduces as m is varied from 2 to 10, i.e. the average node degree $\hat{k}$ varies between 4 and 20 as it is equal to $2m$. Moreover, the NMFPT obtained by the persistent random walk is significantly lower than the NMFPT of the classical random walk. In Fig. 2b we can also see how the average clustering coefficient $\hat{C}$ and the average shortest path length $\hat{l}$ as we vary m. In Fig. 3 it is shown how the NMFPT changes in a small world network as the number of neighbors per node K is varied from 4 to 20, which also represents the average node degree $\hat{k}$. The NMFPT for the random network are given in Fig. 4 where the link probability p is varied from 0.1 to 0.2, which translates into a change of the average node degree from 10 to 20. In both networks the persistent random walk have lower NMFPT than the classical one. The corresponding network characteristics of these two network types for the different parameters are given in Figs. 3b and 4b.

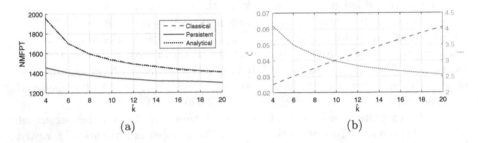

Fig. 2. (a) NMFPT in Barabasi-Albert networks with different average node degrees $\hat{k}$; and (b) the corresponding average clustering coefficient $\hat{C}$ and average shortest path length $\hat{l}$.

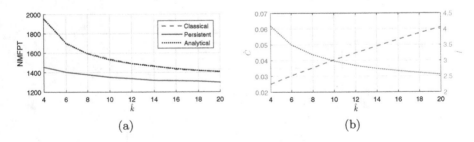

Fig. 3. (a) NMFPT in Watts-Strogatz networks for different average node degrees $\hat{k}$; and (b) the corresponding average clustering coefficient $\hat{C}$ and average shortest path length $\hat{l}$.

The scale free and the small world network have the same number of links, as well as nodes. However, it turns out that the small world network achieves a higher benefit from the persistence and the NMFPT is reduced to a very low value, even though it has a higher NMFPT than the scale free network with

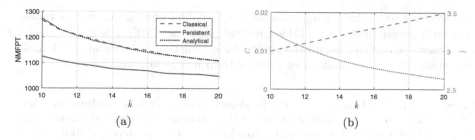

(a)   (b)

Fig. 4. (a) NMFPT in Erdős-Rényi networks for different average node degree $\hat{k}$; and (b) the corresponding average clustering coefficient $\hat{C}$ and average shortest path length $\hat{l}$.

the classical random walk. On the other hand, we had to use larger number of links in the random networks as otherwise it is extremely hard to generate such networks with the same number of links using the Erdős-Rényi model. Even though the persistence decreases the NMFPT also in these networks, they see the least amount of reduction in the NMFPT in comparison with the others.

We have also examined the searching performance of the persistent random walk on a real data of the Internet on autonomous systems level from 02.01.2000 [22], shown in Fig. 5, which is well known to be scale free. This dataset consists of 6474 nodes and 13233 edges, and it is available at [23]. In this case we have only one network instance and we have averaged the results over 50000 node pairs. The analytical method yields a NMFPT of 19312, while numerical simulations of random paths have resulted in a similar value of 19297. Using the persistent random walk with $\alpha = 1$ and different values of β we have obtained the values given in Table 1, where one can see how the NMFPT is reduced as β is decreased.

Fig. 5. The Internet topology [22] at autonomous system level visualized in Gephi, where the larger size and darker color indicates a higher node degree.

On the other hand, for a fixed value of $\beta = 0.01$ and several different α values, the results shown in the same Table 1 indicate how the NMFPT becomes smaller as we increase α. Once again we can conclude that both $\alpha > 1$ and $\beta < 1$ make the persistent random walk more efficient than its classical counterpart.

Table 1. NMFPT obtained using the persistent random walk in the Internet where the analytical calculations indicate a NMFTP $= 19312$ and the classical random walk yields a NMFPT $= 19297$. Upper table provides the effects of varying β while $\alpha = 1$, and the lower one shows the effects of varying α while $\beta = 0.01$

β	1	0.5	0.25	0.125	0.0625	0.03125
NMFPT	19384	18339	17959	17676	17479	17335

α	1	2	4	8	16
NMFPT	17353	16604	16493	16247	16132

4   Conclusions

In this work we studied random searching in complex networks using a biased random walk that prefers moving to "more distant nodes" and "hesitates to return" to the previously visited node. This biasing is quantified with two parameters α and β, and the influence of these parameters on the efficiency of the algorithm was studied. The numerical simulations of such a biased random walk have shown that the searching is faster than with the classical approach. Since the presented algorithm uses memory to store only the last visited node (besides the current) one can try to generalize the approach for keeping record of more nodes. Also, since this is not some direct generalization of the mentioned persistent random walk on complex networks, there might be another alternatives that could provide even better searching. One can also try to make a related Markov model that would use the pair of the current and the past node as a state. Then, the MFPT could be calculated analytically by using the adjacency matrix elements.

Acknowledgment. This research was partially supported by the Faculty of Computer Science and Engineering at Ss Cyril and Methodius University in Skopje. LK thanks DFG for support.

References

1. Pearson, K.: The problem of the random walk. Nature **72**(1865), 294 (1905)
2. Kac, M.: Random walk and the theory of brownian motion. Am. Math. Mon. **54**(7), 369–391 (1947)
3. Codling, E.A., Plank, M.J., Benhamou, S.: Random walk models in biology. J. R. Soc. Interface **5**(25), 813–834 (2008)

4. Doyle, P.G., Snell, J.L.: Random Walks and Electric Networks. Mathematical Association of America, Washington, DC (1984)
5. Grady, L.: Random walks for image segmentation. IEEE Trans. Pattern Anal. Mach. Intell. **28**(11), 1768–1783 (2006)
6. Brin, S., Page, L.: The anatomy of a large-scale hypertextual web search engine. Comput. Netw. ISDN Syst. **30**(1), 107–117 (1998)
7. Cochrane, J.H.: How big is the random walk in gnp? J. Polit. Econ. **96**(5), 893–920 (1988)
8. Pólya, G.: Über eine aufgabe der wahrscheinlichkeitsrechnung betreffend die irrfahrt im straßennetz. Math. Ann. **84**(1–2), 149–160 (1921)
9. Grinstead, C.M., Snell, J.L.: Introduction to Probability. American Mathematical Society, Providence (2012)
10. Viswanathan, G.M., Buldyrev, S.V., Havlin, S., Da Luz, M.G.E., Raposo, E.P., Stanley, H.E.: Optimizing the success of random searches. Nature **401**(6756), 911–914 (1999)
11. Tejedor, V., Voituriez, R., Bénichou, O.: Optimizing persistent random searches. Phys. Rev. Lett. **108**(8), 088103 (2012)
12. Bell, W.J.: Searching Behaviour: The Behavioural Ecology of Finding Resources. Springer Science & Business Media, New York (2012)
13. Berg, H.C.: E. Coli in Motion. Springer Science & Business Media, New York (2008)
14. Watts, D.J., Strogatz, S.H.: Collective dynamics of small-worldnetworks. Nature **393**(6684), 440–442 (1998)
15. Barabási, A.-L., Albert, R.: Emergence of scaling in random networks. Science **286**(5439), 509–512 (1999)
16. Noh, J.D., Rieger, H.: Random walks on complex networks. Phys. Rev. Lett. **92**(11), 118701 (2004)
17. Fronczak, A., Fronczak, P.: Biased random walks in complex networks: the role of local navigation rules. Phys. Rev. E **80**(1), 016107 (2009)
18. Tejedor, V., Bénichou, O., Voituriez, R.: Global mean first-passage times of random walks on complex networks. Phys. Rev. E **80**(6), 065104 (2009)
19. Levin, D.A., Peres, Y., Wilmer, E.L.: Markov Chains and Mixing Times. American Mathematical Society, Providence (2009)
20. Networkx: Software package for complex networks in python language. https://networkx.github.io/
21. Erdos, P., Rényi, A.: On the evolution of random graphs. Publ. Math. Inst. Hung. Acad. Sci **5**(1), 17–60 (1960)
22. Leskovec, J., Kleinberg, J., Faloutsos, C.: Graphs over time: densification laws, shrinking diameters and possible explanations. In: Proceedings of the Eleventh ACM SIGKDD International Conference on Knowledge Discovery in Data Mining, pp. 177–187. ACM (2005)
23. Leskovec, J., Krevl, A.: SNAP Datasets: Stanford large network dataset collection, June 2014. http://snap.stanford.edu/data

Weed Detection Dataset with RGB Images Taken Under Variable Light Conditions

Petre Lameski[✉], Eftim Zdravevski, Vladimir Trajkovik, and Andrea Kulakov

Faculty of Computer Science and Engineering,
University of Sts. Cyril and Methodius in Skopje,
Ruger Boskovik 16, 1000 Skopje, Macedonia
lameski@finki.ukim.mk

Abstract. Weed detection from images has received a great interest from scientific communities in recent years. However, there are only a few available datasets that can be used for weed detection from unmanned and other ground vehicles and systems. In this paper we present a new dataset (i.e. Carrot-Weed) for weed detection taken under variable light conditions. The dataset contains RGB images from young carrot seedlings taken during the period of February in the area around Negotino, Republic of Macedonia. We performed initial analysis of the dataset and report the initial results, obtained using convolutional neural network architectures.

Keywords: Dataset · Weed detection · Machine learning · Signal processing · Precision agriculture

1 Introduction

The automation of the agricultural food production is gaining in popularity in the scientific communities and in the industry. The main goal of the automation is to reach the agricultural food demand growth, which currently is lower than the growth of the agricultural food production [1]. The automation in agriculture could be increased by introduction of unmanned aerial vehicles (UAVs) and unmanned ground vehicles (UGVs) that can monitor the growth of the crops and automatically react if some factors, such as water shortage, weed or insect infestation or plant illness, is detected. The introduction of robotic systems in the agricultural food or resource production is proven to increase the crops yield per land-unit [2]. There are quite a few challenges that need to be overcome to have a fully automated solution that can be used in the industry. One of the challenges is the sensing of weeds in the fields under variable light conditions from simple RGB images. While multi-spectral solutions have been applied for plant-weed segmentation and other agricultural tasks [3–5], multi-spectral cameras are still too expensive for small farmers. This introduces the need for an efficient and robust system for weed and other anomaly detection based on RGB cameras that would be more affordable for small farmers.

© Springer International Publishing AG 2017
D. Trajanov and V. Bakeva (Eds.): ICT Innovations 2017, CCIS 778, pp. 112–119, 2017.
DOI: 10.1007/978-3-319-67597-8_11

In this paper we present the Carrot-Weed dataset, containing RGB images of young carrot seedlings taken during the period of February in the area around Negotino, Republic of Macedonia. It can be used for evaluating and bench-marking algorithms for crop-weed segmentation and weed detection by building machine learning and deep-learning models. The Carrot-Weed dataset is publicly available on https://github.com/lameski/rgbweeddetection, and the results of our initial analysis of the dataset are presented in this paper.

In Sect. 2 we overview some of the existing datasets for weed detection in the literature. Then, in Sect. 3 we describe the process of generating the dataset, the characteristics of the dataset. After that, in Sect. 4, we provide some initial results obtained by using segmentation with convolutional neural networks, and finally in Sect. 5, we conclude the paper and discuss some ideas for future work.

2   Related Work

There are several available datasets for weed detection. The dataset presented in [6] is a crop-weed segmentation dataset with images taken under constant light conditions with Near Infrared (NIR) and Red (R) channels. Authors in [7] describe a dataset that is also taken with constant light conditions and with RGB+NIR images. In [8] authors automatically generate the weed detection dataset based on 3D models of the plants and use them to train weed detection algorithms. Although there are only a few publicly available datasets for weed detection, there more datasets for plant detection and recognition. For example, in [9] a dataset obtained from Pl@ntNet [10] is used for plant identification task in the LifeClef competition. Other datasets exist for plant identification, such as the leaf shape dataset presented in [11], where the images of the segmented leaf contours of 100 different plants are provided for the task of plant classification based on the contours and a set of previously calculated descriptors for each image.

In this paper we describe the Carrot-Weed dataset that is obtained under natural variable light conditions from RGB camera, which is the main difference from other datasets. In general, the variable light conditions and using a common RGB camera instead of NIR or RGB+NIR cameras should make the segmentation and weed detection tasks more difficult. In the next section, the method that was used to obtain and annotate the dataset is described and some statistics for the dataset are presented.

3   Weed Detection Dataset

The dataset contains RGB images taken from approximately 1 m height. The images are taken using a cell phone camera with 10 mega pixels. After the initial acquisition, the images are processed so that the masks for the weeds, carrot (plant) and ground are generated. Several example images from the dataset are presented in Fig. 1. The variable light conditions can be observed from the presented images.

Fig. 1. Example images from the dataset

In the first step, aiming to initially separate the vegetation pixels from the ground pixels, vegetation indices are generated. The ExGExR [12] index was used to perform the initial vegetation segmentation. The ExG (Excess Green) index is calculated using (2). The ExG index, proposed in [13], gives the difference between the detected light values of green channel and the red and blue channels. Prior to calculating the ExG index, the R (Red), G (Green) and B (Blue) channel values in the images are normalized using (1). The normalization is performed to reduce the influence of the different lighting conditions to the color channels. The result of this normalization is shown in Fig. 2.

$$r = \frac{R}{255}, g = \frac{G}{255}, b = \frac{B}{255} \tag{1}$$

$$ExG = 2g - r - b \tag{2}$$

$$ExR = 1.4r - b \tag{3}$$

$$ExGExR = ExG - ExR \tag{4}$$

Using the ExG index for segmentation the vegetation pixels from the ground or other object pixels in the images yields good results. The ExGExR index,

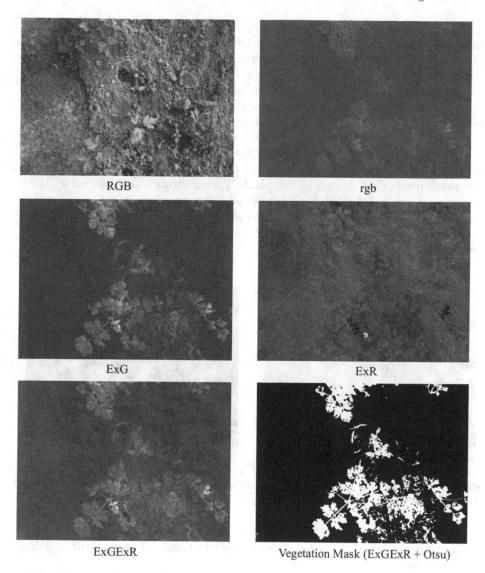

RGB

rgb

ExG

ExR

ExGExR

Vegetation Mask (ExGExR + Otsu)

Fig. 2. Example (RGB) image from the dataset, normalized (RGB) image, ExG result image, ExR result image, ExGExR result image and Vegetation Mask obtained using ExGExR + Otsu (Color figure online)

calculated with (4), improves the detection. Prior to its calculation, the ExR [12] index is calculated with (3).

A normalized gray-scale representation of the ExG and ExGExR images is presented in Fig. 2. One of the best perks of these indices is that after their calculation, the Otsu algorithm for automatic threshold calculation [14], can be applied to segment the vegetation pixels from the other pixels in the image.

The combination of the ExGExR index and the Otsu thresholding of the obtained image, yields a good result when used for segmentation of the vegetation pixels. An example of a segmented image using the described approach is presented in Fig. 3.

After the initial segmentation, morphological operations are applied on the resulting vegetation mask, such as opening and closure, to reduce the noisy mask pixels and fill in the empty parts of the leaves. Finally, the resulting masks are manually inspected and amended to obtain the final dataset masks.

An exemplary image, along with its mask, is shown in Fig. 3. Later, the segmentation of the weed and plant pixels of the images is performed by modifying the mask pixels that belong to the plant. The resulting mask has values 0 for the ground pixels, 1 for the weed pixels and 2 for the carrot plant pixels, which can be used for evaluating segmentation algorithms. In Fig. 3, for visualization purposes the pixels are represented with values of 50 and 100 instead of 1 and 2, respectively.

Fig. 3. Example image with its annotation. The ground is colored black, the weeds are colored with dark gray and the useful plant (the carrot) is colored in light gray

The dataset contains 39 images with the same size of 3264 × 2448. Since the image sizes are large, the dataset can be split to smaller images using a sliding window approach.

The dataset consists of 311, 620, 608 pixels from which 26, 616, 081 carrot plant pixels, 18, 503, 308 weed plant pixels and 266, 501, 219 ground pixels.

In the dataset there are more pixels of carrot plants than of weed plants. In addition to the imbalanced dataset, the weed infestation in the images is quite large and there is high overlap between the weed plants and the carrot plants, making the weed segmentation task very challenging. Most of the classification algorithms would require a balanced distribution of each classes in order to successfully train the classification model. In this paper, the benchmark results of the pixel classification are obtained by a semantic segmentation approach that uses convolutional neural networks. This approach and its application on the dataset is described in the next section.

4   Initial Experiments and Results

We performed initial experiments on the dataset using the SegNet [15] convolutional architecture for semantic segmentation. Other classification approaches to classify patches of the image as weed, plant or land, would require to first segment the weed, plant and land patches from the image. Having these segments would have already solved the problem of weed detection in the image, unless we are interested in detecting different types of weeds.

The SegNet architecture uses a combination of convolutional and deconvolutional layers that allow semantic segmentation of the image. We modified the proposed architecture so that it recognizes only three classes: weed, plant and ground, which are presented using the manually performed annotations in the dataset with the pixel labels 1, 2 and 0 respectively.

Since the dataset was obtained under variable light conditions, we built models using the original RGB images and also using the same images converted to the Lab and HSV color-spaces. The Lab color-space transforms the RGB colors in such way that when compared in distance, it gives smaller distances for similar colors and larger distances for different colors, when observed by a human eye. The HSV color space presents the colors as a degree in a circle using the H value. The V value represents the light value of the pixel. These color spaces give good results for object classification and HSV is also used in [16] for plant pixels detection in combination with other parameters. We used these transformations motivated by their successful application in other approaches for both object recognition tasks and segmentation tasks.

Due to the large size of the images, we used a sliding window approach with size of 256×256 to obtain image patches that can be used in the SegNet architecture. For training purposes, we introduced additional limitation that considers patches for the training set only if they contain at least 5% of plant pixels and 5% of weed pixels. This reduced the available patches, but was expected to make the training model to converge faster. This expectation, however was only based on experimental observation and was not measured.

Further, we experimented with two scenarios. In the first scenario, we only used one randomly selected image for the training and the rest of the images for testing. In the second scenario, we used the leave-one-image-out approach for training and testing. The initially obtained results are presented in Table 1.

Similar results were also reported in [8] when using SegNet for plant/weed segmentation from RGB images. It can be observed that SegNet over-fits on the data because the trained model, when tested on the training data, fit the data successfully with over 95% correctly segmented pixels.

As it can be observed from the results, even tough we expected the Lab and HSV color spaces to improve the segmentation accuracy, this was not the case. The usage of the HSV and Lab transformations did not improve the accuracy, but instead resulted in similar or even worse accuracy compared to using unmodified RGB images.

Table 1. Results obtained with RGB, HSV and Lab conversion of the dataset using 1 image for train and 38 for test and 38 for train and one image for test

(Train/test) and Type	Vegetation, Land	Weed, Plant, Land
SegNet (1/38) RGB	0.694	0.590
SegNet (1/38) HSV	0.429	0.481
SegNet (1/38) Lab	0.534	0.521
SegNet (38/1) RGB	0.713	0.641
SegNet (38/1) HSV	0.601	0.609
SegNet (38/1) Lab	0.71	0.630

5   Conclusion and Future Work

In this paper we presented the Carot-Weed dataset for weed detection and segmentation using RGB images. The dataset contains 39 images taken from approximate height of 1 m with 10 MP camera under variable lighting conditions.

The dataset is a good starting point for initial development of machine vision algorithms for weed segmentation or weed detection from RGB images and is one of the rarely available public datasets for weed detection in plant images taken from close proximity. The initial results are promising and give good directions for further development of algorithms for weed detection.

The convolutional neural network approach used in the experiments reported in this paper could be modified and further simplified in order to avoid the reported over-fitting on the data.

Further research is needed to design an adequate deep learning architecture that could be successfully applied for the purpose of weed detection from RGB images. Some conventional approaches, such as unsupervised segmentation and patch classification should also be investigated.

A model with a good weed detection accuracy from RGB images could pave the way to a cheap and robust weed detection sensor that could be used by both farmers with large plantations and also small farmers that cannot afford the significantly more expensive multi-spectral recording equipment.

Acknowledgments. The work presented in this paper was partially financed by the University of Sts. Cyril and Methodius in Skopje, Macedonia, Faculty of Computer Science and Engineering.

References

1. Ray, D.K., Mueller, N.D., West, P.C., Foley, J.A.: Yield trends are insufficient to double global crop production by 2050. PloS One **8**(6), e66428 (2013)
2. Cosmin, P.: Adoption of artificial intelligence in agriculture. Bull. Univ. Agric. Sci. Vet. Med. Cluj-Napoca. Agriculture 68(1) (2011)

3. Mulla, D.J.: Twenty five years of remote sensing in precision agriculture: key advances and remaining knowledge gaps. Biosyst. Eng. **114**(4), 358–371 (2013). Special Issue: Sensing Technologies for Sustainable Agriculture

4. de Castro, A.I., Jurado-Expósito, M., Peña-Barragán, J.M., López-Granados, F.: Airborne multi-spectral imagery for mapping cruciferous weeds in cereal and legume crops. Precis. Agric. **13**(3), 302–321 (2012)

5. Torres-Sánchez, J., López-Granados, F., De Castro, A.I., Peña-Barragán, J.M.: Configuration and specifications of an unmanned aerial vehicle (UAV) for early site specific weed management. PLOS ONE **8**(3), 1–15 (2013)

6. Haug, S., Ostermann, J.: A Crop/Weed field image dataset for the evaluation of computer vision based precision agriculture tasks. In: Agapito, L., Bronstein, M.M., Rother, C. (eds.) ECCV 2014. LNCS, vol. 8928, pp. 105–116. Springer, Cham (2015). doi:10.1007/978-3-319-16220-1_8

7. Potena, C., Nardi, D., Pretto, A.: Fast and accurate crop and weed identification with summarized train sets for precision agriculture. In: Chen, W., Hosoda, K., Menegatti, E., Shimizu, M., Wang, H. (eds.) IAS 2016. AISC, vol. 531, pp. 105–121. Springer, Cham (2017). doi:10.1007/978-3-319-48036-7_9

8. Di Cicco, M., Potena, C., Grisetti, G., Pretto, A.: Automatic model based dataset generation for fast and accurate crop and weeds detection. arXiv preprint (2016). arXiv:1612.03019

9. Goëau, H., Joly, A., Bonnet, P., Selmi, S., Molino, J.F., Barthélémy, D., Boujemaa, N.: Lifeclef plant identification task 2014. In: CLEF2014 Working Notes. Working Notes for CLEF 2014 Conference, Sheffield, UK, September 15–18, 2014, CEUR-WS, pp. 598–615 (2014)

10. Barthélémy, D., Boujemaa, N., Mathieu, D., Molino, J., Bonnet, P., Enficiaud, R., Mouysset, E., Couteron, P.: The pl@ ntnet project: a computational plant identification and collaborative information system. Technical report, XIII World Forestry Congress (2009)

11. Mallah, C., Cope, J., Orwell, J.: Plant leaf classification using probabilistic integration of shape, texture and margin features. Signal Process. Pattern Recogn. Appl. **5**(1) (2013)

12. Meyer, G.E., Neto, J.C.: Verification of color vegetation indices for automated crop imaging applications. Comput. Electron. Agric. **63**(2), 282–293 (2008)

13. Woebbecke, D., Meyer, G., Von Bargen, K., Mortensen, D., et al.: Color indices for weed identification under various soil, residue, and lighting conditions. Trans. ASAE Am. Soc. Agric. Engineers. **38**(1), 259–270 (1995)

14. Otsu, N.: A threshold selection method from gray-level histograms. IEEE Trans. Syst. Man Cybern. **9**(1), 62–66 (1979)

15. Badrinarayanan, V., Kendall, A., Cipolla, R.: Segnet: A deep convolutional encoder-decoder architecture for image segmentation. arXiv preprint (2015). arXiv:1511.00561

16. Moorthy, S., Boigelot, B., Mercatoris, B.: Effective segmentation of green vegetation for resource-constrained real-time applications. In: Precision agriculture 2015, pp. 93–98. Wageningen Academic Publishers (2015)

Influence of Algebraic T-norm on Different Indiscernibility Relationships in Fuzzy-Rough Rule Induction Algorithms

Andreja Naumoski[✉], Georgina Mirceva, and Kosta Mitreski

Faculty of Computer Science and Engineering, Ss. Cyril and Methodius University in Skopje, Skopje, Macedonia
{andreja.naumoski,georgina.mirceva,kosta.mitreski}@finki.ukim.mk

Abstract. The rule induction algorithms generate rules directly in human-understandable if-then form, and this property is essential of successful intelligent classifier. Similar as crisp algorithms, the fuzzy and rough set methods are used to generate rule based induction algorithms. Recently, a rule induction algorithms based on fuzzy-rough theory were proposed. These algorithms operate on the well-known upper and lower approximation concepts, and they are sensitive to different T-norms, implicators and more over; to different similarity metrics. In this paper, we experimentally evaluate the influence of the T-norm Algebraic norm on the classification and regression tasks performance on three fuzzy-rough rule induction algorithms. The experimental results revealed some interesting results, moreover, the choice of similarity metric in combination with the T-norm on some datasets has no influence at all. Based on the experimental results, further investigation is required to investigate the influence of other T-norms on the algorithm's performance.

Keywords: Fuzzy tolerance relationship · Fuzzy rough sets · Rule induction algorithms · Classification · Regression

1 Introduction

As a major step towards building intelligent information system that produce knowledge in human understandable form (if-then rules), the fuzzy rule induction algorithms form a large group of methods and techniques that constantly evolve in more and more accurate algorithms. The use of these algorithms in data analysis, allows users to enhance the transparency in the learned models compared to other methods. Many of these rule induction methods and techniques derive a concise and human understandable set of rules for tasks at hand like classification and prediction of nominal or numeric values. Several known methods can be included in this category; first-order fuzzy rule induction [1, 2], fuzzy associate rule mining [3, 4], and semantics-preserving modelling [5, 6]. Beside many advantages of these methods compared to the crisp methods, still there are some disadvantages that need to be addressed. One such drawback is that the accuracy and efficiency of these methods decreases as the dimensionality of the dataset increases. One way to avoid this is to use feature selection as pre-processor [7]. Therefore, tighter integration of feature selection and rule induction is desirable. Before such method was

© Springer International Publishing AG 2017
D. Trajanov and V. Bakeva (Eds.): ICT Innovations 2017, CCIS 778, pp. 120–129, 2017.
DOI: 10.1007/978-3-319-67597-8_12

derived, several research papers try to integrate the rough set theory (RST) in rule induction.

RST in the last decade [8] has become a research discipline of great interest of many researchers. The application is very wide and applied in many practical domains. However, because this theory has been recently introduced, there have been few attempts to build fuzzy-rough methods for rule induction. Many research works are focused on using crisp rough theory to generate fuzzy rule set [9, 10], but no fuzzy-rough concepts are used in these algorithms, and therefore the RST advantages were not taken into account. One of the first steps to avoid fuzzy logical connectives, one of the important steps towards integration of RST in rule induction was introducing lower and upper approximation concepts as authors stated in [11]. This method is known as gradual decision ruling, and uses lower and upper approximations in order to define positive and negative relationship between credibility of premises and conclusions. More recently, a fuzzy-rough approach for rule induction was presented in [12]. In this research, fuzzy reducts in combination with fuzzy-rough feature selection pre-processing step are introduced to generate rules from the data at hand. It is interesting to note that fuzzy-rough concepts are used for building decision trees and afterwards into rules. Such method for generating certain and possible rule sets from hierarchical data is introduced in [13].

One of the many advantages of the RST applicability is the alternative approach that preserves the underlying semantics and in the same time allowing generality. The methods based on RST can build models where no parameters are required and thus eliminating the need of human interfering in the process of model induction. Two main areas where RST is very successful are feature selection and rule induction [7]. These two approaches involve analysis of equivalence classes generated from the partitioning of the universe of discourse by sets of features. Therefore, it is natural to integrate these two parts in a hybrid feature selection/rule induction algorithm that combines the advantages of both theories. Most recent attempt to do this is done in [7] by introducing QuickRules algorithm. The QuickRules algorithm is based on greedy hill-climbing strategy, but any search method can be implemented. In this direction, another approach using vaguely quantified rough set (VQRS) measure [7, 14] which can be used to replace the traditional dependency measure in order to better handle noise and uncertainty. This is very practical when applying these algorithms to real-world problems. These two approaches are very flexible and can be defined with different similarity metrics, T-norm and Implicator combinations or VQRS model. That's why the aim of this paper is to investigate the influence of four fuzzy similarity metrics and two T-norms; namely KD and Algebraic, using various datasets on the two rule induction algorithms: QuickRules and VQRS. Evaluation is done using AUC-ROC metric to estimate the accuracy of the obtained models.

The rest of the paper is organised as follows. Section 2 provides necessary details for RST, while Sect. 3 presents the definition of two fuzzy-rough rule induction algorithms. Section 4 summarizes the experimental approach and presents experimental results on a series of classification problems, while Sect. 5 concludes the paper.

2 Fuzzy Rough Set Theory

In order to proceed with the used terms of rule induction algorithm in terms of fuzzy rough data analytics, we first need to define what information system is. Information system is defined as couple (X, A), where $X = \{x_1, x_2, ..., x_n\}$ and $A = \{a_1, a_2, ..., a_n\}$ are finite non-empty sets of objects and attributes that can be discrete-valued (qualitative) or real-values (quantitative). U is a universe of discourse. Each attribute can take value from a finite set, and comparison between the values is done with strict equality basis. This is done by using so-called a-indiscernibility relationship (fuzzy similarity metrics), defined as:

$$R_a(x, y) = \{(x, y) \in U^2 \,|\, \forall a \in B, a(x) = a(y)\} \tag{1}$$

When a is quantitative, its values are drawn from a closed interval of real numbers, and compared by means of R_a. Here we present four different a-indiscernibility relationships that are used in the experiments:

$$\text{Sim1} = R_a^1(x, y) = 1 - \frac{|a(x) - a(y)|}{a_{\max} - a_{\min}} \tag{2}$$

$$\text{Sim2} = R_a^2(x, y) = \exp(-\frac{(a(x) - a(y))^2}{2\sigma_a^2}) \tag{3}$$

$$\text{Sim3} = R_a^3(x, y) = \max(\min(\frac{(a(y) - (a(x) - \sigma_a))}{(a(x) - (a(x) - \sigma_a))}, \frac{((a(x) + \sigma_a) - a(y))}{((a(x) + \sigma_a) - a(x))}), 0) \tag{4}$$

$$\text{Sim4} = R_a^4(x, y) = \max(0, \min(1, \frac{\beta - \alpha^* a(x) - a(y)}{\sigma_a})), \quad \alpha = 0.5, \beta = 1.0 \tag{5}$$

for each $x, y \in X$, while σ_a denotes the standard deviation of a. The four fuzzy a-indiscernibility relationships will be further tested and we will examine their influence on the rule induction algorithms. Each fuzzy a-indiscernibility relationships will be referred with the upper index numbers given within the equations. For a given subset B of A, the fuzzy B-indiscernibility relationship R_B is defined as:

$$R_B(x, y) = T(R_a(x, y)) \tag{6}$$

For each $x, y \in X$ and $a \in B$. T represents the selected t-norm and set B may contain both quantitative and qualitative attributes. R_B is a fuzzy tolerance that means it is a reflective and symmetric relationship, and used to approximate fuzzy sets in X. In this direction, we can define the upper and lower approximations for a given fuzzy set A, R_B is defined by:

$$\underline{R_B}(y) = \inf_{x \in X} I(R_B(x, y), A(x)) \tag{7}$$

$$\overline{R_B}(y) = \sup_{x \in X} T(R_B(x, y), A(x)) \tag{8}$$

For all $y \in X$, I represent implicator operator. An implicator I is a $[0, 1]^2 \to [0, 1]$ mapping that is decreasing in its first and increasing in its second argument, satisfying $I(0, 0) = I(0, 1) = I(1, 1) = 1$ and $I(1, 0) = 0$. On the other hand, T-norm T is an increasing, commutative, associative $[0, 1]^2 \to [0, 1]$ mapping that satisfies $T(x, 1) = x$ for x in $[0, 1]$. In this paper, we will use two T-norms: Kleene-Dienes (KD) and the Algebraic T-norm. And for implicator operator we will use the KD Implicator, which is defined as $I(x, y) = \max(1 - x, y)$.

Now, in order to use these fuzzy-rough concepts for knowledge discovery or in our case rule induction we first need to define the concept of decision system. A decision system $(X, A \cup \{d\})$ is an information system in which d $(d \notin A)$ is selected attribute called decision. Decision systems are often used in classification, but nonetheless they can be used also for solving regression data mining problems. Based on the decision values, X is partitioned into number of non-overlapping decision classes X_k $(k = 1, ..., p)$. In order to make a decision based on the given information system, we will use decision reducts. Given $B \subseteq A$, the fuzzy B-positive region is a fuzzy set in X that contains each object y to the extent that all objects with approximately equal values for the attributes in B, that have equal decision values d [15]. In order to calculate the decision values, B-positive region is defined for each value y and its predictive ability by calculating the degree of dependency of d on B. The B-positive region is defined as:

$$POS_B(y) = \left(\bigcup_{i=1}^{p} \underline{R_B} \right)(y) = \max_{i=1}^{p} \inf_{x \in X} I(R_B(x, y), A_k(x)) \tag{9}$$

The predictive ability of d of attributes in B is reflected by the ratio γ_B (degree of dependency) is defined as:

$$\gamma_B = \frac{|POS_B|}{|POS_A|} = \frac{\sum\limits_{x \in X} POS_B(x)}{\sum\limits_{x \in X} POS_A(x)} \tag{10}$$

If $\gamma_B = 1$, i.e. B preserves the decision-making power of A, and in [16], the subset B of A is called superreduct. If it cannot be further reduced, there exist no proper subset B' of B that satisfies $POS_B' = POS_A$, it is called a decision reduct. Computing all decision reducts is NP-hard problem, and from a practical purpose, it often suffices to generate a single decision superreduct. QuickReduct is a heuristic hill-climbing search algorithm used in [16] to calculate all the decision superreducts. So, in [7] the authors adopt the QuickReduct algorithm to rule induction method of QuickRules.

3   Rule Induction Algorithms

QuickRule as we mentioned earlier, is based on hybrid fuzzy rule induction that combines the fuzzy RST and the crisp rule induction procedures. This is done with simple strategy; generate (induce) fuzzy rules by overlaying decision reducts over the original (training) decision systems and reading off the values. Using recursively partitioning the universe of discourse of the features present in the decision reduct, each equivalence class forms a single rule. Because the partitioning is produced by reduct, it is guaranteed that each equivalence class is a subset of, or equal to, a decision concept, meaning that the attribute values are good predictors of the decision concept [7]. The implementation and use of the reduct concept ensured that each element of the data is covered by the set of rules that are produced. However, this approach has some disadvantages, namely, it generates rules that are often too specific, as each rule precursor always includes every feature appearing in the final reduct. For this purpose, authors in [7] integrated the rule induction step directly into the feature selection process, and thus generating rules on the fly. So, authors in [7] adapt the QuickReduct algorithm, so at each step, fuzzy rules maximally cover the training objects, with a minimal number of attributes. In this way, the authors in [7] combined rule induction and feature selection concepts, by constructing rules from tolerance classes and corresponding decision concepts.

3.1   QuickRules Algorithm

The QuickRules algorithm proceeds in a similar fashion as the QuickReduct by extending this algorithm with rule induction in the induction phase. The rule set is maintained in separate set, while the fuzzy set C in X set the current degree of coverage of each object in the training data by the current set of rules. In the initial stage of the algorithm, all sets are empty. Then the function $covered(C)$ returns the set of objects that are maximally covered in C defined as:

$$covered(C) = \{x \in X \mid Cov(x) = POS_A(x)\} \tag{11}$$

According to Eq. 11, an object is considered to be included in the set of rules if its membership to $Cov()$ is equal to that of the positive region of the full feature set [7]. A rule for an object y is constructed only when it has not been covered maximally yet – $y \notin covered\,(Cov)$. This is also true for an attribute subset $B \cup \{a\}$, which means the object y should belong maximally to POS $_{B \cup_{\{a\}}}$. This insures that the rule created for object y is included in the final decision rule set when $y's$ tolerance class $R_B(y)$ is fully included in a decision concept. Consequently, the attribute values generated this tolerance class are good indicators of the concept.

The QuickRules algorithm also checks that the new created rules are only added if it has the same or higher coverage than the existing rules [7]. If an existing rule has a coverage that is strictly smaller than the new rules, it is deleted [7]. Finally, if the new rule has fulfilled the required criteria, the rule set is updated together with the coverage. The coverage value is calculated by taking the union of the rule's tolerance class with the current coverage. If all objects are fully covered, no further rules are added to the decision rule set. The process of rule induction terminates when each object belongs to the positive region to the maximum extend, which is same condition for the rule set to cover all objects maximally. Thus, when the QuickRules algorithms has finished, the resulting rule set will cover all objects.

3.2 VQRS Algorithm

The VQRS algorithm revises the hybridization process by introducing vague quantifiers like "*some*" or "*most*" into the definition of the upper and lower approximations. In RST, an object belongs to the upper approximations when it is related to one of the elements in the set, while in lower approximation only retains the object related to all the elements of the set. This is because the methods based on upper and lower approximations use external quantifiers for upper approximation (*sup*), and universal quantifier for lower approximation (*inf*). When using rule induction methods dealing with real-values that comes with property of noise and thus the methods are prone to errors and inconsistency. Therefore, relaxing the definition of the upper and lower approximations, by introducing vague quantifiers like *most* and *some*, the authors in [14] can improve the ability of the method to deal with these problems.

The vague quantifiers is an a object y that belongs to the lower approximation of a set A to the extent the most elements related to y are in A. Equally, the object y belongs to the upper approximations to the extent that some elements related to y are in A [14]. The VQRS algorithm uses linguistic quantifiers to decide to what extend an object belongs to the lower and upper approximation. In VQRS algorithm the couple (Q_1, Q_2) represents the fuzzy quantifiers that model the linguistic quantifiers, then the lower and upper approximation of A by R_a are defined as follows:

$$\underline{R_{Q1}}(y) = Q_1(\frac{|R_B \cap A|}{R_B}) = Q_1(\frac{\sum\limits_{x \in X} \min(R_A(x, y), A(x))}{\sum\limits_{x \in X} R_A(x, y)}) \tag{12}$$

$$\overline{R_{Q2}}(y) = Q_2(\frac{|R_B \cap A|}{R_B}) = Q_2(\frac{\sum\limits_{x \in X} \min(R_B(x, y), A(x))}{\sum\limits_{x \in X} R_B(x, y)}) \tag{13}$$

where the fuzzy set intersection is defined by the min T-norm and the fuzzy set cardinality by the sigma-count operation [14].

4 Experimental Setup and Results

4.1 Experimental Setup

This section presents the experimental setup of the fuzzy and fuzzy-rough classification algorithms: QuickRules and VQRS, over fourteen benchmark datasets from [17, 18].

Classification datasets are used to experimentally evaluate the influence of the methods on classification tasks. Detail description for each dataset can be found in Table 1.

Table 1. Characteristics of the datasets used for experimental evaluation

Dataset for classification	Classification		Dataset for classification	Classification	
	Objects	Attributes		Objects	Attributes
Glass	214	10	Waterclass1	218	117
Heart-Statlog	270	14	Waterclass2	218	117
Ionosphere	351	35	Waterclass3	218	117
Iris	150	5	Waterclass4	218	117
Mushroom	8124	23	Waterclass5	218	117
Vehicle	946	19	Waveform	5000	41
Vote	435	17	Wine	178	14

As we mentioned, we evaluate algorithms based on two T-norms: Algebraic T-norm $(T(x, y) = x * y)$ and the KD T-norm $(T(x, y) = \max(0, x + y - 1))$. One implicator is used for all experiments: KD implicator $(I(x, y) = \max(1 - x, y))$. The VQRS rule indication algorithm as defined in [14] is used with the following settings: $Q_1 = Q(0.1; 0.6)$ and $Q_2 = Q(0.2; 1.0)$, according to Eq. 14. All other algorithms are set to their default values as they are proposed in their original papers.

$$Q_{(\alpha,\beta)}(x) = \begin{cases} 0, & x \leq \alpha \\ \dfrac{2(x-\alpha)^2}{(\beta-\alpha)^2}, & \alpha \leq x \leq \dfrac{\alpha+\beta}{2} \\ 1 - \dfrac{2(x-\alpha)^2}{(\beta-\alpha)^2}, & \dfrac{\alpha+\beta}{2} \leq x \leq \beta \\ 1, & \beta \leq x \end{cases} \tag{14}$$

The experiments dedicated to evaluate the classification tasks, use train (descriptive) and test (predictive) dataset for testing the performance of the algorithm. Standard 10-fold cross validation is used to evaluate the predictive performance of the algorithms (test dataset used for testing – See Table 3.).

4.2 Experimental Results

In both tables, in which we present the results from the experiments, all results for each algorithm are given in the following form: Method[Similarity Metric]. Method set contains the two algorithms: QuickRules (QR) and VQRS (VQ) algorithms, while similarity metric set contains four elements {Sim1 - Eq. 2, Sim2 - Eq. 3, Sim3 - Eq. 4, Sim4 - Eq. 5}. We use the standard evaluation metrics for analysing the obtained results: AUC-ROC metric for both descriptive and predictive performance of the obtained models.

The influence of the two T-norms in combination of four similarity metrics on the QR and VQ rule induction algorithms is depicted in Table 2. Comparing the results between the similarities metrics, it is easy to note that Sim3 metric doesn't have any influence on the T-norm change. In contrast, the other metrics have positive effect on the T-norm change in both algorithms by increasing the accuracy of the algorithm. We must also note that there is no difference between the QR and VQ algorithms improvements when applying T-norms Algebraic compared to KD T-norm (compare rows 1–4 with rows 5–8). It is interesting to note that for some datasets (Mushroom, Vote and Waveform) no influence have been detected by the experiments no matter what similarity metric or T-norm is used. This is interesting since these datasets, especially Mushroom and Waveform are datasets with relatively large number of objects compared to the other datasets. Similar results are obtained when estimating the predictive performance. Comparing the results between descriptive and predictive results (between Tables 2 and 3), it easy to note that Sim3 metric in Table 3 in some cases achieves better results when using VQRS algorithm compared to the QR algorithm when using T-norm Algebraic. In this part of the experiment, metrics Sim1 and Sim4 have lower number of experiments where they achieve better results, compared with the train experiments. The trend of non-influenced datasets is still present in the test experiments as we noticed in the descriptive experiments (Mushroom, Vote and Waveform).

Table 2. Results of evaluation using AUC-ROC metric (difference between T-norm Algebraic and T-norm KD) using the train dataset for testing (descriptive performance). Bolded results show higher performance for T-norm Algebraic, underlined results show lower performance for T-norm KD and N/A denotes that for some T-norm there is no rules in the rule set.

Dataset/ Algorithm	QR^1	QR^2	QR^3	QR^4	VQ^1	VQ^2	VQ^3	VQ^4
Glass	**0.39**	**0.17**	**0.08**	**0.39**	**0.71**	**0.09**	0.06	0.71
Heart-Stat	**0.10**	**0.04**	0.00	**0.10**	**0.04**	**0.50**	0.00	0.40
Ionosphere	**0.08**	**0.01**	0.00	**0.08**	0.35	N/A	0.01	0.34
Iris	**0.15**	**0.16**	0.00	**0.15**	0.94	0.47	0.06	0.94
Mushroom	0.00	0.00	0.00	0.00	0.00	0.00	0.00	0.00
Vehicle	**0.35**	**0.17**	**0.01**	**0.35**	0.75	0.50	0.04	0.75
Vote	0.00	0.00	0.00	0.00	0.00	0.00	0.00	0.00
Waterclass1	**0.09**	<u>0.01</u>	0.00	**0.09**	0.36	0.07	0.00	0.36
Waterclass2	**0.17**	<u>0.01</u>	0.00	**0.17**	0.32	0.05	0.00	0.32
Waterclass3	**0.15**	**0.01**	0.00	**0.15**	0.31	0.07	0.00	0.31
Waterclass4	**0.22**	**0.02**	0.00	**0.22**	0.38	0.05	0.00	0.38
Waterclass5	**0.14**	**0.01**	0.00	**0.14**	0.14	0.01	0.00	0.14
Waveform	0.00	0.00	0.00	0.00	0.00	0.00	0.00	0.00
Wine	**0.02**	0.00	0.00	**0.02**	N/A	N/A	0.00	N/A

Table 3. Results of evaluation using AUC-ROC metric (difference between T-norm Algebraic and T-norm KD) using the test dataset for testing (predictive performance). Bolded results show higher performance for T-norm Algebraic, underlined results show lower performance for T-norm KD and N/A denotes that for some T-norm there is no rules in the rule set.

Dataset/ Algorithm	QR^1	QR^2	QR^3	QR^4	VQ^1	VQ^2	VQ^3	VQ^4
Glass	**0.30**	**0.12**	**0.07**	**0.30**	**0.17**	**0.10**	**0.05**	**0.17**
Heart-Stat	**0.13**	**0.02**	0.03	**0.13**	**0.05**	**0.38**	0.00	**0.05**
Ionosphere	**0.15**	**0.01**	0.02	**0.15**	**0.31**	N/A	0.00	**0.05**
Iris	**0.14**	**0.15**	**0.01**	**0.14**	**0.44**	**0.46**	**0.07**	**0.44**
Mushroom	0.00	0.00	0.00	0.00	0.00	0.00	0.00	0.00
Vehicle	**0.32**	**0.19**	**0.01**	**0.32**	**0.76**	**0.42**	**0.01**	**0.76**
Vote	0.00	0.00	0.00	0.00	0.00	0.00	0.00	0.00
Waterclass1	0.10	**0.04**	**0.01**	**0.10**	**0.13**	**0.05**	**0.03**	**0.13**
Waterclass2	0.01	0.01	0.02	0.01	**0.04**	**0.02**	0.01	**0.04**
Waterclass3	0.01	**0.04**	0.01	0.01	**0.06**	**0.05**	**0.01**	**0.06**
Waterclass4	0.06	**0.03**	0.01	0.06	0.08	0.03	**0.02**	0.08
Waterclass5	0.04	**0.04**	0.03	0.04	**0.01**	**0.00**	0.01	**0.01**
Waveform	0.00	0.00	0.00	0.00	0.00	0.00	0.00	0.00
Wine	**0.01**	0.00	0.00	**0.01**	N/A	N/A	**0.00**	N/A

5   Conclusion

In this paper, we investigated the influence of Algebraic T-norms compared with the T-norm KD in combinations with four fuzzy similarity metrics on two rule induction algorithms. The two rule induction algorithms QuickRules and VQRS are based on fuzzy-rough concepts of upper and lower approximations. The experimental results presented in the paper, present some interesting results in both training and testing experiments. In most of the cases, the implementation of the Algebraic T-norm increased the values of the classification performance towards positive trend. There are also datasets that are no influenced by the T-norm, not even when we change the similarity metrics. This is an interesting finding, and our future plans include investigating the impact of implicators on the rule induction methods accuracy, as well as other T-norms and similarity metrics on method accuracy, not only on these types of datasets, but other also.

Acknowledgement. This work was partially financed by the Faculty of Computer Science and Engineering at the Ss. Cyril and Methodius University in Skopje.

References

1. Drobics, M., Bodenhofer, U., Klement, E.P.: FS-FOIL: an inductive learning method for extracting interpretable fuzzy descriptions. Internat. J. Approx. Reason **32**, 131–152 (2003)
2. Prade, H., Richard, G., Serrurier, M.: Enriching relational learning with fuzzy predicates. In: Proceedings of Principles and Practice of Knowledge Discovery in Databases, pp. 399–410 (2003)
3. Cloete, I., Van Zyl, J.: Fuzzy rule induction in a set covering framework. IEEE Trans. Fuzzy Syst. **14**(1), 93–110 (2006)
4. Xie, D.: Fuzzy associated rules discovered on effective reduced database algorithm. In: Proceedings of the 14th IEEE International Conference on Fuzzy Systems, pp. 779–784 (2005)
5. Marin-Blazquez, J.G., Shen, Q.: From approximative to descriptive fuzzy classifiers. IEEE Trans. Fuzzy Syst. **10**(4), 484–497 (2002)
6. Qin, Z., Lawry, J.: LFOIL: linguistic rule induction in the label semantics framework. Fuzzy Sets Syst. **159**(4), 435–448 (2008)
7. Jensen, R., Cornelis, C., Shen, Q.: Hybrid fuzzy-rough rule induction and feature selection. In: IEEE International Conference In Fuzzy Systems (FUZZ-IEEE 2009), pp. 1151–1156 (2009)
8. Pawlak, Z.: Rough Sets: Theoretical Aspects of Reasoning About Data. Kluwer Academic Publishing, Boston (1991)
9. Hsieh, N.-C.: Rule extraction with rough-fuzzy hybridization method. In: Washio, T., Suzuki, E., Ting, K.M., Inokuchi, A. (eds.) PAKDD 2008. LNCS, vol. 5012, pp. 890–895. Springer, Heidelberg (2008). doi:10.1007/978-3-540-68125-0_89
10. Shen, Q., Chouchoulas, A.: A rough-fuzzy approach for generating classification rules. Pattern Recogn. **35**(11), 2425–2438 (2002)
11. Greco, S., Inuiguchi, M., Slowinski, R.: Fuzzy rough sets and multiple-premise gradual decision rules. Int. J. Approximate Reasoning **41**, 179–211 (2005)
12. Wang, X., Tsang, E.C.C., Zhao, S., Chen, D., Yeung, D.S.: Learning fuzzy rules from fuzzy samples based on rough set technique. Inf. Sci. **177**(20), 4493–4514 (2007)
13. Hong, T.P., Liou, Y.L., Wang, S.L.: Learning with hierarchical quantitative attributes by fuzzy rough sets. In: Proceedings Joint Conference on Information Sciences. Advances in Intelligent Systems Research (2006)
14. Cornelis, C., De Cock, M., Radzikowska, A.: Vaguely quantified rough sets. In: Proceedings 11th International Conference on Rough Sets, Fuzzy Sets, Data Mining and Granular Computing (RSFDGrC 2007), pp. 87–94 (2007)
15. Jensen, R., Shen, Q.: Computational Intelligence and Feature Selection: Rough and Fuzzy Approaches. Wiley-IEEE Press, Hoboken (2008)
16. Blake, C.L., Merz, C.J.: UCI Repository of machine learning databases. Irvine, University of California (1998). http://archive.ics.uci.edu/ml/
17. Jensen, R.: Fuzzy-Rough Datasets. Richard Jensen Datasets (2017). http://users.aber.ac.uk/rkj/site/?page_id=81
18. Naumoski, A., Mirceva, G., Mitreski, K.: A novel fuzzy based approach for inducing diatom habitat models and discovering diatom indicating properties. Ecol. Inform. **7**(1), 62–70 (2012)

An Investigation of Human Trajectories in Ski Resorts

Boris Delibašić ⓘ, Sandro Radovanović(✉) ⓘ, Miloš Jovanović ⓘ,
Milan Vukićević ⓘ, and Milija Suknović

Faculty of Organizational Sciences, University of Belgrade,
Jove Ilića 154, 11000 Belgrade, Serbia
{boris.delibasic,sandro.radovanovic}@fon.bg.ac.rs

Abstract. Analyzing human trajectories based on sensor data is a challenging
research topic. It has been analyzed from many aspects like clustering, process
mining, and others. Still, less attention has been paid on analyzing this data based
on hidden factors that drive the behavior of people. We, therefore, adapt the
standard matrix factorization approach and reveal factors which are interpretable
and soundly explain the behavior of a dynamic population. We analyze the motion
of a skier population based on data from RFID-recorded ski entrances of skiers
on ski lift gates. The approach is applicable to other similar settings, like shopping
malls or road traffic. We further applied recommender systems algorithms for
testing how well we can predict the distribution of ski lift usage (number of ski
lift visits) based on hidden factors, but also on other benchmark algorithms. The
matrix factorization algorithm showed to be the best recommender score predictor
with an RMSE of 2.569 ± 0.049 and an MAE of 1.689 ± 0.019 on a 1 to 10 scale.

Keywords: Ski lift transportation patterns · Principal component analysis ·
Matrix factorization · Recommender systems

1 Introduction

The problem of skier movements has been previously analyzed with clustering algorithms [6]. It has been identified that skiers tend to group their ski trajectory around a
specific ski lift which best determines other lifts they might choose based on vicinity
and skillfulness level. In [4] authors use several fuzzy algorithms for analyzing ski
trajectories. They propose two type of skier behavior: variety seeking and loyal. In this
paper, we expand these researches by identifying hidden factors that drive such behavior.

In order to fulfill skiers' needs, ski resorts strive to understand patterns of skiers'
transportation. Radio frequency identification (RFID) collected data can help in understanding those patterns, as most ski resorts are equipped with such technology. Usually,
ski resorts keep track of all ski lift gate entrances. This generates massive data which is
highly underutilized.

B. Delibašić and S. Radovanović—Two authors contributed equally.

© Springer International Publishing AG 2017
D. Trajanov and V. Bakeva (Eds.): ICT Innovations 2017, CCIS 778, pp. 130–139, 2017.
DOI: 10.1007/978-3-319-67597-8_13

In this paper, we analyzed the 2009 season skiing population from Mt. Kopaonik and we set the following research questions:

– Can we reveal factors that drive the choice of ski lifts? and
– Can we predict for each skier the distribution of ski lift visits per day?

With first research question, we wanted to investigate and reveal hidden factors that drive decisions on spatial trajectories of skiers. Additionally, we show that, based on these factors, predictions on the distribution of visits to ski lifts can be made for each skiing day. We observe that the hidden factors enable recommender algorithms to make sound predictions. Besides that, hidden factors allow the ski resort to understand the drivers for ski lift choice in ski resorts, and can help in planning of future capacities of the resort, having in mind the satisfaction of the skiing population.

The problem of predicting the skier distribution of ski lift visits is challenging since there are many ski lifts (multi-label classification), and there are no additional features about skiers (lack of input features). Traditional data mining and machine learning algorithms perform poorly or cannot even be applied in these settings. As a suitable method for prediction, we, therefore, apply recommender systems [13]. For this purpose, we created and evaluated several rating prediction recommender system algorithms. In order to evaluate the performance of the algorithm, we used root mean squared error (RMSE) and mean absolute error (MAE). Every obtained result is validated using 10-fold cross validation.

These analyses are of crucial importance for ski resort management because, currently, there is no control for mismatches between the available ticket formats and skiers' transportations. Therefore, benefits from this approach can be viewed with the redesign of the existing ski lift tickets formats. This research demonstrates that not all skiers' transportation needs are satisfied and with an understanding of skiers' behavior and with proper prediction model. Ski resort management can fill the gap between mismatch and skiers' transportation. The technology for coding new ski lift formats already exists at a ski resort, so providing new ski lift tickets formats is feasible. Besides adapting to skiers' needs, the new ski lift ticket formats could also provide ski resorts a tool for better utilization of ski lifts by providing more affordable ski lifts tickets on less occupied lifts or less occupied time periods. With this approach ski resort can (1) provide a larger variety of ski tickets, (2) produce better pricing models and consequently improve income, (3) motivate skiers to search for the best ski ticket price, and (4) improve utilization of ski lifts.

The remainder of this paper is structured as follows. In Sect. 2, background on similar research is presented. Section 3 presents materials and methods. Section 4 present results, while Sect. 5 concludes the paper.

2 Background

Data analysis in ski resorts management is often used as a decision support tool for better planning of operations and budgets. In paper [10] skier-days prediction was performed using ensemble methods. Their study yields at providing ski resorts trustworthy predictions for efficient planning of operations in a ski resort and planning offers for skiers. Ski lift ticket pricing was also studied [15], as well as pricing models for hotels [16].

On the other hand, analyzing RFID transportation data in order to get patterns of usage can be seen in the case of London metro stations in the paper [11]. Human trajectories have been analyzed using clustering methods, where authors identify 15 temporal clusters in the data.

Most of the papers regarding ski data discuss the risk of ski injuries [14] and more recently also the prediction of ski injuries [3]. To the best of our knowledge, there are only several papers which use RFID data for skiing pattern analysis. In [4] four modifications of the fuzzy k-medoids algorithm were used to identify clusters in a skier day in the Val Gardena, Italy. The original dataset was re-coded in order to discard information about ski lifts, i.e. only transitions were recorded. Several cluster models were tested with a different number of clusters and the finding was that there are two types of skiers, variety seekers, who tend to change ski lifts often during a skiing day, and loyal skiers, who tend to stick to one or few ski lifts. In paper [6] a clustering approach was used on Mt. Kopanik, Serbia, for analyzing the RFID data. Interesting clusters such as nine temporal clusters and seventeen spatial clusters were identified. It was concluded that ski patterns most often have one ski lift as a dominant on a day. The choice of this lift also influences the choice of other ski lifts which are combined with the most dominant lifts. The dominated ski lifts are usually close to the dominant one and have similar skillfulness level requirements for skiers.

With clustering methods, one can obtain a snapshot situation in data and identify representatives (centroids) in data. However, one cannot make predictions with clustering methods. But, revealing hidden factors from data can be used for making predictions. The motivation for application of recommender systems came from [5] where they predicted the occurrence of disease. In has been shown that recommender systems are more suitable compared to traditional both single and multi-output machine learning methods. Therefore, recommender systems provided better results in terms of classification accuracy.

Human motion prediction is commonly applied in surveillance [1, 17] and they use algorithms with the same goal as machine learning, i.e. predict a posteriori probability where a person will go. In paper [7] intention of the person is predicted. They used time series analysis and achieved satisfactory results. The main conclusion is that simpler models provide better performance. However, our opinion is that performance can be improved especially since we don't observe time when skier checked ski lift but how many times a skier will visit various ski lift.

3   Materials and Methods

3.1   Data

Kopaonik is the largest ski resort in Serbia where skiing season starts from December 1st and lasts until May 4th, depending on weather conditions.

In order to obtain plausible predictive models, we used and analyzed only skiers' transportation with 6-days tickets. The dataset was prepared so a row presents a skier skiing on a certain day, and the columns present the ski lifts used. The table content presents the frequency of ski lift usage by skiers on certain days.

We analyze only the subset of the most crowded season periods (1.1.2009.–31.3.2009). This period represents season peak and it covers 89.27% of the overall season ski lift transportations. During this period, a total of 17 ski lifts were operating on the ski resort and they provide access to 32 ski slopes of which 15 are blue-easy, 10 are red-medium, and 7 are black-difficult slopes.

The data set used for this research contains 1,248,755 ski lift transportation records that correspond to 21,121 skiers. On average, there are 14,031 ski lift transportations per day. The dataset has been filtered for infeasible ski lift transitions, such as records with travel time equal to zero or unrealistically low or high. This dataset was further processed in order to be usable for PCA and recommender systems. The dataset was prepared so a row presents a skier skiing on certain day (skier day), and the columns are ski lifts. There are over 400,000 rows and 17 columns. Each column presents an output which should be predicted. Values in matrix present number of ski lift visits. Skiers had at average 2.714 (±2.484) ski lift visits per lift where the maximum value for one ski lift is 51 visits. More details about data are presented in Sect. 4.1 Descriptive Data Analysis.

3.2 Experimental Setup

First, we employed descriptive statistics and data visualization in order to understand data. Namely, we used pivot tables and histograms. This allowed us to understand the problem. This information will also be important to the readers since ski lift visits have unique characteristics.

In order to get answers to the first research question (Can we reveal factors that drive the choice of ski lifts), we created and evaluated several matrix factorization techniques. Here we present only principal component analysis results as the results obtained with PCA were most informative. We selected six principal components since they cover more than 75% of the variation in data. With those components we were able to understand typical behavior of skiers. In other words, we found latent factors which are easily interpreted and present useful information for ski resorts.

In order to get an answer to the second research question (Can we predict for each skier the distribution of ski lift visits per day), we utilized recommender system algorithms. The reason why we used recommender system is that we have no input features besides skier and ski lift, which mean that traditional data mining and machine learning algorithms are not suitable. Additionally, we have multiple outputs. Namely, we created and evaluated random prediction, global average, user item baseline (which were used as baseline method), Slope One algorithm [12], BiPolar Slope One [12], Matrix Factorization rating prediction [9], Biased Matrix Factorization [8] and Matrix factorization with factor-wise learning [2]. Each recommender system was created and evaluated using 10-fold cross validation in order to prevent overfitting. In order to perform model selection, we measured mean absolute error (MAE) and root mean squared error

(RMSE). A number of factors and iterations are learned using inner 10-fold cross validation which selected the best combination using grid search over parameter space optimizing MAE. Experiments were conducted using the RapidMiner advanced analytics environment.

4   Results

In this section, we present the results and answers for research questions presented in the Experimental setup section.

4.1   Descriptive Data Analysis

In Table 1 one can observe ski lifts and their distribution of ski lift transportations. We can notice that three ski lifts are most popular. Namely, Pancicev vrh with 17.18% of ski lift transportations, Mali Karaman with 17.97% and Karaman greben with 21.49%. It is worth to notice that 46.58% of skiers starts from Mali Karaman.

Table 1. Ski lifts and their basic characteristics

Ski-lift	Code	Difficulty	% of ski lift visits
Suncana dolina	SUN	Blue	1.82%
Malo jezero	JEZ	Blue	4.14%
Masinac	MAS	Blue	4.58%
Centar	CEN	Blue-Red	0.90%
Suvo rudiste	SUV	Blue-Red	1.18%
Pancicev vrh	PAN	Red	17.18%
Duboka 1	DB1	Red-Black	7.20%
Krcmar	KRC	Red	0.88%
Duboka 2	DB2	Blue	6.53%
Gvozdac	GVO	Black	2.43%
Karaman greben	KGB	Blue	21.49%
Knezevske bare	KNE	Blue	2.41%
Mali Karaman	MAK	Blue	17.97%
Marine vode	MAR	Blue	4.18%
Karaman	KAR	Blue	3.51%
Jaram	JAR	Blue	0.90%
Gobelja relej	GOR	Red-Black	2.43%

We can also observe on Fig. 1 that ski lifts have small or non-existing linear relationship among each other with the majority of ski lifts having covariance near to zero. However, we can notice that some popular ski lifts have a negative relationship between each other (darker color). Also, we can notice that there is some negative relationship between Karaman greben and Duboka 1 and Duboka 2 ski lifts and between Mali Karaman and Masinac. On the other side, we can observe a positive relationship between

Mali Karaman and Marine vode and between Pancicev vrh and Duboka 1 ski lifts. This has to do with the fact that these lifts cover areas with different skillfulness level, and therefore don't attract the same population of skiers.

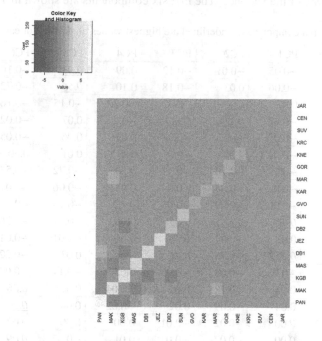

Fig. 1. Covariance plot

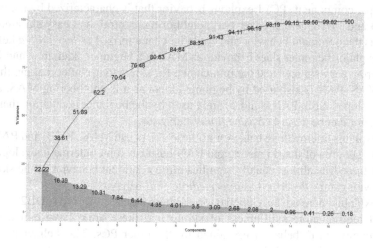

Fig. 2. Percentage of explained variance in PCA

Further, we performed principal components analysis in order to get reduced data set and to provide interpretation. Percentage of explained variance is presented in

Fig. 2. From Fig. 2 we notice that the "elbow" of the curve is with 6 principal components, i.e. after the sixth component, the increase of variance explained by adding new components is not huge, as it was the sixth component. The six principal components explained 76.48% of the variance. The first six components are shown in Table 2.

Table 2. Principal components. Underlined are highest values of ski lifts for each component

Ski-lift	PC1	PC2	PC3	PC4	PC5	PC6
SUN	−0.05	−0.01	−0.12	0.09	_0.23_	−0.16
JEZ	−0.06	0.03	−0.18	0.10	_0.56_	−0.72
MAS	0.07	_0.28_	−0.32	−0.87	−0.15	−0.16
PAN	−0.82	_0.11_	_0.51_	−0.22	0.07	−0.02
SUV	−0.01	0.00	−0.03	0.02	0.06	−0.05
CEN	−0.01	0.00	−0.01	0.01	0.01	0.00
DB1	−0.19	−0.10	−0.11	_0.20_	−0.72	−0.53
KRC	−0.02	−0.03	−0.04	0.04	−0.06	−0.01
DB2	−0.06	−0.09	−0.09	_0.12_	−0.19	−0.05
KGB	_0.46_	_0.52_	_0.63_	0.06	−0.11	−0.26
KNE	0.01	−0.07	0.00	0.00	−0.01	−0.01
MAK	_0.25_	−0.76	_0.41_	−0.34	0.03	−0.22
GVO	−0.02	−0.08	−0.05	0.04	−0.14	−0.03
MAR	0.04	−0.12	0.04	−0.04	0.02	0.06
KAR	0.04	−0.09	0.02	−0.02	0.04	_0.10_
JAR	0.00	−0.02	−0.01	0.00	0.00	0.02
GOR	0.00	−0.07	−0.03	0.01	−0.02	_0.09_

Principal component (PC) 1 explains behavior that is characterized by the usage of lifts KGB and MAK, which cover two neighboring, central, and easy ski areas. This behavior is the dominant behavior and could be characterized as *less-skilled* behavior.

PC2 explains beginner skier behavior as MAS is a beginner skier area, and KGB is a neighboring easy ski area and the first choice for skiers to use after making their first steps at MAS. PAN is also used by beginner skiers as it is neighboring MAS, and has some blue slopes. Still, PAN is more rarely used by the beginner population than KGB. This behavior can be described as *beginner* behavior.

PC3 explains intermediate behavior as these skiers use MAK, KGB, and PAN. This covers a huger area of the ski resort, and PAN has also some intermediate slopes. Still using this three lifts almost equally signifies more confident behavior in the ski resort. This behavior can be described as *intermediate skill* behavior.

PC4 explains more advanced behavior, as these skiers cover DB1, DB2, which are areas that require more skillfulness of skiing and these areas have higher vertical descents compared to behavior explained by the other PCs. This behavior could be characterized as *advanced* behavior.

PC5 explains skiing behavior that is located in the southern areas of the ski resort. These are skiers that know the ski resort better (they need to know how to get to these slopes), like easy slopes for skiing, and like less crowded areas. This behavior could be labeled as *no-stress* behavior.

PC6 explains behavior that is located in the northern parts of the ski resort (KAR, GOR). These skiers know the ski resort very well, like more difficult slopes, and less crowded areas. This behavior could be labeled as *skilled-and-lonely* behavior.

4.2 Recommender Systems

Performances of recommender systems are presented in Table 3. Our goal was to predict for each ski lift on a 1 to 10 scale the number of ski lift visits. If skier had more than 10 ski lifts visits (a rather rare event) were also regarded as having 10 ski lift visits.

Table 3. Performances of recommender systems

Algorithm	MAE	RMSE
Random	27.228 ± 0.088	23.047 ± 0.087
Global average	5.486 ± 0.053	2.700 ± 0.016
User item baseline	5.388 ± 0.052	2.569 ± 0.014
Slope one	2.878 ± 0.048	1.831 ± 0.019
BiPolar slope one	2.841 ± 0.049	1.744 ± 0.018
Matrix factorization	**2.569 ± 0.049**	**1.689 ± 0.019**
Biased matrix factorization	3.162 ± 0.076	1.946 ± 0.047
Factor wise matrix factorization	6.568 ± 0.731	2.827 ± 0.293

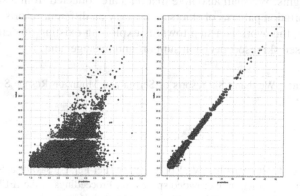

Fig. 3. (a) User item baseline and (b) Matrix factorization predictions vs. real values

We can observe that recommender systems have rather good performances with MAE less than 2. This means that on average we miss 2 ski lift visits from 10. Also, it is worth to notice that baseline methods (Random guessing, Global average and User Item Baseline) have higher MAE. The best performing algorithm was Matrix Factorization which uses singular value decomposition, i.e. hidden factors that drive ski decisions. Further, inspection shows that this algorithm also has the lowest RMSE which means that it does not have large misses. On Fig. 3 we present performance of User Item

Baseline (on the left side) Matrix Factorization model (on the right side). Predictions are presented on the x-axis, while real values are presented on the y-axis.

5   Conclusions

Usage of RFID technologies to monitor and control skier transportation through ski lifts leads to massive data acquisition. However, data obtained from this is still underused. This paper deals with the problem of studying skiers' transportation patterns recognition and prediction in order to gain insight into skiers' behavior and for revealing information that could be used for optimization of skiers' satisfaction in ski resorts.

In this paper we gained insight into skier behavior through descriptive statistics (Table 1), creation and interpretation of covariance plot and PCA model (Figs. 1, 2 and Table 2). Using this we were able to find interesting patterns in ski lift visits which can be used in ski resort decision making. Further, we created and evaluated recommender systems in order to predict the number of ski visits (Table 3 and Fig. 3). Obtained results show that we can accurately predict with MAE less than 2 how many ski visits skier will have. We note that this prediction is made solely based on the behavior of other skiers without other additional information about skier (demographic, behavior, the level of knowledge etc.), ski lift or other.

A major limitation of this paper is lack data about skier, ski lifts or other useful data, so analysis and prediction could be improved. Authors are aware that, although the sample represents the biggest group of the population, using the whole dataset could yield richer insights. We shall also note that data are collected from a single ski resort and that transportation patterns of skiers could vary on different resorts.

To conclude, this work is a step towards exploiting existing, underused resources for evidence-based ski resort decision support and management.

Acknowledgement. We thank Ski resorts of Serbia and Mountain Rescue Service Serbia for allowing us to use data.

References

1. Alahi, A., Goel, K., Ramanathan, V., Robicquet, A., Fei-Fei, L., Savarese, S.: Social LSTM: human trajectory prediction in crowded spaces. In: Proceedings of the IEEE Conference on Computer Vision and Pattern Recognition, pp. 961–971 (2016)
2. Bell, R., Koren, Y., Volinsky, C.: Modeling relationships at multiple scales to improve accuracy of large recommender systems. In: Proceedings of the 13th ACM SIGKDD International Conference on Knowledge Discovery and Data Mining, pp. 95–104. ACM, August 2007
3. Bohanec, M., Delibašić, B.: Data-mining and expert models for predicting injury risk in ski resorts. In: Delibašić, B., Hernández, Jorge E., Papathanasiou, J., Dargam, F., Zaraté, P., Ribeiro, R., Liu, S., Linden, I. (eds.) ICDSST 2015. LNBIP, vol. 216, pp. 46–60. Springer, Cham (2015). doi:10.1007/978-3-319-18533-0_5
4. D'Urso, P., Massari, R.: Fuzzy clustering of human activity patterns. Fuzzy Sets Syst. **215**, 29–54 (2013)

5. Davis, D.A., Chawla, N.V., Christakis, N.A., Barabási, A.L.: Time to CARE: a collaborative engine for practical disease prediction. Data Min. Knowl. Discov. **20**(3), 388–415 (2010)
6. Delibašić, B., Marković, P., Delias, P., Obradović, Z.: Mining skier transportation patterns from ski resort lift usage data. IEEE Trans. Hum. Mach. Syst. **47**(3), 417–422 (2016)
7. Ferrer, G., Sanfeliu, A.: Bayesian human motion intentionality prediction in urban environments. Pattern Recognit. Lett. **44**, 134–140 (2014)
8. Gemulla, R., Nijkamp, E., Haas, P.J., Sismanis, Y.: Large-scale matrix factorization with distributed stochastic gradient descent. In: Proceedings of the 17th ACM SIGKDD International Conference on Knowledge Discovery and Data Mining, pp. 69–77. ACM, August 2011
9. Jamali, M., Ester, M.: A matrix factorization technique with trust propagation for recommendation in social networks. In: Proceedings of the Fourth ACM Conference on Recommender Systems, pp. 135–142. ACM, September 2010
10. King, M.A., Abrahams, A.S., Ragsdale, C.T.: Ensemble methods for advanced skier days prediction. Expert Syst. Appl. **41**(4), 1176–1188 (2014)
11. Lathia, N., Smith, C., Froehlich, J., Capra, L.: Individuals among commuters: building personalised transport information services from fare collection systems. Pervasive Mobile Comput. **9**(5), 643–664 (2013)
12. Lemire, D., Maclachlan, A.: Slope one predictors for online rating-based collaborative filtering. In: Proceedings of the 2005 SIAM International Conference on Data Mining, pp. 471–475. Society for Industrial and Applied Mathematics, April 2005
13. Lu, J., Wu, D., Mao, M., Wang, W., Zhang, G.: Recommender system application developments: a survey. Decis. Support Syst. **74**, 12–32 (2015)
14. Ruedl, G., Kopp, M., Sommersacher, R., Woldrich, T., Burtscher, M.: Factors associated with injuries occurred on slope intersections and in snow parks compared to on-slope injuries. Accid. Anal. Prev. **50**, 1221–1225 (2013)
15. Schamel, G.: Weekend vs. midweek stays: modelling hotel room rates in a small market. Int. J. Hosp. Manage. **31**(4), 1113–1118 (2012)
16. Wolff, F.C.: Lift ticket prices and quality in French ski resorts: Insights from a non-parametric analysis. Eur. J. Oper. Res. **237**(3), 1155–1164 (2014)
17. Xie, D., Shu, T., Todorovic, S., Zhu, S.C.: Modeling and inferring human intents and latent functional objects for trajectory prediction. *arXiv* preprint arXiv:1606.07827 (2016)

Courses Content Classification
Based on Wikipedia and CIP Taxonomy

Atanas Dimitrovski$^{(\boxtimes)}$, Ana Gjorgjevikj$^{(\boxtimes)}$, and Dimitar Trajanov

Faculty of Computer Science and Engineering,
Ss. Cyril and Methodius in Skopje, Skopje, Macedonia
a.dimitrovski5@gmail.com, ana.gorgevic@gmail.com,
dimitar.trajanov@finki.ukim.mk

Abstract. The amount of online courses and educational content available on the Internet is growing rapidly, leaving students with large and diverse number of choices for their areas of interest. The educational content is spread into diverse e-learning platforms, making its search and comparison even more challenging. Classifying educational content into a standardized set of academic disciplines or topics can improve its search, comparison and combination to better meet students' inquiries. In this paper we make use of well-known techniques from Information Retrieval to map course descriptions into two common sets of topics, one manually created and well-controlled, i.e. CIP, and one collaboratively created, i.e. Wikipedia. We then analyze and compare the results to see how the size of the topic schemes and their associated data, such as textual descriptions, affect the accuracy of the end results.

Keywords: E-learning · Text classification · Wikipedia · CIP

1 Introduction

The Internet has revolutionized today's education. We have seen a rapid growth in the number of online courses available from different platforms. Sites like Coursera[1], Udacity[2] and edX[3] offer courses from number of educational fields. The easy accessibility, the big number of choices and the fact that most are free of charge draw a lot of users. Lots of students who cannot attend college because of distance, price or some other problem, can now easily get education through this kind of courses. With the number of massive open online courses (MOOC) increasing, searching for the right course, comparison and combination of courses from different providers, identifying prerequisites and designing curricula that fit user's needs, becomes a nontrivial task. The use of different classification taxonomies for description of the academic fields the courses belong to or folksonomies to specify what the courses are about, which differ between

---

[1] https://www.coursera.org/.
[2] https://www.udacity.com/.
[3] https://www.edx.org/.

© Springer International Publishing AG 2017
D. Trajanov and V. Bakeva (Eds.): ICT Innovations 2017, CCIS 778, pp. 140–153, 2017.
DOI: 10.1007/978-3-319-67597-8_14

the platforms and are sometimes rather shallow or rather noisy make the afore-mentioned problem even worse. The need for an automated solution to the course classification problem and the necessity for a comprehensive dataset of academic disciplines against which courses can be classified was the main motivation for this work.

In this paper we propose a system which classifies courses into a set of academic disciplines, utilizing courses short textual descriptions. The main goal of this work is comparison of the classification results using two sources of academic disciplines, one manually created, controlled, but more restrictive and more slowly evolving, i.e. the Classification of Instructional Programs (CIP) taxonomy[4], and one collaboratively created, containing the most current information, but a lot of noise as well, i.e. Wikipedia[5]. Using standard techniques from Information Retrieval, the vector space model and TF-IDF weighting scheme, we measure the similarity of the course description and the description of the academic disciplines to classify the course with the most similar disciplines. We assume that the most similar academic description to the course description is in fact the best candidate for the course to be classified with, where the similarity is measured as the cosine of the angle between the vectors. Analysis of the results, supported by examples, is provided.

2   Related Work

A lot of work on classifying courses and building systems for recommending courses has been done over the years. A system for recommending courses based on students' historical grades was proposed by Apaza et al. [1]. It is based on Latent Dirichlet Allocation (LDA) and recommends courses from the Massive Online Open Courses (MOOCs). LDA is used as feature descriptor of courses, whereas machine learning is used to predict user preferences based on grading information in college, creating a user vector. Based on this information, a rating is given to a specified MOOC course. POS (part of speech) weighted algorithm has been proposed by Xu [11] for easier searching of MOOC courses. POS algorithm is based on the tf-idf algorithm and tries to expand on it by taking into consideration the meaning of the terms. Most weight is given to verbs and nouns. Adverbs and adjectives are given less weight than verbs and nouns, but more weight from other types of terms. The tf-idf formula is changed by changing the tf part where instead of simply calculating the ratio of term frequency and document length, a weight value is assigned to each term according to its POS and then the ratio of the weighed term frequency and weighed document length is found. Based on the proposed POS Weighted TF-IDF algorithm, a vertical MOOC search engine COURSES was developed. Zhuhadar et al. [12] envision a different approach in building MOOCs platforms. The classic hierarchical organization of courses used by most MOOCs, is replaced with semantic relationships

---

[4] https://nces.ed.gov/ipeds/cipcode/Default.aspx?y=55.
[5] https://en.wikipedia.org.

with learning objects using Semantic Web and social networks. Various Collaborative Semantic Filtering technologies can be used for building semantically Enriched MOOCs management systems. They use the Social Semantic Collaborative Filtering model to create a prototype of an enriched MOOCs' platform called JeromeDL. They claim that the overall social network becomes better informed when using the social semantic collaborative filtering technique to disseminate information, and manage to prove it with an experiment which constructs a social network model that corresponds to the small world phenomena. Shatnawi et al. [8] try to improve some of the problems surrounding MOOCs (quality of individual tutoring, communication and feedback) using data mining algorithms. They go over different text mining, text-clustering and topic modeling algorithms and propose an intelligent MOOCs feedback management system. Feedbacks from the instructors are stored in the system. Students interact with the system and produce large amounts of streaming data. This data is labeled with entity names from a domain ontology created from the courses, and then, according to the labels, it is clustered. In the cluster, new text data is compared to the existing, to find the most similar text, and the feedback for that text is returned to the student. If no similar text is found, a message to the instructor is send, to provide feedback for that particular text.

Course builder using Wikipedia has been proposed by Limongelli et al. [6], which tends to help teachers create courses and share knowledge using Wikipedia. A query is constructed in a module called "Terms Manage" from a user input, and the Wikipedia search engine is queried. The retrieved pages are filtered by means of the cosine similarity between the query formed by the Terms Manager module and the HTML pages, with a tf-idf terms-weighting technique. A module called "The Pages Manager" represents the results To the user, in a suitable graphic way. SemCCM [3] is a system that tries to utilize Semantic Web tools, methods and datasets for automatic semantic annotation of Learning Management Systems (LMS) courses. The system is envisioned as a complement to an existing LMS, retrieving courses content and user data from the existing LMS. It identifies the DBpedia resources which are most relevant to a given course, as well as the Wikipedia categories, used as more general areas covered by the courses. Users' competencies are inferred based on their completed courses and the ranked DBpedia resources for each course. Relevance of the DBpedia resources for the courses is calculated using a variation of tf-idf. Gasparetti et al. [2] use the Wikipedia content as source of information for identifying prerequisite relations between learning objects. After identifying the topics, i.e. Wikipedia articles that appear in the learning content, they evaluate a set of hypothesis that allow automatic inference of prerequisite relations. Tam et al. [10] propose a complete e-learning system network that tries to automate/semi-automate the process from searching for any possible relationship among involved concept or modules to the ultimate generation and optimization of the resulting learning paths, especially in the presence of incomplete information stored in the form of learning object. Performing explicit semantic analysis (ESA) on the course materials has been proposed, additionally using relevant Wikipedia articles to fill in any missing information. This is followed by using a heuristic based concept clustering algorithm to group relevant concepts before finding their relationship measures. Lastly an evolutionary optimizer is used to return the optimal learning sequence, and later on it is improved by using hill-climbing heuristic. This represents an attempt to provide more personalized and systematic advice through optimizing the learning paths as more suitable and viable solution for individual student or group of students, thus allowing the instructor to adopt a different learning path of concepts that may help the students to achieve better academic performance.

3   Course Classification

This section describes the classification datasets which were used for classifying the courses, the system's architecture and its implementation, as well as the algorithm used for classifying the courses.

3.1   Datasets

The development of the course classification system required definition of a set of classes against which the courses will be classified. The classes should represent instructional programs and have some meaningful textual description available, needed for the classification method. Two available datasets that fulfilled the requirements were selected, i.e. the Classification on Instructional Programs (CIP) and Wikipedia. The purpose of using different datasets (significantly in size) was to compare and analyze results from same input and see how much corpus size affects end results.

The Classification of Instructional Programs (from now on referred as CIP) is a taxonomic coding scheme of instructional programs. Its purpose is to facilitate the organization, collection, and reporting of fields of study and program completions. The CIP was originally developed in 1980 by the National Center for Education Statistics (NCES) in the U.S. Department of Education, with revisions occurring in 1985, 1990, 2000 and 2010. CIP classes are divided into three levels. The first level consists of more general classes, while the third has more specific classes. The CIP classification scheme, which can be downloaded from the National Center for Education Statistics (NCES)[6], contains CIP title, code, definition, examples and some other categories. In this paper we will use CIP classes from the third level and their description, to classify given courses with CIP title and CIP code.

Wikipedia is a free online encyclopedia that has over 40 million articles in over 250 different languages. Because of this nature, Wikipedia has been used as a data source for many applications, analytics and experiments. DBpedia is a crowd-sourced community effort to extract structured information from Wikipedia and make this information available on the Web [5]. DBpedia also

---

[6] https://nces.ed.gov/ipeds/cipcode/resources.aspx?y=55.

provides different datasets for downloading, which cover most of the Wikipedia content. In this paper, we will use the dataset composed from the abstracts of the Wikipedia articles in English. The abstracts represent a brief summary of the articles. The dataset can be downloaded from the Dbpedia downloads page[7]. We will use this dataset to classify a course with a Wikipedia article title, using the abstract.

The size of the Dbpedia dataset was approximately 3 gigabytes, while the size of the CIP database was approximately 500 kilobytes.

3.2 Document Similarity

Each of the classes in the dataset has a description which will be used for classification of the courses. By calculating the similarity between the class description and the courses, we can classify the course with the class which description had the highest similarity with the given course. This reduces our goal to a document similarity problem. The variety of algorithms addressing this problem, makes choosing one, a non trivial task.

We will discuss some of the most used measures for calculating the similarity of two objects (in our case the class description and the course). The similarity or the distance between the two objects is represented as a single numeric value, which states the closeness the two objects.

Euclidean distance is a standard distance measure, which calculates the distance between two points in a euclidean space. If we represent two documents with a weighted term vectors a and b, we can calculate the euclidean distance as follows:

$$ED(a, b) = \sqrt{\sum_{t=1}^{n}(a_t - b_t)^2} \tag{1}$$

where ED is Euclidean distance and $T = \{t_1, t_2, ..., t_n\}$ is the term set. This measurement doesn't take into consideration the length of the documents, so two documents similar in content, but with big difference in length, can have a great distance in the euclidean space resulting in a small similarity value for the two documents.

Jaccard Coefficient is a measure, which calculates similarity by measuring the ratio of the number of shared elements between the two documents and the number of all the elements of the two documents, or in other words, it is the intersection of the documents, divided by the union. If we represent two documents with a weighted term vectors a and b, we can calculate the Jaccard coefficient as follows:

$$JC(a, b) = \frac{\sum_{t=1}^{n} a_t b_t}{\sum_{t=1}^{n} a_t^2 + \sum_{t=1}^{n} b_t^2 - \sum_{t=1}^{n} a_t b_t} \tag{2}$$

where JC is Jaccard Coefficient and $T = \{t_1, t_2, ..., t_n\}$ is the term set.

---

[7] http://downloads.dbpedia.org/2016-04/core-i18n/en/long_abstracts_en.tql.bz2.

Cosine similarity is a measure which defines similarity by measuring the angle between two vectors. If we represent the documents as vectors in the vector space, we can measure their similarity by measuring the angle between them. If two documents have a 0° angle, they can be treated as identical. Cosine similarity is independent from the document length, meaning that the similarity value won't be affected from the documents length. Cosine similarity will be discussed more in Sect. 3.4.

Locally sensitive hashing is a technique from the Data Mining branch for finding similar documents. The idea behind this algorithm is that similar documents will have similar hash-codes. This algorithm is mainly used for detecting plagiarism.

A lot of work has been done on comparing similarity measures in the field of document clustering. Huang [4] expands the work done by Strehl et al. [9] in which similarity measures are discussed and compared how they perform in document clustering. Both papers conclude that euclidean distance is an ineffective metric, whereas both jaccard and cosine similarity produce similar results. In our paper, we decided to use cosine similarity as a similarity measure for our classification algorithm. In our future work, we plan to also use other similarity measures and make a comparative analysis to conclude which measure performs the best in our classification algorithm.

3.3   Classification System Architecture

Figure 1 represents the architecture of the course classification system. The system can be separated in two parts. First part of the system is the creation of inverted indexes for both datasets, using the Apache Lucene library[8]. For the CIP inverted index, we used the description of the CIP titles, whereas for the Wikipedia inverted index we used the abstract from the given article. From now on, we will refer to CIP descriptions and Wikipedia abstracts as documents, using the Apache Lucene terminology. For both datasets, data pre-processing was done consisting of tokenization, stop words removal and stemming using the Porter stemmer to reduce words to their root form. In the CIP inverted index, CIP code and title were stored. CIP description was left out from the inverted index as it is not needed for the final result and it will consume valuable memory. Only the Wikipedia article title was stored for the Wikipedia inverted index. The abstract was not stored for the same reason the CIP description was not stored in the CIP inverted index. For every document, a term vector is stored in the index. Term vectors consists of every word that the document contains. It also stores information about the number of times the words occurred in the document. The inverted indexes represent a base on which the classification system is build.

Input in the classification system are PDF documents from the courses, which are uploaded through a REST endpoint in the system. Using the PDFBox java library[9], the content is extracted from the document. The system, based on the

---

[8] https://lucene.apache.org/core/.
[9] https://pdfbox.apache.org/.

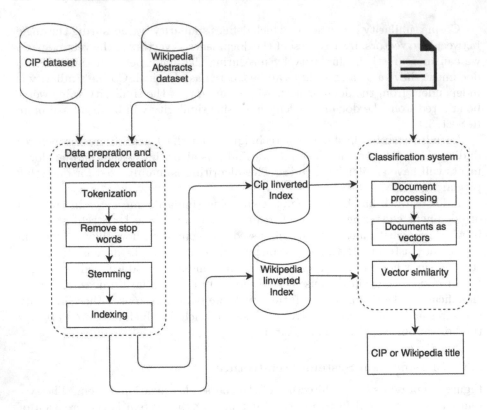

Fig. 1. Course classification system architecture

content, classifies the course with a CIP or Wikipedia title. The processes of deciding what class to be assigned to the document is described in Sect. 3.4.

3.4   Classification Algorithm

After the text extraction from the course document, the system classifies the text with a class. The process of deciding which class to be assigned to the extracted text starts with adding the text in to the inverted indexes. Before adding the text, the same data pre-processing that was done on the CIP and Wikipedia datasets, is performed on the extracted text. This process is done to keep the consistency in the inverted indexes. The algorithm for classifying courses consists of ranking the documents from the classification schemes, based on their similarity with the given course, and assigning the class from the best ranked document. For every comparison made between the course and documents, we emulated the vector space model by representing both text documents with weighted vectors. The axis of the vector space model were constructed of the whole set of words of the two documents included in the comparison. Similarity between the documents is measured by calculating the cosine of the angle between the vectors.

TF-IDF (term frequency - inverse document frequency) weighting was used for constructing the document vectors. Its formula can be seen in Eq. 4. tf represents the number of times the word occurs in the given document, whereas idf (Eq. 3) represents the number of documents in the corpus (N) divided by the number of documents in which the word occurs (df). Logarithm function is used on the idf to normalize the value. The purpose of using tf-idf weighting is to give more value to words that occur many times within a small number of documents [7].

$$idf = log(\frac{|N|}{df})$$ (3)

$$tf\text{-}idf = tf \times idf$$ (4)

Term frequency, document frequency and corpus size values can all be retrieved from the created inverted indexes. When comparing a course with a document, the system constructs tf-idf vectors with the size of the set of words from both text documents. For every word in the vector, tf-idf calculation is done using the values retrieved from the inverted index. The angle between the vectors is measured by calculating the cosine similarity between the two. The value obtained represents the similarity between the two documents. Given two vectors a and b, cosine similarity is calculated as follows:

$$\cos(a, b) = \frac{a \cdot b}{||a||\,||b||} = \frac{\sum_{i=1}^{n} a_i b_i}{\sqrt{\sum_{i=1}^{n} a_i^2}\sqrt{\sum_{i=1}^{n} b_i^2}}$$ (5)

Because of the small size of the CIP dataset, when classifying a course, comparison is made between the course and every document from the CIP, while when using the Wikipedia abstract, we first use the Lucene library to query the inverted index for best results by converting the course in a query, constructed of the course words separated by the OR operator and then we perform the comparison between the course and the results obtained from the Lucene query.

4   Discussion

The course classification system presented in Sect. 3 was ran over two datasets containing courses from the computer science field. The first dataset was obtained from MIT OpenCourseWare (OCW)[10] and contains 265 courses from the Electrical Engineering and Computer Science department. The second dataset was obtained from the Faculty of Computer Science and Engineering (FCSE)[11] and contains 35 courses from the Computer Science and Engineering program. Courses from the MIT OCW dataset are represented with their title

---

[10] https://ocw.mit.edu/.
[11] http://finki.ukim.mk.

and course description, whereas courses from the FCSE dataset, are represented as PDF documents, giving different information about the course. In this paper, we used the "Course Content" part from the document which describes what the student will learn if he takes that course. Both datasets were classified over the CIP and Wikipedia inverted indexes using the classification system. For every course, we gathered the best ranked CIP and Wikipedia classes. The ranking algorithm was described in Sect. 3.4 of this paper. Best ranked classes represent greatest similarity between its CIP definition or Wikipedia abstract (depending on the used inverted index) and course description. In Tables 1 and 2 some of the results obtained from running the system over the CIP inverted index are given. Table 1 represents the results for the courses "Robotics" and "Fundamentals of Computer Graphics" from the FSCE dataset. The best matches for both courses, "Robotics Technology/Technician" for "Robotics", "Computer Graphics for Fundamentals" of "Computer Graphics", have high values of cosine similarity with 0.6636958166 and 0.497835655 respectively. The system matched the courses with classes representing the same field as the courses. We can see that the system accurately classified these courses with an appropriate CIP class. If we compare the two texts, both from the course and CIP description, we can see why the system chose these results. For the course "Robotics", both texts contain the words: "robot", "use" and "control". The words "use" and "control" are more frequent in the CIP corpus, while the word "robot" is not. The rarity of the word "robot" contributes for big idf value. Combined with the tf values for the words, it contributes for the CIP class "Robotics Technology/Technician" to have a high cosine similarity value. The same can be found for the course Fundamentals of Computer Graphics. The best ranked CIP class "Computer Graphics" has seven words in common with the course. High term frequency of the words in the texts, and high idf values for some of the words, were the reason why the system classified the course with this CIP class. In Table 2, results for two MIT courses are given. Same characteristics as the ones described for the results in Table 1, for why the system chose the given CIP classes, can be detected for the results in Table 2. MIT course "Introduction to Numerical Methods" and CIP class "Algebra and Number Theory" have three words in common: "matrices", "linear", "algebra". All three words have low document frequency, with "matrices" appearing in only 3 documents, "linear" in 6 and "algebra" in 5. This contributes for high idf scores resulting with high tf-idf values which results in high cosine similarity value. The second course in Table 2 has more words in common (program, programming, software, design, include, testing) with the best ranked CIP class, which even though most the words are common in the corpus (program is included in 1710 documents, software in 46, include in 1226), the frequency of the words in the two texts contributes for high cosine similarity value.

The results that were described suggest that the system managed to correctly classify the courses, but if we look at the results in Table 3 we can see that is not true for every course. Moreover, if we look after the best ranked class in Tables 1 and 2, we can see a significant drop in the cosine similarity

Table 1. Cosine similarity results for FSCE courses run over CIP classification scheme

FCSE course title	CIP title	Cosine similarity
Robotics	Robotics Technology/Technician	0.6636958166
	Undersea Warfare	0.2797877673
	Automation Engineer Technology/Technician	0.2562566923
	Mechatronics, Robotics, and Automation Engineering	0.2436635451
	Surgical Technology/Technologist	0.1585793418
Fundamentals of Computer Graphics	Computer Graphics	0.497835655
	CAD/CADD Drafting and/or Design technology/Technician	0.381013755
	Graphic Communications, Other	0.337921630
	Graphic and Printing Equipment Operator, general Production	0.2552102418
	Medical Illustration/Medical Illustrator	0.2437037558

Table 2. Cosine similarity results for MIT courses run over CIP classification scheme

MIT Course Title	CIP Title	Cosine Similarity
Introduction to Numerical Methods	Algebra and Number Theory	0.4039465832
	Mathematics and Statistics	0.2918484769
	Computational Mathematics	0.2315138222
	Mechatronics, Robotics, and Automation Engineering	0.2212023272
	Laser and Optical Engineering	0.2210414242
Software Construction	Computer Programming/Programmer, General	0.3011612213
	Computer Software and Media Applications, Other	0.2789394801
	Computer Programming, Specific Applications	0.2483644444
	Locksmithing and Safe Repair	0.2168353175
	Computer Software Technology/Technician	0.2161809295

Table 3. Cosine similarity results for FCSE/MIT courses run over CIP classification scheme

FCSE/MIT course title	CIP title	Cosine similarity
Calculus2	Advanced Legal Research/Studies, General	0.3001409335
Data Mining	Mining and Petroleum Technologies/Technicians, Other	0.3031102324
Mobile Platforms and Programming	Vehicle Maintenance and Repair Technologies, Other	0.4954344611

results. This suggests that there were not many candidates in the classification process. Classifying "Calculus 2" with "Advanced Legal Research/Studies" is clearly wrong, and the classification of "Data mining" with "Mining and Petroleum Technologies/Technicians" suggests that the word "mining" was used in a different context. We suspect that the reason for this behavior is the size of the corpus that was used for classifying the courses. CIP classification scheme only contains 1719 documents, and its inverted index takes 500 kb of memory. Because of the small corpus, many of the words that we know of today won't be included in the inverted index, resulting in a small vocabulary. Input text can only be compared with the 1719 documents, which represents a small number for the system to chose from, and resulting with wrong results. The words that are not included in the inverted index, but occur in the input text, won't be taken in consideration. Small vocabulary and choice of documents, contributes small amount of words to have a big impact in the final result. This is the case for the FCSE course "Mobile Platforms and Programming", where only one word is responsible for determining the CIP class. The word "mobile" occurs in only 9 documents in the corpus, which contributes to a high idf value, which results in a high cosine similarity between the course "Mobile Platforms and Programming" and the CIP class "Vehicle Maintenance and Repair Technologies". Because of the small corpus and the small amount of documents to chose from, the system was not able to find a document which contains more common and more important words.

To see how the results will deffer when using a bigger corpus, we repeated the same process over the Wikipedia abstracts dataset. The size of in this dataset is much bigger with 4,847,276 different documents contained in the corpus. Cosine similarity values between three courses and best ranked Wikipedia abstracts for each of them is given in Table 4. If we compare the results obtained from the CIP inverted index we can see a significant difference. Course "Data Mining" which was classified with "Mining and Petroleum Technologies/Technicians" is now classified with "Data Mining" which represent a straight match. Also, for the course "Mobile Platforms and Programming", where only one word was responsible for classifying it with a CIP class, is now classified with "Mobile

Table 4. Cosine similarity results for FSCE/MIT courses run over Wikipedia classification scheme

FCSE/MIT course title	Wikipedia title	Cosine similarity
Data Mining	Data mining	0.5793786778
	Data analysis	0.5468895059
	Data pre-processing	0.4925073853
Mobile Platforms and Programming	Mobile application development	0.692758599
	Mobile Web	0.6075720147
	Remote mobile virtualization	0.4036265023
Software Construction	Software development	0.4647611515
	Software factory	0.4480353994
	Software analytics	0.4297361022

application development" with whom they have 11 words in common (mobile, system, web, applications, conceptual, development, focus, infrastructure, user, interface, platform). We can also see, that the difference in cosine similarity between best ranked documents is not as significant as in the results obtained from the CIP classification scheme. In Figs. 2 and 3, we can see the average cosine similarity of the best five ranked documents for every course, from both datasets. We can see that in almost every case for the MIT courses, and in every case for the FCSE courses, the cosine similarity is bigger for the results obtained from the Wikipedia classification scheme.

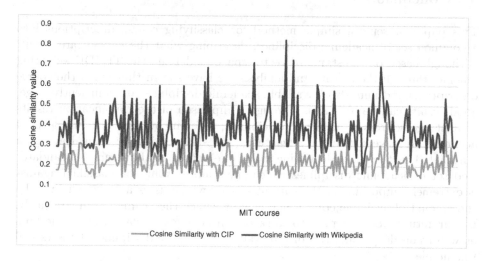

Fig. 2. Average cosine similarity of the 5 best ranked results for every MIT course for both Wikipedia and CIP classification schemes

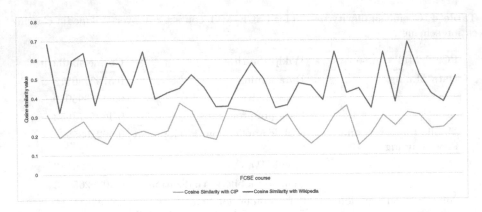

Fig. 3. Average cosine similarity of the 5 best ranked results for every FCSE course for both Wikipedia and CIP classification schemes

Common for all the courses when classifying them over the Wikipedia classification scheme was the big number of words they had in common with the Wikipedia documents, whereas the number of common words with the CIP documents when the classification was done over the CIP classification scheme was much smaller. We can conclude, that the size of the corpus, plays a significant role in classifying text documents when using tf-idf vectors to represent the documents, and cosine similarity to calculate the similarity between those documents.

5   Conclusion

This paper presents a simple method for classifying course descriptions into a common set of academic disciplines, describing what the courses are about. Although based on the standard vector-space model and the TF-IDF weighting scheme, through the testing and analysis we have shown that when the course descriptions and number of available academic disciplines are sufficiently large, good results can be obtained. Through comparison of the classification results against two sources of academic disciplines, different by their size and level of accuracy, we have shown the advantages and shortcomings of each. Having an automated method for identifying what one course is about and a standardized set of academic disciplines which the courses would be classified against is extremely important in the current setting where users are overwhelmed by the amount of available educational content which they cannot compare easily. In our future work we plan to work on improving the classification method, as well as on deeper analysis of the available academic disciplines datasets and taxonomies.

References

1. Apaza, R.G., Cervantes, E.V., Quispe, L.C., Luna, J.O.: Online courses recommendation based on lda. In: SIMBig, pp. 42–48 (2014)
2. Gasparetti, F., Limongelli, C., Sciarrone, F.: Exploiting wikipedia for discovering prerequisite relationships among learning objects. In: 2015 International Conference on Information Technology Based Higher Education and Training (ITHET), pp. 1–6. IEEE (2015)
3. Gjorgjevik, A., Stojanov, R., Trajanov, D.: Semccm: course and competence management in learning management systems using semantic web technologies. In: Proceedings of the 10th International Conference on Semantic Systems, pp. 140–147. ACM (2014)
4. Huang, A.: Similarity measures for text document clustering. In: Proceedings of the Sixth New Zealand Computer Science Research Student Conference (NZCSRSC2008), Christchurch, New Zealand, pp. 49–56 (2008)
5. Lehmann, J., Isele, R., Jakob, M., Jentzsch, A., Kontokostas, D., Mendes, P.N., Hellmann, S., Morsey, M., Van Kleef, P., Auer, S., et al.: Dbpedia–a large-scale, multilingual knowledge base extracted from wikipedia. Semant. Web 6(2), 167–195 (2015)
6. Limongelli, C., Gasparetti, F., Sciarrone, F.: Wiki course builder: a system for retrieving and sequencing didactic materials from wikipedia. In: 2015 International Conference on Information Technology Based Higher Education and Training (ITHET), pp. 1–6. IEEE (2015)
7. Manning, C.D., Raghavan, P., Schütze, H., et al.: Introduction to Information Retrieval, vol. 1. Cambridge University Press, Cambridge (2008)
8. Shatnawi, S., Gaber, M.M., Cocea, M.: Text stream mining for massive open online courses: review and perspectives. Syst. Sci. Control Eng. Open Access J. 2(1), 664–676 (2014)
9. Strehl, A., Ghosh, J., Mooney, R.: Impact of similarity measures on web-page clustering. In: Workshop on Artificial Intelligence for Web Search (AAAI 2000), vol. 58, p. 64 (2000)
10. Tam, V., Lam, E.Y., Fung, S.: A new framework of concept clustering and learning path optimization to develop the next-generation e-learning systems. J. Comput. Educ. 1(4), 335–352 (2014)
11. Xu, R.: Pos weighted tf-idf algorithm and its application for an mooc search engine. In: 2014 International Conference on Audio, Language and Image Processing (ICALIP), pp. 868–873. IEEE (2014)
12. Zhuhadar, L., Kruk, S.R., Daday, J.: Semantically enriched massive open online courses (moocs) platform. Comput. Hum. Behav. 51, 578–593 (2015)

Tendencies and Perspectives of the Emotions Usage in Robotics

Vesna Kirandziska$^{(\boxtimes)}$ and Nevena Ackovska

Faculty for Computer Science and Engineering, Skopje, Macedonia
{vesna.kirandziska,nevena.ackovska}@finki.ukim.mk

Abstract. Emotions are psychological phenomena present in the living beings. However, there is still no consensus for a more precise definition of the emotions. As a complex concept, the emotions can be part of a robot model for different usage in variety of robots. This paper presents the state of the art of robot behavior models that use emotions in a chronological order. The aim of this paper is to formalize the usage of emotion in robotics. Thus, a definition and explanation of a set of distinct parts in robotic models influenced by emotions is proposed. On the other hand, the properties of robots that use emotions, the so-called emotional robots, are also explored. As a result, the most important different properties that emotional robots should possess are retrieved and explained in detail. The aim of the formalism presented in this paper is to give tribute to the work already done and to give possible directions for the future work with emotional robots.

Keywords: Robotics · Emotion model · Human-robot interaction · Social robotics · Robotic behavior

1 Introduction

The definition, the description as well as the origin of emotions is constantly under debate. Some say that emotions are the result of brain processing, while others state that emotions are the result of physiological modifications and the environment. One aspect of emotions is the way emotions are represented in respect to their differences. Usually, in the robotics researches, emotions are represented using different emotion models: a discrete or a continuous space model. In the discrete model, emotions are represented as categories of emotions, which can be basic and complex. One widely used discrete model is the one proposed by Ekman [1] who distinguished six basic emotions (anger, disgust, fear, happiness, sadness, and surprise). In the continuous space model, emotions are points or areas in the multidimensional space. One popular three-dimensional continuous emotion model is the arousal-valance-stance space by Braezael [2].

From a historical point of view, until 1990s, emotions were not considered as important in intelligent robotics behavior models. Still, some revolutionary visionaries have included emotions in learning models. One is Mowrer who in 1960 [3] improved the classical conditioning technique for learning that is based on stimulus-response associations by associating stimulus with emotions. An alternative to the famous reinforcement learning mechanism that takes inspiration from emotions was proposed in 1982 [4].

© Springer International Publishing AG 2017
D. Trajanov and V. Bakeva (Eds.): ICT Innovations 2017, CCIS 778, pp. 154–164, 2017.
DOI: 10.1007/978-3-319-67597-8_15

In this proposal, the concept of state evaluation was introduced. Some states were evaluated a priori with pleasant and/or unpleasant emotional state and these labels were propagated to all other states while the robot explores its environment. The proposed architecture, named Crossbar Adaptive Array (CAA), actually is a solution to the delayed reinforcement problem [5].

After 1990s, the great potential of the usage of emotions started to be recognized and consequently researchers started building computational models of human emotions [1, 2]. As a result, the emotions were used in creating interactive systems that are able to recognize the user's emotion, express emotions and in robots that use emotions [6]. Artificial emotions can be used as control mechanism that reflects how the robot is affected by and adapts to different factors over time [7]. A robot can express emotion by facial expression, body movement, sound production etc. and it can recognize the user's emotions by processing video or sound signals.

One simple control model was proposed in [8] where a fuzzy state machine that controls emotion expression and robot action selection was designed. The control model in [9] uses a multilevel hierarchy of emotions where emotions are used to modify active behaviors at the sensory-action level. A hybrid architecture that uses emotions to improve the general adaptive performance is given in [10]. Here, both reactive and deliverable emotions exist: reactive emotions can change the model parameter in response to appraisals from the next environmental data, while deliberative emotions present the learned associations between current state and actions. More complex control mechanisms are the ones that enable learning and one of the most popular learning techniques is reinforcement learning. A modification of the reinforcement learning model, proposed by Lin [11], that uses neural networks was upgraded by using emotion in the model [7, 12]. A recurrent network that imitates a simplified version of a human hormone system represents the emotion system of this model. Emotions were included in the reinforcement function, and they helped in event detection and influenced on the learning of the meta-control values. Another example where an emotion model was integrated in a reinforcement learning framework was built in 2001 [13]. This model has improved the performances of the robots learning task.

EARL (emotion, adaptation and reinforcement learning) framework was built by Broekens in 2007 [8] to model the relation between emotions and learning in emotional robots. The emotions in this framework are used as social reinforcement (from human emotion perception), as meta-parameters (parameters in the robot model) and in emotion expression (robot's body expresses its current state). In [6] the intrinsically motivated reinforcement-learning framework was used as a base to build a robotic learning model. A new approach for emotion-based reward design that is inspired by the way humans and other animals appraise their environment in nature is presented in [6]. With the model built, survival and adaptive skills were introduced to the robot. An affect-based behavior model named TAME (Traits, Attitudes, Moods and Emotions) was built in 2003 [14], where moods and emotions represent long-term and short-term emotional states, accordingly. Personality and affect module was designed to add affect in robotic behavior. A behavior selection system built in 2013 has four modules: Cognition, Emotion, Behavior-selection and Behavioral-making [15, 16]. The Cognition module calculates emotion-inducing factors (four parameters), the Emotional module calculates the next emotional state based on its previous state, the Behavior-selection module determines the probability distribution for all possible behaviors given the current emotional state.

In our previous work, a classifier for emotion detection was done, where the Ekman basic emotions were successfully detected using well-chosen sound features [17]. Later, facial data were included as input data to the classifier [18]. The experimental data have shown that the created robot emotion classifier and human recognition on other people's emotions have similar power [17]. This kind of emotion classifier can be used in a robot to improve social human-robot interaction. In [19] we have discussed the ethical consequences on both humans and robots from the involvement of an emotion classifier in a social robot. In the taxonomy of learning agents presented in [20] it was stated that this architecture follows the self-learning paradigm. In this architecture, emotions are used as the robots' internal evaluation mechanism on what the learning was based.

As seen in the introduction, varieties of models that use emotions exist. The aim of this paper is to categorize different aspects in which the emotions are used in the models and to distinguish the characteristics a robot would obtain if it uses and/or possesses emotions. The following section provides the elements of the robotic model where emotions can be included. The Sect. 3 describes the characteristics specifics for an emotional robot. In Sect. 4, some examples of robots that implement emotion-based model are described and in the end, the conclusions are emphasized.

2   Emotions as Parts of Different Components of the Robotic Models

In the humans, the emotion represents a psychological energy [21]. The emotional state is the internal manifestation of this energy, while the human behavior is its external manifestation. Beside the influence that emotions have on human physical actions, emotions influence several cognitive processes, among which are problem-solving, decision making, thinking, and perception [22, 23]. Human memory, that represents the accumulated knowledge taken from the experience, is also under the influence of emotions. Considering all impact of emotions on humans, it was proven in [24] that intelligence in general is much impacted by emotions.

Emotion usage in human behavior implies that there could be varieties of ways they can be used in robots. According to [25], an artificial emotion has two aspects in application: as the carrier link that links the external environment and the robot internal state and as a part of the evaluation mechanism of robots learning process.

If we look deeper in the emotion usage in a robot, emotions can influence many different components in a robotic model. In the Introduction we have made an overview of existing robotic models that incorporate emotions. By analyzing them from the perspective of where exactly in the model emotions are used we have made a set of different components or modules in a generic robotic model that use emotions (Fig. 1). Note that, to our knowledge, this kind of analysis that includes all the aspects of emotions in robotics has not been published. Some specific robotic models can contain some or all of the components presented in Fig. 1 and usually they give a detailed explanation of the relationship between the components. For example, in the model proposed in [11], emotions were used in all components, while the model in [16] does not use emotions for behavior execution nor for the robot's goals.

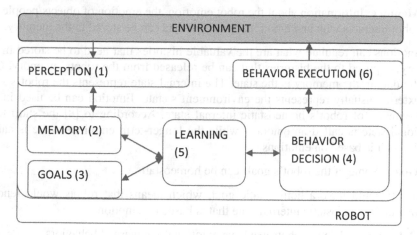

Fig. 1. Schematic view of a generic robotic model with the components in a robotic model that can use emotions.

The perception module (1) receives raw signal data from the environment and transfers valuable information to the robot that outputs the actions, which are executed by the robot actuators and effectors (Behavior Execution (6)). The brain of the robot is the processor(s), and it represents the relation between the perception and behavior execution module. An important part of the brain is the memory component (2) that stores all the valuable information for the robot. Some robots have a goal component (3) that contains information about their goals, which can be changed during the robot's lifetime. The robotic behavior decision module (4) is the center of the brain that has the task to choose the best behavior given all information from the environment and given the goals of the robot. The ability of the robot to learn is enabled with the learning module (5) which is responsible to change the robot model in a way it would improve the interaction between the robot and its environment considering the past events and histories. In the sequel, a definition and explanation of the usage of the emotions for each component is given.

(1) **Perception**: Emotions can be used in the affective world perception and/or emotion perception of people that interact with the robot.

Emotions can reflect how the robot is affected by different perceptual information it gathers in the world [23]. For example, emotions can be used for valuable information selection: some information from the environment can be ignored, while some can be considered very important. From another perspective a parameter, representing the emotional valance, can be added to each data perceived from the environment. Positive valance can be attached to an event, object or situation that have brought or will bring positive emotion to the robot, and negative valance can be attached in all other cases. From the aspect of emotion perception, the perception module could perceive emotions of the people it interacts with. Information from human facial expression, voice, and body gesture can be used to distinguish human emotion.

(2) **Memory**: Information about the robot emotion, the emotion of objects/people the robot interacts with and emotionally driven data can be stored in the memory.

Emotions can regulate what are the valuable histories that need to be stored in the memory, opposed to the histories that can be released from the robots' memory. One specific data in the memory is the state. The internal state represents the robot's state and external usually represents the environment's state. Emotion can be used in the representation of robot's homeostatic internal state. According to [8] the short-term emotional state is called an emotion, while the longer-term emotional state is called mood, and it is based on emotions.

(3) **Goals**: Some of the robot's goals can be homeostatic.

Robots could have a homeostatic goal, which means that robots would tend to improve their homeostatic internal state that is based on emotions.

(4) **Behavior decision**: Robots can have emotionally governed behaviors.

An emotion can influence the specific reactive actions robots make, but also the long term robotic behavior selection that consists of a plan of actions that should be performed in some order [8, 26]. For example, if the robot is happy than its next reactive behavior could be to smile. On the long run, the robot could choose a behavior that increases the probability that the robot's emotional goal is accomplished. Every robot can have a decision-making module that suits the robot's personality, which is something that the robot is "born" with or it has adapted to during its lifetime.

(5) **Learning**: Different learning methods can be influenced by emotions.

The robot learning can be teacher-based, where the robot learns from a teacher, or self-learning, if the robot learns by itself [20]. In teacher-based learning (supervised learning), the robot learns from advice or by reinforcement. The kind of advice or reinforcement, that is an evaluation of the robot's current state that includes information about emotion, is called emotional feedback [23, 27]. For example, if a robot learns from imitating another object or human, a human teacher can present emotional feedback to the robot by expressing his/her emotions after realizing how the robot performs on its imitation [28]. Self-learning can be enabled by introducing generically built evaluation function that could evaluate emotions [5].

(6) **Behavior execution**: Some robotic actions directly express emotions, while others incorporate the robot's emotional state.

One special behavior execution that can occur especially in humanoid robot is the emotion expression behavior. In this kind of behavior, the actions are closely connected to express the robot internal emotional state. The most famous way robots' express emotion is by facial expression, but also they could be expressed in the robot's voice and body gesticulations. Beside this, the robot's current emotional state could influence on the way every other behavior or action is executed. For example, a happy robot may speak faster than a robot that is not happy.

3   Emotions in Robot's Properties or Characteristics

Having emotions in a robotic behavior model brings benefits for the robot to have different characteristics. In this overview, we have distinguished some important characteristics of a robot that is influenced by emotions, and they are presented on Fig. 2. These robot's characteristics have been observed in past studies where emotional robots are created or robotic models for emotional robots have been proposed. Some of them are explained in the sequel.

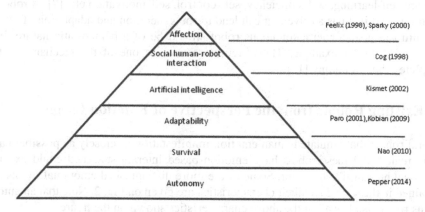

Fig. 2. Important characteristics of robots that benefit from the usage of emotions and samples of robots that have these characteristics.

By definition, the affection is the act or process of affecting or being affected. One kind of manifestation of affection is through emotion. A robot that has emotions might be affected, while a robot that expresses emotions is affecting the environment.

In social interaction, a robot explicitly communicates with and learns from all parties in their society of robots or humans and this can be improved with emotions [29]. The emotions can facilitate believable human-robot interaction in the sense that the information about the emotions of interacting parties is interchanged and the communication itself depends on this information [27]. The emotion perception and emotion expression enables a higher level of social human-robot interaction. Social human-robot interaction finds its application in personal service robots that usually interact with people with no special skills or training to operate with robots [22].

Artificial intelligence is the simulation of human intelligence processed by a machine, and in this case in a robot in particular. Intelligent processes include learning, reasoning and self-correction. Some researches argue that a robot with no emotion could not be intelligent [23]. Traditional artificial intelligence control in a robot uses variety of existing popular machine learning methods, but ignores the concept of emotion.

Introducing emotions to all intelligence processes in a robot may solve this problem and enable a true intelligence control.

As Darwin argued, emotions have been a great psychological factor in the evolutionary adaptation of humans [1]. The characteristic to adapt is valuable for robots. Adaptation of a robot to a new environment situations and scenarios is closely connected to the learning module of a robot. The role of artificial emotions in a robot is to enhance the adaptability by regulating the robot's emotion [23].

Darwin postulated that emotions in humans serve to increase the survivability of the system. Emotions in robots provide support for a robot's survival in a complex world, since they modulate the behavior response of the robot directly [27].

A robot is autonomous if the system can fulfill its goal without human intervention or intervention from any other system. Some properties an autonomous robot should have are self-learning, self-sufficiency, self-control, self-motivation etc. [7]. A robot is self-motivated if it has a drive that can lead to the generation and adaptation of goals. The ultimate goal for an autonomous robot should be of a homeostatic nature, like emotional state for example. Thus, emotions serve as one of the mechanisms for complete robot autonomy [12].

4   Existing Robots from the Perspective of Emotion Usage

Several robots that emulate human emotion manifestation as closely as possible have been created. All these robots have emotion-based internal state and could execute emotional human-like behavior. Some of the more distinguished emotional robots are explained in the sequel and their characteristics are given on Fig. 2. Note that an autonomous robot also has all of the above characteristics shown on the figure.

In 1998 Cog was built with a goal to evaluate the models of human social behaviors and to enable social human-robot interaction [30]. The robots Sparky [31] and Feelix [30] present emotional facial expression using their 4 degrees of freedom on their faces. FACS (Facial Action Coding System) [32] system was used to generate facial emotional behavior. The results in [30] show that humans can recognize Feelix's emotions: anger, sadness, fear, happiness, surprise and neutral.

Some emotional robots have been created for the purpose of robot-assisted therapy. For example, the robot Paro is such a robot [33]. Paro has the appearance of a baby harp seal. Paro's behavior tends to improve the patients' emotional state and it is shown that it increases the patients' relaxation and motivation, and decreases their stress [33]. One of the best properties of Paro is that it can adapt to its patient. This robot has been build in 2001, and since 2003 have been used in the market. Another example is KOBIAN that assists elderly and disabled people during their daily activities. KOBIAN can express emotions through both facial expressions and body movement. With a high recognition rate, the emotions it expresses are recognized by the people it interacts with and enables social human-robot interaction [34].

The robot Kismet [35] is one complex intelligent robot that uses emotions. Its model is behavior-based and its goal is to have homeostatic middle-ground state that represents its inner state. Its emotional state is represented as point in a three-dimensional emotional space with dimensions: arousal, valance and stance. Kismet can express its emotional state by face expression and by its voice. The facial expressions are generated using an interpolation-based technique and the expressive utterances are generated by assembling strings phonemes with pitch accents. Kismet can express six emotions: anger, distrust, fear, happiness, sorrow and surprise.

Since the creation of Nao, a popular programmable humanoid robot, many researchers have used this robot to implement emotion-based robotic behavior [2, 36]. In some, Nao has an emotional inner state where different emotion models have been used. One example the Pleasure-Arousal-Dominance model used to represent Nao's emotion in [37] where fuzzy control system controls the emotion state. Nao has various sensors that enable valuable sensor data from the environment. Nao can potentially express emotions through its voice or movement, but has limited facial expressions [38]. In [39] several emotional postures for anger, sadness and happiness for the humanoid robot Nao were developed. In our past work, we have also used Nao as a teacher in the education of children. Our research shows impressive results with the Nao robot in its social interaction with children and especially with children with special needs. Note that we have used emotion evaluation from the children as a feedback that influences the next robotic action [40, 41].

The greatest achievement with the Nao robot was done in 2010 that was a result of the project called FEELIX Growing, where a group of scientist led by Lola Canamero modeled Nao to behave as a young child [42]. The properties of the control model are: Nao can express emotions, it can detect human emotions, it interacts socially and emotionally with all people, it can learn from the people it interacts with, it has an affective inner state that tends to be positive, its experiences are labeled as good or bad etc. With this work, Nao became the first robot capable of developing emotions and forming bonds with humans [42]. The follower of Nao is the robot called Pepper that was developed by Aldebaran SoftBank Robotics in 2014. Pepper is autonomous robot used on the marked as a social companion for humans. Pepper has a built in emotion engine that gives it the ability to understand people's feelings and it is the first robot that responds to emotional signature as laughing or frowning. Based on the feeling of other people the robot is controlled [43].

5   Conclusion

In this paper an overview of the concepts of emotions in robotics is given. The global shift in the usage of emotions for different tasks in robotics is explored and exposed. The overview presents a variety of models that use emotions and can serve as a concise review of researches done in the area of emotions in robotics.

Today, emotional robots can be used for assistive therapy for elderly or disabled people or for education, but also these can be used as personal robots. In social robots, emotions are used to facility believable human-robot interaction, but emotions are also important in the robot autonomy. Building autonomous intelligent robot is the goal in the creation of novel personal or social robots and these must be able to represent emotions and act emotionally so they could survive in an environment with humans, other living creatures, and other robots. Following this reasoning, a formalization of different components of a robotic model that could be implemented to use emotions is also presented in this paper, and to our knowledge, this was not previously done. These components are Perception, Memory, Goals, Behavior Decision, Learning and Behavior Execution. Following the need to build such a system, many characteristics of the robots are also differentiated. Some of them are recognized as affection, social human-robot interaction, artificial intelligence, adaptability, survival and autonomy. In the future, all robots that possess some or all of these characteristics should incorporate the concept of emotion.

References

1. Ekman, P.: Basic emotions. In: Dalgeleish, T., Power, M. (eds.) Handbook of Cognition and Emotion. Wiley, New York (1999)
2. Braezael, C.: Function meets style: insights from emotion theory applied in HRI. IEEE Trans. Syst. Man Cybern. Part C Appl. Rev. 2(34), 187–194 (2004)
3. Mowrer, O.H.: Learning Theory and Behavior. Wiley, New York (1960)
4. Bozinovski, S.: A self-learning system using secondary reinforcement. In: Trappl, R. (ed.) Cybernetics and System Research, pp. 397–402. Elsevier, North-Holland (1982)
5. Bozinovski, S.: Crossbar adaptive array: the first connectionist network that solved the delayed reinforcement learning problem. In: Dobnikar, A., Steele, N., Pearson, D., Albert, R. (eds.) Artificial Neural Networks and Genetic Algorithms. Spriger, Vienna (1999)
6. Sequeira, P.: Socio-emotional reward design for intrinsically motivated learning agents. Ph.D. thesis. Instituto Superior Técnico, Universidade de Lisboa (2013)
7. Gadino, S.P.: Reinforcement learning in autonomous robots: an empirical investigation of the role of emotions. Ph.D. thesis. University of Edinburg (1999)
8. Broekens, J.: Emotion and reinforcement: affective facial expressions facilitate robot learning. In: Huang, Thomas S., Nijholt, A., Pantic, M., Pentland, A. (eds.) Artifical Intelligence for Human Computing. LNCS, vol. 4451, pp. 113–132. Springer, Heidelberg (2007). doi: 10.1007/978-3-540-72348-6_6
9. Murphy, R.R., Lisetti, C., Tardiff, R., Irish, L., Gage, A.: Emotion based control of cooperating heterogeneous mobile robots. IEEE Trans. Robot. Autom. 18(5), 744–757 (2002)
10. Hollinger, G., Georgiev, Y., Manfredi, A., Maxwell, B.A., Pezzementi, Z.A., Mitchell, B.: Design of a social mobile robot using emotion-based decision mechanisms. In: 2006 IEEE/RSJ International Conference on Intelligent Robots and Systems, pp. 3093–3098 (2006)
11. Lin, L.J.: Reinforcement learning for robots using neural networks. Ph.D. thesis. Carnegie Mellon University (1993)
12. Gadanho, S.S., Hallam, J.: Emotion-triggered learning for autonomous robots. In: Workshop at the 5th International Conference of the Society for Adaptive Behavior, Zurich (1998)
13. Gadanho, S., Hallam, J.: Emotion-triggered learning in autonomous robot control. Cybern. Syst. 32(5), 531–559 (2001)
14. Moshkina, L., Arkin, R.C.: On TAMEing robots. In: Proceedings of IEEE International Conference on Systems, Man and Cybernetics, Washington, DC, USA (2003)
15. Jitviriya, W., Koike, M., Hayashi, E.: Emotional model for robotic system using a self-organizing map combined with Markovian model. J. Robot. Mech. 5(27), 563–570 (2015)
16. Watada, S., Obayashi, M., Kuremoto, T., Mabu, S.: A decision making system of robots introducing a re-construction of emotions based on their own experiences. J. Robot. Netw. Artif. Life 1(1), 27–32 (2014)

17. Kirandziska, V., Ackovska, N.: A robot that perceives human emotions and implications in human-robot interaction. In: Proceedings of IEEE RO-MAN 2014, Edinburgh, pp. 495–498 (2014)

18. Kirandziska, V., Ackovska, N., Madevska Bogdanova, A.: Comparing emotion recognition from voice and facial data using time invariant features. Int. J. Comput. Electr. Autom. Control Inf. Eng. World Acad. Sci. Eng. Technol. 5(10), 737–741 (2016)

19. Kirandziska, V., Ackovska, N.: A concept for building more humanlike social robots and their ethical consequence. In: Proceedings of the International Conferences, ICT, Society and Human Beings (MCCSIS), 15–19 July, Lisbon, Portugal, pp. 37–44 (2014). (Best Paper)

20. Ackovska, N.: Taxonomy of learning agents. In: Didactical Modeling Annals, vol. 4 (2010/2011)

21. Xin, L., Lun, X., Zhi-Iang, W., Dong-mei, F.: Robot emotion and performance regulation based on HMM. Int. J. Adv. Robot. Syst. 10, 1–6 (2013)

22. Bartneck, C., Forlizzi, J.: A design-centered Framework for social-human interaction. In: Ro-Man 2004, Kurashiki, Okayama Japan, pp. 591–594. IEEE (2004)

23. Yang, F., Zhen, X.: Research on the Agent's behavior decision-making based on artificial emotion. J. Inf. Comput. Sci. 8(11), 2722–2723 (2014)

24. Damasio, A.R.: Descartes' Error: Emotion, Research and Human Brain. Penguin Publ. Group, New York (2005)

25. Wang, Z.I.: Artificial psychology and artificial emotions. CAAI Trans. Intell. Syst. 1, 38–43 (2006)

26. Scheutz, M.: Using roles of emotions in artificial agents: a case study from artificial life. In: Proceedings AAAI, San Jose, California, pp. 42–48 (2004)

27. Arkin, R.C.: Moving up the food chain: motivation and emotion in behavior-based robots. In: Fellous, J. (ed.) Who Needs Emotions: The Brain Meets the Robot. Oxford University Press, Oxford (2005)

28. Rao, R.P.N., Shon, A.P., Maltzoff, A.N.: A Bayesian model of imitation in infants and robots. In: Nehaniev, C.L., Dautenhahn, K. (eds.) Imitation and Social Learning in Robots, Humans and Animals, pp. 217–247. Camridge University Press, New York (2007)

29. Fong, T., Nourbakhsh, I., Dautenhahn, K.: A Survey of socially interactive robots: concepts, design and applications. Technical report CMU-RI-TR, no. 29 (2002)

30. Brooks, R.A., Breazeal, C., Marjanović, M., Scassellati, B., Williamson, M.M.: The cog project: building a humanoid robot. In: Nehaniv, C.L. (ed.) CMAA 1998. LNCS(LNAI), vol. 1562, pp. 52–87. Springer, Heidelberg (1999). doi:10.1007/3-540-48834-0_5

31. Scheeff, M., et al.: Expressions with sparky: a social robot. In: Proceedings of the Workshop Interactive Robot Entertainment (2000)

32. Ekman, P., Freisen, W.: Measuring facial movement with Facial Action Coding System. In: Ekman, P. (ed.) Emotion in the Human Face. Cambridge University Press, Cambridge (1982)

33. Shibata, T., et al.: Mental commit robot and its application to therapy of children. In: Proceedings of the International Conference on AIM (2001)

34. Zecca, M., Yu, M., Endo, K., Iida, F., Kawabata, Y., Endo, N., Itoh, K., Takanishi, A.: Whole body emotion expressions for KOBIAN humanoid robot - Preliminary experiments with different emotional patterns. In: IEEE RO-MAN (2009)

35. Braezal, C.: Designing Sociable Robots. MIT Press, Cambridge (2002)

36. Ackovska, N.: System software in minimal biological systems (in Macedonian). Ph.D. thesis, St. Cyril and Methodius University, Skopje (2008)

37. Nanty, A., Gelin, R.: Fuzzy controlled PAD emotional state of a NAO robot. In: Proceedings of Conference on Technologies and Applications of Artificial Intelligence, Washington, pp. 90–96 (2013)

38. Manohar, V., Crandall, J.W.: Programming robots to express emotions: Interaction paradigms, communication modalities and context. IEEE Trans. Hum. Mach. Syst. **44**(3), 362–373 (2014)
39. Erden, M.S.: Emotional postures for the humanoid robot Nao. Int. J. Soc. Robot. **5**(4), 441–456 (2013)
40. Tanevska, A., Ackovska, N., Kirandziska, V.: Robot-assisted therapy: considering the social and ethical aspects when working with autistic children. In: Proceedings of the 9th International Workshop on Human-Friendly Robotics - HFR 2016, Genova, pp. 57–60 (2016)
41. Tanevska, A., Ackovska, N., Kirandziska, V.: Assistive robotics as therapy for autistic children. In: International Conference for Electronics, Telecommunications, Automation and Informatics, Struga (2016)
42. Nao, The robot able develop emotions form bond humans. Daily Mail Reporter, 13 August 2010
43. Guizzo, E.: Meet Pepper, Aldebaran's New Personal Robot With an "Emotion Engine". IEEE Spectrum, 5 June 2014

Image Retrieval for Alzheimer's Disease Based on Brain Atrophy Pattern

Katarina Trojacanec[✉], Slobodan Kalajdziski, Ivan Kitanovski, Ivica Dimitrovski,
Suzana Loshkovska, for the Alzheimer's Disease Neuroimaging Initiative*

Faculty of Computer Science and Engineering, Ss. Cyril and Methodius University,
Rugjer Boshkovik 16, PO Box 393 Skopje, Macedonia
{katarina.trojacanec,slobodan.kalajdziski,ivan.kitanovski,
ivica.dimitrovski,suzana.loshkovska}@finki.ukim.mk

Abstract. The aim of the paper is to present image retrieval for Alzheimer's Disease (AD) based on brain atrophy pattern captured by the SPARE-AD (Spatial Pattern of Abnormality for Recognition of Early Alzheimer's Disease) index. SPARE-AD provides individualized scores of diagnostic and predictive value found to be far beyond standard structural measures. The index was incorporated in the image signature as a representation of the brain atrophy. To evaluate its influence to the retrieval results, Magnetic Resonance Images (MRI) provided by the Alzheimer's Disease Neuroimaging Initiative (ADNI) were used. For this research, baseline images of the patients with diagnosed AD and normal controls (NL) were selected from the dataset, including 416 subjects in total. The obtained experimental results showed that the approach used in this research provides improved retrieval performance, by using semantically precise and powerful, yet low dimensional image descriptor.

Keywords: Image retrieval · Alzheimer's Disease · MRI · SPARE-AD · ADNI

1 Introduction

Alzheimer's Disease (AD) is the most common form of dementia in elderly people with extremely high tendency to prevalence. Despite of the active and continuous research, there is still no cure for AD [1]. The ability to provide early diagnosis or even predict the disease are crucial challenges for physicians and researchers in this domain [2]. By addressing them, better patient prognosis, reduced costs for long-term care, as well as easier and more efficient conducting of the clinical trials, might be provided.

---

*Data used in preparation of this article were obtained from the Alzheimer's Disease Neuroimaging Initiative (ADNI) database (adni.loni.usc.edu). As such, the investigators within the ADNI contributed to the design and implementation of ADNI and/or provided data but did not participate in analysis or writing of this report. A complete listing of ADNI investigators can be found at: http://adni.loni.usc.edu/wp-content/uploads/how_to_apply/ADNI_Acknow ledgement_List.pdf; e-mail: edrake@bwh.harvard.edu.

© Springer International Publishing AG 2017
D. Trajanov and V. Bakeva (Eds.): ICT Innovations 2017, CCIS 778, pp. 165–175, 2017.
DOI: 10.1007/978-3-319-67597-8_16

Magnetic Resonance Imaging (MRI) is found to be an important diagnostic technique, noninvasively providing efficient extraction of imaging markers sensitive to the disease [3]. Enormous number of scans full of rich information, suitable for understanding and detecting disease pathology, arise this way. Efficient storage and organization in the medical databases is essential to provide precise and semantically relevant retrieval. Being able to retrieve images in the database with the similar brain condition, pathology or disease, might be very useful in the clinical centers to support the diagnosis process [4], or for educational purposes [5].

Research have shown that medical image retrieval systems have pure searching capabilities, despite some reported attempts for improvement [6, 7]. The main challenging part is the image representation. It should properly reflect clinically relevant information to provide precise and semantically meaningful retrieval. Traditional techniques for feature extraction directly represent the visual image information [8–12]. They have several limitations including lack of semantical relevance, incompleteness by excluding possibly relevant spatial information and inefficiency regarding the high dimension of the feature vector. On the other side, wide range of research have shown the power of the imaging markers with respect to the disease, such as entorhinal thickness, volume of the ventricular structures: left land right lateral ventricle, third and fourth ventricle, and the volume of the hippocampus and amygdala [13–16]. Considering this, the image description can be raised to a higher level by involving the domain knowledge. Hence, the MRIs might be represented by the measurements of the brain regions of interest (ROI), namely cortical thickness of the separate parts of the brain cortex and volume of the brain regions. This strategy is used in [6, 17] to overcome the limitations of the traditional approach for feature. This way, the process is subtly transformed from the retrieval of images with similar visual characteristics, to the retrieval of images with similar structural appearance/change with the query.

However, research demonstrates the capability of even more sophisticated analysis of brain atrophy, namely, Spatial Pattern of Abnormality for Recognition of Early Alzheimer's Disease (SPARE-AD) index [18–20]. According to the research, it is superior over the ROI measurements in early diagnosis of AD and powerful distinction between different phases of the disease, and even conversion from one stage to another [21–23].

Thus, we propose to base the image retrieval on the SPARE-AD index, by using it as subject/image signature or incorporate it into the representation consisting of brain structural measurements. We hypothesize that this kind of representation will improve the precision and relevance of the retrieval results, while remaining highly efficient by summarizing the high dimensional rich image information within a single index.

2   Materials and Methods

2.1   Participants and Inclusion Criteria

Data used in the preparation of this study were obtained from the Alzheimer's Disease Neuroimaging Initiative (ADNI) database. The ADNI was launched in 2003 as a public-private partnership, led by Principal Investigator Michael W. Weiner, MD. Investigation on whether serial MRI, positron emission tomography (PET), other biological markers, such as cerebrospinal fluid markers, APOE status and full-genome genotyping via blood sample, as well as clinical and neuropsychological assessments can be combined to measure the progression of mild cognitive impairment and early AD has been the primary goal of ADNI. The database contains data about cognitively normal individuals, adults with early or late Mild Cognitive Impairment, and people with early AD with different follow up duration of each group, specified in the protocols for ADNI-1, ADNI-2, and ADNI-GO (see http://www.adni-info.org).

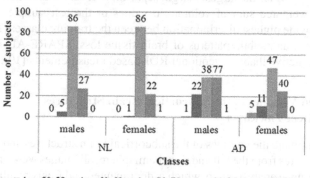

Fig. 1. Demographic information such as gender and age per class for the Screening 1.5T subset. The horizontal axis denotes gender per class, whereas the vertical one – the number of subjects. In each class, the number of subjects grouped by age range and gender is represented.

In this research, MRI at 1.5T from ADNI-1 standardized screening dataset were selected by using the criteria that each subject belongs to AD or NL group. A total of 416 subject were selected, 188 AD and 228 NL. Demographic information such as gender and age is represented on Fig. 1.

2.2   Spatial Pattern of Abnormality for Recognition of Early Alzheimer's Disease

In this research, the image retrieval is based on a pattern learned from the reach information contained in the images. The Spatial Pattern of Abnormality for Recognition of Early Alzheimer's Disease (SPARE-AD) maximally captures and reflects spatial pattern of brain atrophy caused by AD [24]. Statistical parametric maps reflecting spatial patterns of brain atrophy and their visualizations are commonly applied in neuroimaging research. Quantifying such patterns and their representation as indices is of crucial importance in this domain [18]. Basically proposed for quantifying spatial pattern of abnormality for recognition of early AD [19, 25], it is found to be a powerful AD indicator providing optimal distinction between AD and normal controls (NL) at similar age, as well as conversion from NL to Mild Cognitive Impairment (MCI), and from MCI to AD [18]. Initially calculated by voxel-wise tissue density mas [26], it is further enhanced with calculation on a ROI-based manner using the state-of-the-art multi-atlas segmentation method [27] and available for ADNI-based examinations on [28].

Main steps for calculating SPARE-AD starts with brain extraction from the available images using a multi-atlas based algorithm MASS [29]. Hierarchical parcellation using the MUSE method for multi-atlas ROI parcellation [27] is then performed and followed by the additional pattern analysis. For that purpose, SVM classifier for high-dimensional space is trained to separate AD and NL subjects. To provide unbiased results, the pattern of atrophy is learned from a training set and applied to a test set using leave-one-out cross-validation [18, 19]. The SPARE-AD index represents the amount of atrophy patterns in specific brain regions, indicating the distance of a test sample from the hyperplane that provides the largest margin separation. It can be assumed that this score represents the weighted sum of volumetric values of the test sample in those brain regions that provide highest discrimination between the training classes [18]. Detecting spatially complex and subtle patterns of brain tissue loss, SPARE-AD is found to be much more distinctive than conventional ROI-based measurements [19].

2.3 Estimated Volumes of the Subcortical Brain Structures and Cortical Thickness

To be able to estimate the volumes of the subcortical brain structures and cortical thickness of the structures from the left and right hemisphere, all images were processed using the Freesurfer image analysis suite, which is documented and freely available for download online [30]. The Freesurfer's processing pipeline includes removal of non-brain tissue using a hybrid watershed/surface deformation procedure, automated Talairach transformation, segmentation of the subcortical white matter and deep grey matter volumetric structures, intensity normalization, tessellation of the grey matter white matter boundary, automated topology correction, and surface deformation following intensity gradients. After the completion of the cortical models, registration to a spherical atlas follows, which utilizes individual cortical folding patterns to match cortical geometry across subjects. Then the parcellation of the cerebral cortex into units based on gyral and sulcal structure is performed. Estimations of the volume and cortical thickness of the brain structures are then calculated.

2.4 Image Retrieval

When a query subject is given to the image retrieval system, it makes a comparison between this subject and all other already stored in the database. The result is a sorted list of patients based on the similarity to the query, with the most similar one at the top. In fact, the comparison is performed between the subjects' representation, making it one of the crucial parts of the retrieval pipeline. In this SPARE-AD score is used as image signature. Several scenarios regarding are examined in this context. The first one (SC 1) is solely based on SPARE-AD index. The others are extension of the previous research based on regional volumes of subcortical brain structures as well as cortical thickness of the regions from the left and right hemisphere [6]. Namely, the second scenario (SC2) uses estimated volumes of the subcortical structures (55 in total) combined with the SPARE-AD. The third scenario (SC3) combines cortical thickness of the brain structures

(68 in total) with SPARE-AD score, while the forth (SC4) is based on a combination of volumes and cortical thickness together with the SPARE-AD.

Regarding the scenarios SC2-SC4, feature subset selection was also applied to select the most relevant features and reduce the potential noise using the Correlation-based Feature Selection (CFS) method. According to the previous research [6] it significantly improved the retrieval performance while reducing the dimension of the image descriptor. The algorithm evaluates subsets of features regarding the usefulness of individual features for predicting the class along the degree of intercorrelation among them. Valuable and relevant feature subsets contain features highly correlated with the class, yet uncorrelated to each other [31].

Considering the small number of subjects in the examined cohort, leave-one-out strategy was used. Thus, each subject representation was used as a query against the representations of the rest of the subjects. Three methods were used as similarity measurements: Euclidean distance (1), Manhattan distance (2) and Canberra distance (3) [32]. In these equations, d is the distance between the subjects' representation i and j, k is the feature index, whereas n is the total number of features in the feature vector.

$$d^{euc}(i,j) = \sqrt{\sum_{k=0}^{n-1} |y_{i,k} - y_{j,k}|^2} \tag{1}$$

$$d^{man}(i,j) = \sum_{k=0}^{n-1} |y_{i,k} - y_{j,k}| \tag{2}$$

$$d^{can}(i,j) = \sum_{k=0}^{n-1} \frac{|y_{i,k} - y_{j,k}|}{|y_{i,k}| + |y_{j,k}|} \tag{3}$$

To evaluate the output of the retrieval, standardized evaluation metrics were used:

- Mean Average Precision (MAP) – the mean of the average precision scores for each query, evaluation metric for the general performance of the retrieval system;
- Precision at first 1 (P1) – precision of the first (top) returned subject;
- Precision at first 5 (P5) – precision of the first (top) 5 returned subjects;
- Precision at first 10 (P10) – precision of the first (top) 10 returned subjects;
- Precision at first 20 (P20) – precision of the first (top) 20 returned subjects;
- Precision at first 30 (P30) – precision of the first (top) 30 returned subjects;
- R-precision (RP) – precision at first (top) R returned subjects, where R is the total number of relevant subjects.

The retrieved subject is assumed to be relevant if it belongs to the same category (AD/NL) as the query one.

3 Experimental Results and Discussion

Evaluation of the proposed image representation based on SPARE-AD index for image retrieval in AD context is presented in this section. The retrieval based only on the regional measurements, which are subject of the research in [6], is run through the same experimental setup as in this research. Those results are also presented here for comprehensiveness and clearance, to be able to directly compare the output of the performed retrieval. In this context, the scenario that uses volumetric measurements of subcortical structures is denoted as SC5, whereas SC6 is the scenario based on the cortical thickness estimates. The scenario that uses a combination of volumetric measurements and cortical thickness of the brain regions is labeled as SC7. The results are grouped by the similarity measurement used to calculate distance between the subjects' representation. Table 1 provides the results for all scenarios in the case of Euclidean distance. The results regarding Manhattan distance are presented in Table 2, whereas the evaluation when the comparison is performed using the Canberra distance is given in Table 3. Evan though for the SC1 Euclidean and Manhattan distance are the same metric, both distances are shown for consistency in the representation of the results.

Table 1. Evaluation of the retrieval performance for all scenarios - Euclidean distance.

Scenario	MAP	RP	APat1	APat5	APat10	APat15	APat20	APat30
SC1	0.81	0.76	0.83	0.83	0.83	0.83	0.84	0.84
SC2	0.74	0.68	0.84	0.82	0.82	0.81	0.81	0.81
SC3	0.77	0.71	0.84	0.83	0.83	0.83	0.82	0.82
SC4	0.77	0.72	0.85	0.84	0.84	0.83	0.83	0.82
SC5	0.69	0.65	0.77	0.74	0.74	0.74	0.74	0.73
SC6	0.73	0.68	0.79	0.78	0.77	0.77	0.77	0.77
SC7	0.75	0.71	0.84	0.81	0.81	0.81	0.80	0.80

According to the performed evaluation, the power of the SPARE-AD index is evident. Table 1 shows that the best overall retrieval performance on the bases of MAP score is related to SC1. The situation is similar regarding the R-precision as well. Considering Table 2, again SC1 outperforms all other scenarios regarding MAP and RP. In this cases the results were similar in all scenarios except on SC1. The reason is that in this scenario they are the same metric, so that the results are equal. The case where Euclidean distance was used provided slightly better precision at the level of the firs retrieved image, whereas Manhattan distance led slightly smoother decrease of the precision through the other levels, keeping higher precision at higher levels. In both cases, the combination of the ROI-based measurements combined with SPARE-AD index outperformed the ROI-based scenarios used in [6], respectively. However, the best scenario remains the SC1 based solely on SPARE-AD index. This means that adding additional information, not always improves the results.

Table 2. Evaluation of the retrieval performance for all scenarios - Manhattan distance.

Scenario	MAP	RP	APat1	APat5	APat10	APat15	APat20	APat30
SC1	0.81	0.76	0.83	0.83	0.83	0.83	0.84	0.84
SC2	0.74	0.69	0.83	0.83	0.82	0.81	0.81	0.81
SC3	0.77	0.72	0.84	0.84	0.83	0.83	0.82	0.81
SC4	0.78	0.73	0.83	0.84	0.84	0.83	0.83	0.82
SC5	0.70	0.65	0.75	0.75	0.75	0.74	0.74	0.73
SC6	0.73	0.68	0.77	0.79	0.77	0.77	0.77	0.76
SC7	0.75	0.71	0.81	0.81	0.80	0.80	0.80	0.80

Regarding the case in which Canberra distance was used to compare the images, the influence of the SPARE-AD index used in the image signature is also evident (Table 3). However, in this case the scenario based on the combination of the cortical thickness measurements and the SPARE-AD index provides the best retrieval results considering the MAP value and RP. Moreover, the MAP value obtained in this case (0.85) was the best overall result. Considering the results based on Canberra distance, which is known to be very sensitive to small changes near zero [32], the improvements were even more emphasized. In this case, the MAP value is significantly better in all scenarios SC1-4 in comparison to the results of the previous research [6].

Table 3. Evaluation of the retrieval performance for all scenarios - Canberra distance.

Scenario	MAP	RP	APat1	APat5	APat10	APat15	APat20	APat30
SC1	0.84	0.80	0.83	0.83	0.83	0.83	0.84	0.84
SC2	0.83	0.77	0.85	0.84	0.85	0.85	0.85	0.85
SC3	0.85	0.81	0.86	0.86	0.85	0.85	0.85	0.86
SC4	0.84	0.79	0.86	0.86	0.86	0.86	0.86	0.86
SC5	0.69	0.65	0.74	0.75	0.74	0.74	0.73	0.73
SC6	0.73	0.68	0.79	0.79	0.78	0.77	0.77	0.77
SC7	0.75	0.69	0.83	0.80	0.80	0.79	0.79	0.79

Regarding the dimension of the feature vector, the most efficient scenario was definitely the SC1, with only one feature. In all other cases, feature subset selection was applied to further reduce the descriptor dimension and select the most relevant features. It is important to emphasize that in all cases where the combination of ROI-based measurements SPARE-AD index was part of the most relevant feature subset. Among the other features selected in most of the cases, well known AD indicators in the literature such as hippocampal, amygdala, inferior lateral ventricle volumes, as well as entorhinal thickness were distinguished. When volume estimates were combined with SPARE-AD index, only five features were selected in most of the cases. In the case of a combination of the cortical thickness and SPARE-AD score, eight features were selected in the most of the cases, whereas in the SC4 scenario, 14 features were selected as the most relevant and distinctive in most of the cases.

Considering the good retrieval performance and low dimension of the descriptor, it can be recommended to use SPARE-AD as powerful image signature for image retrieval for AD together with Canberra distance for comparison. Slightly better retrieval performance can be achieved with a combination of SPARE-AD and cortical thickness estimate of the brain regions, with few more features in the image representation.

Image retrieval based on screening images has one crucial importance from the clinical point of view. That is finding relevant information about similar images to the query one at the moment of the first visit to the hospital. However, extending this research in a longitudinal manner, which will be a part of our future work, will add additional value from the clinical and scientific point of view.

4   Conclusion

Research in this domain is actively moving from the traditional feature extraction strategies towards novel techniques that very often use the domain knowledge aiming to overcome the limitations of the general-purpose retrieval systems in the medical domain. In this paper, efficient and powerful image representation based on the SPAER-AD index was proposed, considering the effectiveness of this score in recognition of early AD, and distinction between AD and normal controls at similar age. The evaluation of the image retrieval using the proposed strategy was performed. In this context, four scenarios were evaluated. In the first one, only SPARE-AD index was used as image representation. The other three scenarios were based on a combination of this score with ROI-based measurements.

The evaluation showed that involving the SPARE-AD index in the image representation significantly improve the retrieval results in all cases. The applications in which the dimension of the feature vector is critical, the image signature based only on SPARE-AD index is recommended. Canberra distance was identified as the most appropriate similarity measurement in this case. However, the best overall retrieval results were obtained by using the combination of SPARE-AD and cortical thickness measurements, with applied feature selection to select the most relevant feature subset. In this case the descriptor dimension is few more features higher than in the scenario based only on SPARE-AD index.

The performed research is very beneficial considering the improved, yet efficient image retrieval, which is of crucial importance for the application domain.

Acknowledgement. Data collection and sharing for this study was funded by the Alzheimer's Disease Neuroimaging Initiative (ADNI) (National Institutes of Health Grant U01 AG024904) and DOD ADNI (Department of Defense award number W81XWH-12-2-0012). ADNI is funded by the National Institute on Aging, the National Institute of Biomedical Imaging and Bioengineering, and through generous contributions from the following: AbbVie, Alzheimer's Association; Alzheimer's Drug Discovery Foundation; Araclon Biotech; BioClinica, Inc.; Biogen; Bristol-Myers Squibb Company; CereSpir, Inc.; Cogstate; Eisai Inc.; Elan Pharmaceuticals, Inc.; Eli Lilly and Company; EuroImmun; F. Hoffmann-La Roche Ltd and its affiliated company Genentech, Inc.; Fujirebio; GE Healthcare; IXICO Ltd.; Janssen Alzheimer Immunotherapy Research & Development, LLC.; Johnson & Johnson Pharmaceutical Research & Development

LLC.; Lumosity; Lundbeck; Merck & Co., Inc.; Meso Scale Diagnostics, LLC.; NeuroRx Research; Neurotrack Technologies; Novartis Pharmaceuticals Corporation; Pfizer Inc.; Piramal Imaging; Servier; Takeda Pharmaceutical Company; and Transition Therapeutics. The Canadian Institutes of Health Research is providing funds to support ADNI clinical sites in Canada. Private sector contributions are facilitated by the Foundation for the National Institutes of Health (www.fnih.org). The grantee organization is the Northern California Institute for Research and Education, and the study is coordinated by the Alzheimer's Therapeutic Research Institute at the University of Southern California. ADNI data are disseminated by the Laboratory for Neuro Imaging at the University of Southern California.

Authors also acknowledge the support of the European Commission through the project MAESTRA - Learning from Massive, Incompletely annotated, and Structured Data (Grant number ICT-2013-612944).

References

1. Doré, V., Villemagne, V.L., Bourgeat, P., Fripp, J., Acosta, O., Chetélat, G., Zhou, L., Martins, R., Ellis, K.A., Masters, C.L., Ames, D.: Cross-sectional and longitudinal analysis of the relationship between Aβ deposition, cortical thickness, and memory in cognitively unimpaired individuals and in Alzheimer disease. JAMA Neurol. **70**(7), 903–911 (2013)

2. Nho, K., Risacher, L.S., Crane, P.K., DeCarli, C., Glymour, M.M., Habeck, C., Kim, S., et al.: Voxel and surface-based topography of memory and executive deficits in mild cognitive impairment and Alzheimer's disease. Brain Imaging Behav. **6**(4), 551–567 (2012)

3. Zhang, D., Wang, Y., Zhou, L., Yuan, H., Shen, D.: Multimodal classification of Alz-heimer's disease and mild cognitive impairment. Neuroimage **55**(3), 856–867 (2011)

4. Oliveira, M.C., Cirne, W., de Azevedo Marques, P.M.: Towards applying content-based image retrieval in the clinical routine. Future Gener. Comput. Syst. **23**(3), 466–474 (2007)

5. Rosset, A., Muller, H., Martins, M., Dfouni, N., Vallée, J.-P., Ratib, O.: Casimage project - a digital teaching files authoring environment. J. Thorac. Imaging **19**(2), 1–6 (2004)

6. Trojacanec, K., Kitanovski, I., Dimitrovski, I., Loshkovska, S., Alzheimer's Disease Neuroimaging Initiative: Medical image retrieval for Alzheimer's Disease using structural MRI measures. In: Fred, A., Gamboa, H., Elias, D. (eds.) BIOSTEC 2015. CCIS, vol. 574, pp. 126–141. Springer, Cham (2015). doi:10.1007/978-3-319-27707-3_9

7. Muller, H., Kalpathy-Cramer, J., Kahn Jr., J.C.E., Hersh, W.: Comparing the quality of accessing medical literature using content-based visual and textual information retrieval. In: SPIE Medical Imaging, International Society for Optics and Photonics, p. 726405 (2009)

8. Akgül, C.B., Ünay, D., Ekin, A.: Automated diagnosis of Alzheimer's disease using image similarity and user feedback. In: Proceedings of the ACM International Conference on Image and Video Retrieval, p. 34 (2009)

9. Agarwal, M., Mostafa, J.: Content-based image retrieval for Alzheimer's disease detection. In: 2011 9th International Workshop on Content-Based Multimedia Indexing (CBMI), pp: 13–18 (2011)

10. Mizotin, M., Benois-Pineau, J., Allard, M., Catheline, G.: Feature-based brain MRI retrieval for Alzheimer disease diagnosis. In: 19th IEEE International Conference on Image Processing (ICIP), pp. 1241–1244 (2012)

11. Liu, S., Cai, W., Song, Y., Pujol, S., Kikinis, R., Feng, D.: A bag of semantic words model for medical content-based retrieval. In: MICCAI Workshop on Medical Content-Based Retrieval for Clinical Decision Support (2013)

12. Liu, S., Liu, S., Zhang, F., Cai, W., Pujol, S., Kikinis, R., Feng, D.: Longitudinal brain MR retrieval with diffeomorphic demons registration: What happened to those patients with similar changes? In: 2015 IEEE 12th International Symposium on Biomedical Imaging (ISBI), pp. 588–591. IEEE (2015)

13. Lötjönen, J., Robin, W., Juha, K., Valtteri, J., Lennart, T., Roger, L., Gunhild, W., Hilkka, S., Daniel, R.: Fast and robust extraction of hippocampus from MR images for diagnostics of Alzheimer's disease. Neuroimage 56(1), 185–196 (2011)

14. Sabuncu, M.R., Desikan, R.S., Sepulcre, J., Yeo, B.T.T., Liu, H., Schmansky, N.J., Reuter, M., et al.: The dynamics of cortical and hippocampal atrophy in Alzheimer disease. Arch. Neurol. 68(8), 1040–1048 (2011)

15. Farag, A.A., Ahmed, M.N., El-Baz, A., Hassan, H.: Advanced segmentation techniques. In: Suri, J.S., Wilson, D.L., Laxminarayan, S. (eds.) Handbook of Biomedical Image Analysis, pp. 479–533. Springer, Boston (2005)

16. Nestor, S.M., Raul, R., Michael, B., Matthew, S., Vittorio, A., Jennie, L.W., Jennifer, F., Robert, B.: Ventricular enlargement as a possible measure of Alzheimer's disease progression validated using the Alzheimer's disease neuroimaging initiative database. Brain 131(9), 2443–2454 (2008)

17. Trojacanec, K., Kitanovski, I., Dimitrovski, I., Loshkovska, S.: Medical image retrieval for Alzheimer's Disease using data from multiple time points. In: Loshkovska, S., Koceski, S. (eds.) ICT Innovations 2015. AISC, vol. 399, pp. 215–224. Springer, Cham (2016). doi: 10.1007/978-3-319-25733-4_22

18. Habes, M., Erus, G., Toledo, J.B., Zhang, T., Bryan, N., Launer, L.J., Rosseel, Y., Janowitz, D., Doshi, J., Van der Auwera, S., Von Sarnowski, B.: White matter hyperintensities and imaging patterns of brain ageing in the general population. Brain 139, 1164–1179 (2016). aww008

19. Davatzikos, C., Xu, F., An, Y., Fan, Y., Resnick, S.M.: Longitudinal progression of Alzheimer's-like patterns of atrophy in normal older adults: the SPARE-AD index. Brain 2009(132), 2026–2035 (2009)

20. Davatzikos, C., Bhatt, P., Shaw, L.M., Batmanghelich, K.N., Trojanowski, J.Q.: Prediction of MCI to AD conversion, via MRI, CSF biomarkers, and pattern classification. Neurobiol. Aging 32(2322), 2319–2327 (2011)

21. Wang, Y., Fan, Y., Bhatt, P., Davatzikos, C.: High-dimensional pattern regression using machine learning: from medical images to continuous clinical variables. Neuroimage 50, 1519–1535 (2010)

22. Filipovych, R., Davatzikos, C.: Semi-supervised pattern classification of medical images: application to mild cognitive impairment (MCI). NeuroImage 55, 1109–1119 (2011)

23. Misra, C., Fan, Y., Davatzikos, C.: Baseline and longitudinal patterns of brain atrophy in MCI patients, and their use in prediction of short-term conversion to AD: results from ADNI. NeuroImage 44, 1415–1422 (2009)

24. Toledo, J.B., Da, X., Bhatt, P., Wolk, D.A., Arnold, S.E., Shaw, L.M., Davatzikos, C.: Relationship between plasma analytes and SPARE-AD defined brain atrophy patterns in ADNI. PLoS ONE 8(2), e55531 (2013)

25. Fan, Y., Shen, D., Gur, R.C., Gur, R.E., Davatzikos, C.: COMPARE: classification of morphological patterns using adaptive regional elements. IEEE Trans. Med. Imaging 2007(26), 93–105 (2007)

26. Davatzikos, C., Genc, A., Xu, D., Resnick, S.M.: Voxel-based morphometry using the RAVENS maps: methods and validation using simulated longitudinal atrophy. NeuroImage 14(6), 1361–1369 (2001)

27. Doshi, J., Erus, G., Ou, Y., Resnick, S.M., Gur, R.C., Gur, R.E., Satterthwaite, T.D., Furth, S., Davatzikos, C., Initiative, A.N.: MUSE: MUlti-atlas region Segmentation utilizing Ensembles of registration algorithms and parameters, and locally optimal atlas selection. NeuroImage **127**, 186–195 (2016)
28. ADNI. http://adni.loni.ucla.edu. Accessed 5 June 2017
29. Doshi, J., Erus, G., Ou, Y., Gaonkar, B., Davatzikos, C.: Multi-atlas skull-stripping. Acad. Radiol. **20**(12), 1566–1576 (2013)
30. FreeSurfer. http://surfer.nmr.mgh.harvard.edu/. Accessed 21 Nov 2016
31. Hall, M.A., Holmes, G.: Benchmarking attribute selection techniques for discrete class data mining. IEEE Trans. Knowl. Data Eng. **15**(6), 1437–1447 (2003)
32. Cha, S.H.: Comprehensive survey on distance/similarity measures between probability density functions. City **1**(2), 1 (2007)

Addressing Item-Cold Start Problem in Recommendation Systems Using Model Based Approach and Deep Learning

Ivica Obadić, Gjorgji Madjarov[✉], Ivica Dimitrovski, and Dejan Gjorgjevikj

Faculty of Computer Science and Engineering, Ss. Cyril and Methodius University,
Rugjer Boshkovikj 16, 1000 Skopje, Macedonia
`obadic.ivica@gmail.com`,
{`gjorgji.madjarov,ivica.dimitrovski,dejan.gjorgjevikj`}`@finki.ukim.mk`

Abstract. Traditional recommendation systems rely on past usage data in order to generate new recommendations. Those approaches fail to generate sensible recommendations for new users and items into the system due to missing information about their past interactions. In this paper, we propose a solution for successfully addressing item-cold start problem which uses model-based approach and recent advances in deep learning. In particular, we use latent factor model for recommendation, and predict the latent factors from item's descriptions using convolutional neural network when they cannot be obtained from usage data. Latent factors obtained by applying matrix factorization to the available usage data are used as ground truth to train the convolutional neural network. To create latent factor representations for the new items, the convolutional neural network uses their textual description. The results from the experiments reveal that the proposed approach significantly outperforms several baseline estimators.

1 Introduction

Advances of internet and popularity of modern interactive web applications led to creation of massive amounts of data by internet users. This data becomes useful only when it is analyzed and turned out into valuable information that can be used in future. *Big data* emerged as a concept which describes the challenge of smart management of this vast amounts of data. Common for almost all *big data* challenges is that no explicit algorithms that can solve such problems in finite number of steps have yet been discovered. For example, there is not explicit algorithm which can correctly predict customers future behavior from past data. Therefore, in order to tackle *big data* challenges and needs, *artificial intelligence* and it's fields, especially *machine learning* and *data mining* became actual and extremely popular.

The users in 'big data world' are faced with huge amount of information and choice of large sets of items. Only a small subset of those items fit to the interests of the users. Therefore, to attract users and keep their attention, the applications

© Springer International Publishing AG 2017
D. Trajanov and V. Bakeva (Eds.): ICT Innovations 2017, CCIS 778, pp. 176–185, 2017.
DOI: 10.1007/978-3-319-67597-8_17

must provide a **personalized** content to the user. Personalized content reduces the time of search for the users, prunes the large search space and directs users to the content of their interest. These needs brought recommendation systems live on the scene.

Recommendation systems apply machine learning and data mining techniques over past historical data about users interactions with items in the system. They create a model for the users and the items and then generate new recommendations on basis of that model. There are two major approaches for building recommendation systems (achieving this goal) denoted as collaborative filtering approaches and content-based approaches.

1.1 Collaborative Filtering

Collaborative filtering (CF) methods produce user specific recommendations of items based on ratings or usage without any additional information about either item or users [1]. There are two main types of collaborative filtering approaches [2]:

- *Neighborhood-based* methods recommend items based on relationships between items or, alternatively, between users.
  - *User-user based* create recommendations based on similarity between users. Similarity between any two users most often is determined on the basis on similarity of expressed preferences given by both users for same set of items [3].
  - *Item-item based* approaches use known ratings made by the same user on similar items. They recommend items to a user that are similar with the items for which user has expressed positive preference in the past. This approaches offer better scalability and improved accuracy [4].
- *Model-based* methods, such as matrix factorization (aka, SVD, SVD++, Time SVD++) [2,5–7] model latent characteristics of the users and items. These methods transform items and users to the same latent factor space that describes the users and items by inferring user feedback. These methods become very popular since their use for movie recommendation in the Netflix Prize [8].

1.2 Content-Based Approach.

Users and items in recommendation systems that use content-based approaches are modeled using profiles [1] that represent their most significant characteristics and features. Profiles are usually comprised of binary or real values and in literature are often called *feature vectors*. Item profiles can be constructed by using their metadata. For example if items are movies, then the director and the actors of certain movie can be used as components for the items feature vector. In most cases item metadata by itself does not provide enough information to create rich model that generates sensible recommendations. Therefore, additional features are created by analyzing item content and other data provided for the item such as expressed preferences for that item by the users in the system. In the work presented in this paper we are relying on such features by learning *latent factor* models for users and items in the system.

1.3   Cold-Start Problem

The rapid growth of users and content in the applications causes recommendation systems to face common and constant issue of generating recommendations for new users and/or new items that appear in the system. This problem in the literature is known as *cold start* problem [9], and means that the systems do not have enough data to generate personalized recommendation for a new user (that has just entered the system and has no previous history), or for a new item (product) added to the system. Due to the massive amount of content generated each day in the applications, recommendation systems especially struggle to properly recommend the new items. These two problems are known as a user-cold start and item-cold start problem.

Therefore, the goal of this paper is to propose a successful approach for addressing item-cold start problem in recommendation systems. Here, we propose to use a latent factor model for recommendation, and predict the latent factors from item's descriptions when they cannot be obtained from usage data. In particular, our approach consists of learning latent factors for a new item from its reviews by using deep convolutional neural network first and then use those latent factors for rating the new item to each user by using model-based approaches ($SVD++$). We evaluate our approach on a dataset provided by Yelp in 'RecSys2013: Yelp Business Rating Prediction' challenge. The results are compared with multiple baseline estimators.

1.4   Organization

The rest of the paper is organized as follows. Section 2 gives description of the dataset used in experiments of this paper and explains it's suitability for the topic of this research. Section 3 describes the algorithm used for learning item latent factors from item review. Special attention is given to the architecture of the convolutional neural network. Experiments and results are presented and discussed in Sect. 4. In the last section (Sect. 5), we make the conclusions and discuss possible further directions for extending our work.

2   The Dataset

Yelp dataset[1] used in the experiments of this paper provides real-world data about users, businesses, reviews and checkins for Phoenix, AZ metropolitan area. Of interest of our research are the reviews of the businesses in the dataset. There are total *229 907* reviews given by *45 981* distinct users for *11 537* distinct businesses. This dataset properly simulates the cold start problem because there are

---

[1] https://www.kaggle.com/c/yelp-recsys-2013/data.

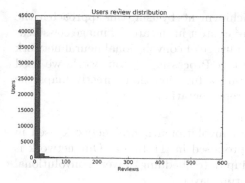

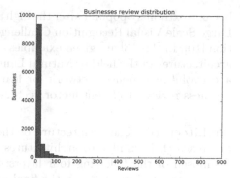

Fig. 1. Users review distribution   **Fig. 2.** Businesses review distribution

known only *0.00043%* of all possible users preferences for businesses. Figures 1 and 2 show the user review distribution and the distribution of reviews for businesses in the dataset respectively. From this figures can be noticed that each user rated only few businesses that leads to a very small number of reviews for majority of businesses in the dataset.

3   Predicting Latent Factors from Item Description

First step in our proposed solution consists of learning latent factors with matrix factorization approach (SVD++ in our case). We represented each user and business with 20 latent factors. Models learned solely with SVD++ are unable to generate predictions in item-cold start problem scenario. Main reason is that latent factors of the new business in the system can not be computed because there are no existing preferences in the database for those businesses. Hence, instead from user preferences data, latent factors for a new business could be learned from other data available for that business. In Yelp dataset, each business has text which represents user textual description about review. Predicting latent factors for a given business from its description is a regression problem. Any traditional regression technique can be used for mapping the feature representation of a business description to latent factors. In our work we use a deep convolutional neural network for mapping the business descriptions directly to latent factors without a need of separate feature extraction approach. Latent factor vectors obtained by applying SVD++ to the available usage data are used as ground truth to train the prediction models. Since there can be multiple reviews for same business, as description for each business, we took the text from the review that has most votes from the other users.

3.1   Deep Convolutional Neural Network

Convolutional neural networks are firstly used in areas of computer vision and speech recognition. Deep learning approaches using convolutional neural networks became extremely popular after the high achievments by using this approach in the Large Scale Visual Recognition Challenge[2] contest in the area of image classification [10]. In [11–13] are given examples on usage of convolutional neural network architectures in the field of Natural Language Processing. In our work, we used a convolutional neural network for learning a function that directly maps the business reviews to latent factor vectors representation.

Architecture. The architecture of the convolutional neural network used in our research is similar to architectures proposed in [11,13,14]. Our network is composed of the following four layers: input (embedding) layer, convolutional layer, max-pooling layer and the final output layer.

The system works by first, constructing a lookup table, where each word is mapped to an appropriate feature vector or word embedding. One way of building the lookup table is by randomly initializing the word vectors. However, previous work showcased that initializing the word embeddings with pre-trained ones provides for better performance as opposed to random initialization. In this work, the input layer of the network maps the business description to a matrix of 300 dimensional Glove word embeddings [15]. Each word in the description is mapped to it's corresponding vector and those vectors are concatenated into a matrix. Alternatively, we can consider our input layer as a function that does the following:

$$f(word_1, word_2, ..., word_n) \mapsto \begin{pmatrix} \overrightarrow{word_1} \\ \overrightarrow{word_2} \\ \vdots \\ \overrightarrow{word_n} \end{pmatrix}$$

Word vectors are additionally updated with backpropagation during the training of the network in order to adapt to the meaning of the corpus in the dataset.

The convolution operation is then applied to the mapped word embeddings and max-over-time pooling in order to get a fixed sized vector. In this step, we apply 50 filters, We use rectified linear units (ReLU) [16] in the convolutional layer and sliding window of size 4. By applying the convolution operation with this setup, input matrix of shape N × 300, where N represents number of words in business description, is transformed into output matrix of shape N × 50. The max-pooling layer subsamples the output of convolutional layer by using region size of N × 1. With this size of region, input to this layer is transformed into a 50 neurons which correspond to the highest values computed by each filter. Neurons from the max-pooling layer are fully connected to the 20 neurons of the output layer which represent the latent factors for the businesses.

3.2 Pre-processing Descriptions

Pre-processing businesses descriptions is necessary step in order to fed the descriptions as an input to the neural network. At first, each description is

---

[2] http://www.image-net.org/challenges/LSVRC.

padded with zeros on the end, so that the length of each description in dataset is equal to the length of the longest description. Then, each word is replaced with it's corresponding Glove word embedding. For the words that were not found in the set of Glove words, the *edit distance* [17] between the corresponding word and each of the words in the Glove set was computed and the word is replaced with the closest word from the glove set if the edit distance is less than or equal to 2, or with a value randomly initialized in the interval $[-0.25, 0.25]$ if no such word exists in the glove set. Random initialization of word embeddings is approach suggested by Kim in [13].

4 Experiments

4.1 Experimental Setup

For our experiments, we took 80% of the businesses and reviews for training, and the other 20% for testing the proposed approach. In order to properly simulate the item-cold start problem, the split that we performed also satisfies the following constraint: each business that occurs in the reviews in the test set, does not occur in any of the reviews in the training set.

The distribution of reviews for businesses in dataset (shown in Fig. 2) shows high imbalance in the number of reviews per business and requires careful decision about the split. Large number of reviews for a business in general allows more accurate and broader description of the business and reduces the outliers effect when training the model. On the other hand, a model learned from businesses with only few reviews in the dataset is much more prone to outliers and user bias effects. In order to test the predictive performance of our model on the two types of businesses according to the number of given reviews, we wanted our split to support two separate evaluations of generated predictions. Separate evaluations need to be performed on predictions for businesses with small number of reviews and for businesses with large number of reviews. Taking this into account, we define two different test sets:

- **Test set 1** consists of reviews from the 15% of businesses with smallest number of reviews in the dataset, which have at least one review with 5 or more votes from other users. This set contains 5 625 review samples in total.
- **Test set 2** consists of reviews from the second half (5%) of the top 10% reviewed businesses which also have at least one review with 5 or more votes from other users. This set contains 45 582 review samples in total.

The training set contains the rest 178 424 reviews (80%). All further experiments described in following subsections are performed over these training set and test sets.

4.2 Parameter Instantiation

In order to properly rate the businesses for each user, we proceed according to the following steps:

1. Users and businesses are modeled with latent factor vectors computed over the training set using *SVD++*.
2. Convolutional neural network is trained to predict businesses latent factors modeled in step 1 using the pre-processed review descriptions in the training set. Following parameters were used in the process of training the network: *learning rate = 0.001*, stochastic gradient descent (*SGD*) as optimization method and *batch size* (number of instances forward-propagated before applying update of the network weights) was set to 64. For the purpose to evaluate the performance of the learned neural network model after each epoch, 10% of the reviews from the training set are used as a validation set. Lowest RMSE is achieved in 23-rd epoch as shown on Fig. 3. Hence, network model computed in the 23-rd epoch is used to obtain business latent factors.
3. As an error function on the output layer we used Root Mean Squared Error (RMSE) between latent factors predicted by the network and the latent factors learned with matrix factorization.

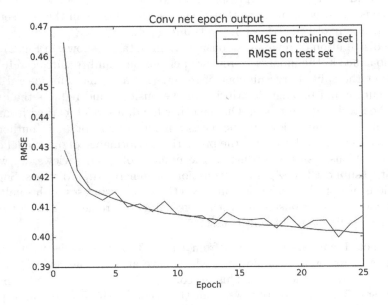

Fig. 3. Convolutional neural network training

4.3   Results and Discussion

We compare our method with baseline approach for overcoming item-cold start problem. The baseline approach estimates latent factor vectors of businesses in the test set by assigning random values for the latent factors. Two different approaches for random initialization are used. As another benchmark, we compare the results of the proposed solution with the results of the predictions obtained only with matrix factorization algorithm (SVD++) under the assumption that when evaluating on each of the test sets separately, the test set is also part of the training set (latent factor vectors are obtained from usage data). This is an upper bound to what is achievable when predicting the latent factors using business reviews. For all comparisons, Root Mean Squared Error (RMSE) is used as an evaluation measure on the predicted business ratings for all the users.

The obtained results from the comparison are given in Table 1. It shows the results for the RMSE scores achieved when the latent factor vectors are randomly generated (*Random 1* and *Random 2*), latent factors are generated by using deep convolutional neural network (*Proposed app.*) and when they are learned from usage data using *SVD++*. *Random 1* initialization assigns random values for the latent factors in the test set using uniform distribution on the interval between the lowest and the highest computed latent factor values from the training set. *Random 2* initialization assigns random values from the interval between the lowest and the highest computed factor values, but only for the corresponding factor. If each business latent factors represent separate row in one matrix, then this kind of initialization assigns a value from the interval $[min, max]$ of its corresponding column to each latent factor. The estimations of the RMSE scores for the both random initializations are performed on 100 runs and the average RMSE score and the variance are reported in the Table 1.

Table 1. Results for RMSE scores achieved when the latent factor vectors are randomized (*Random 1* and *Random 2*), latent factors are generated by using deep convolutional neural network (*Proposed app.*) and when they are learned from usage data using *SVD++* (the upper bound scores)

	Test set 1	Test set 2	Test set 1 + Test set 2
Random 1	2.2070 ± 0.0005	1.8309 ± 9e-05	1.8753 ± 7e-05
Random 2	1.8990 ± 0.0002	1.5659 ± 4e-05	1.6054 ± 4e-05
Proposed app.	1.4094	1.1300	1.1663
SVD++	0.6850	0.5676	0.5877

According to the results show in Table the proposed approach outperforms the baseline approach (*Random 1*) improving the RMSE score for **0.7976** for the test set 1, for **0.7009** for the test set 2 and for **0.7090** for the test set combined from the test set 1 and test set 2.

The proposed approach also shows better results in comparison to the baseline approach that use *Random 2* initialization method with improvement on the RMSE score for **0.4896** on the test set 1, for **0.4359** for the test set 2 and for **0.4391** for the test set combined from the test set 1 and test set 2.

It has been experimentally shown [5] that a small improvement of the RMSE score has high influence on the quality of the top-K recommendations offered to a user. Therefore, from the improvements of the RMSE score that our model shows in comparison to the baseline estimate scores, we can conclude that our proposed algorithm has a significant impact in successfully overcoming the item-cold start problem. In this context, our model achieves highest improvements on the test set 1. This result is very important because test set 1 contains only the least reviewed businesses in our dataset, which best represent item-cold start problem.

There is a large gap between our result and the theoretical maximum (results obtained by using $SVD++$ on the whole dataset), but this is to be expected. Many aspects of the businesses that could influence the user preference cannot be extracted from the businesses reviews only. In our further work, we plan to use other information about the businesses such as description, popularity, type and location.

5   Conclusion

In this paper, we investigated the use of deep learning for successfully addressing the item-cold start problem in recommendation systems. We use latent factor model for recommendation, and predict the latent factors from items' descriptions using deep convolutional neural network when they cannot be obtained from usage data.

We compare our approach to several baseline estimators using RMSE evaluation measure. The results have shown that the proposed approach significantly outperforms the baseline estimators showing a decrease of 0.7090 and 0.4391 of the RMSE score. Also, our results indicate that a lot of research in this domain could benefit significantly from using deep neural networks.

Acknowledgments. We would like to acknowledge the support of the European Commission through the project MAESTRA Learning from Massive, Incompletely annotated, and Structured Data (Grant number ICT-2013-612944). Also, this work was partially financed by the Faculty of Computer Science and Engineering at the Ss. Cyril and Methodius University.

References

1. Rajaraman, A., Ullman, J.D.: Mining of Massive Datasets. Cambridge University Press, New York (2011)
2. Ricci, F., Rokach, L., Shapira, B., Kantor, P.B.: Recommender Systems Handbook, 1st edn. Springer, New York (2010)
3. Segaran, T.: Programming Collective Intelligence, 1st edn. O'Reilly (2007)
4. Bell, R.M., Koren, Y.: Scalable collaborative filtering with jointly derived neighborhood interpolation weights. In: Proceedings of the Seventh International Conference on Data Mining, ICDM 2007, Washington, DC, USA, pp. 43–52. IEEE Computer Society (2007)

5. Koren, Y.: Factorization meets the neighborhood: a multifaceted collaborative filtering model. In: Proceedings of the 14th ACM SIGKDD International Conference on Knowledge Discovery and Data Mining, KDD 2008, New York, NY, USA, pp. 426–434. ACM (2008)

6. Koren, Y.: Collaborative filtering with temporal dynamics. In: Proceedings of the 15th ACM SIGKDD International Conference on Knowledge Discovery and Data Mining, KDD 2009, New York, NY, USA, pp. 447–456. ACM (2009)

7. Oord, A.V.D., Dieleman, S., Schrauwen, B.: Deep content-based music recommendation. In: Proceedings of the 26th International Conference on Neural Information Processing Systems, NIPS 2013, USA, pp. 2643–2651. Curran Associates Inc. (2013)

8. Bennett, J., Lanning, S., Netflix, N.: The netflix prize. In: KDD Cup and Workshop in Conjunction with KDD (2007)

9. Bernardi, L., Kamps, J., Kiseleva, J., Müller, M.J.I.: The continuous cold start problem in e-commerce recommender systems. CoRR, vol. abs/1508.01177 (2015)

10. Krizhevsky, A., Sutskever, I., Hinton, G.E.: Imagenet classification with deep convolutional neural networks. In: Pereira, F., Burges, C.J.C., Bottou, L., Weinberger, K.Q. (eds.) Advances in Neural Information Processing Systems 25, pp. 1097–1105. Curran Associates Inc. (2012)

11. Collobert, R., Weston, J., Bottou, L., Karlen, M., Kavukcuoglu, K., Kuksa, P.: Natural language processing (almost) from scratch. J. Mach. Learn. Res. **12**, 2493–2537 (2011)

12. Kalchbrenner, N., Grefenstette, E., Blunsom, P.: A convolutional neural network for modelling sentences. CoRR, vol. abs/1404.2188 (2014)

13. Kim, Y.: Convolutional neural networks for sentence classification. CoRR, vol. abs/1408.5882 (2014)

14. Stojanovski, D., Strezoski, G., Madjarov, G., Dimitrovski, I.: Twitter sentiment analysis using deep convolutional neural network. In: Onieva, E., Santos, I., Osaba, E., Quintián, H., Corchado, E. (eds.) HAIS 2015. LNCS, vol. 9121, pp. 726–737. Springer, Cham (2015). doi:10.1007/978-3-319-19644-2_60

15. Pennington, J., Socher, R., Manning, C.D.: Glove: Global vectors for word representation (2014)

16. Nair, V., Hinton, G.E.: Rectified linear units improve restricted boltzmann machines. In: Fürnkranz, J., Joachims, T. (eds.) Proceedings of the 27th International Conference on Machine Learning, ICML-2010, pp. 807–814. Omnipress (2010)

17. Cormode, G., Muthukrishnan, S.: The string edit distance matching problem with moves. ACM Trans. Algorithms (TALG) **3**(1), 2 (2007)

Predictive Clustering of Multi-dimensional Time Series Applied to Forest Growing Stock Data for Different Tree Sizes

Valentin Gjorgjioski[1,2,3], Dragi Kocev[2,3], Andrej Bončina[4],
Sašo Džeroski[2,3(✉)], and Marko Debeljak[2,3]

[1] GrabIT, Prilep, Macedonia
[2] Department of Knowledge Technologies, Jožef Stefan Institute, Ljubljana, Slovenia
saso.dzeroski@ijs.si
[3] Jožef Stefan International Postgraduate School, Ljubljana, Slovenia
[4] Biotechnical Faculty, University of Ljubljana, Ljubljana, Slovenia

Abstract. In this paper, we propose a new algorithm for clustering multi-dimensional time series (MDTS). It is based on the predictive clustering paradigm, which combines elements of predictive modelling and clustering. It builds upon the algorithm for predictive clustering trees for modelling time series, and extends it to model MDTS. We also propose adequate distance functions for modelling MDTS. We apply the newly developed approach to the task of analyzing data on forest growing stock in state-owned forests in Slovenia. This task of high importance, since the growing stock of forest stands is a key feature describing the spatio-temporal dynamics of the forest ecosystem response to natural and anthropogenic impacts. It can be thus used to follow the structural, functional and compositional changes of forest ecosystems, which are of increasing importance as the forest area in Europe has been growing steadily for the last 20 years. We have used two scenarios (quantitative and qualitative) to analyze the data at hand. Overall, the growing stock in Slovenian forests has been increasing in the last 40 years. More specifically, the growing stock of the three tree-size has progressive dynamics, which indicates that Slovenian state-owned forests have balanced structure.

Keywords: Time series modelling · Multi-dimensional time series · Forest growing stock · Forest inventory · Predictive clustering trees

1 Introduction

Explanatory analysis of time series data is an important topic in ecological studies. It is used for understanding and predicting the temporal response of ecosystems to variations in ecological, environmental and management factors. The response of ecosystems can be structural, functional and compositional, where the former refers to the spatial arrangement of various components of an ecosystem

© Springer International Publishing AG 2017
D. Trajanov and V. Bakeva (Eds.): ICT Innovations 2017, CCIS 778, pp. 186–195, 2017.
DOI: 10.1007/978-3-319-67597-8_18

(e.g., height of the vegetation, biomass, spatial distribution), the second refers to various ecological processes (e.g., production of organic matter, evapotranspiration) and the latter to the variety of ecosystem components (e.g., species richness and abundance) [1]. Structural, functional and compositional responses are interdependent, so changes of one ecosystem property trigger changes of other ecosystem properties. Scientific evidence shows different responses of ecosystems to global environment changes, such as pollution and climate change.

The most frequently used approach for investigating dynamic changes of forest ecosystem structure is based on the analysis, clustering and modelling of time series data, among which the growing stock data often is the most central. This modelling task is usually based on a mechanistic modelling approach, where the field data are used for calibration and validation of manually constructed models, whose structure is based on existing theoretical knowledge. Such an approach is informative, but includes many parameters, some of which are difficult to set or estimate. It often requires many different types of data that are not always possible to obtain (i.e., crown ratio, regeneration). Due to the large number of parameters that have to be fitted in mechanistic models, it is hard to achieve stability of their output accuracy, and their predictive power is lower [2].

The growing stock is typically modelled as the aggregated value of the volumes of the individual trees and represents the accumulation of wood production of forest trees through the production period of forest stands. With the aggregation of individual tree volumes, a lot of information on its composition is lost. To avoid the limitations of using age structure for larger and more diverse forest areas, the growing stock can be described in more detail by its distribution into diameter at breast height (DBH) classes (e.g., A 10–29 cm, B 30–49 cm, and C 50 cm and more [3]). DBH classes are a good substitute for age classes, which can be used for describing growing stock dynamics in even and uneven-aged forests. The task of modelling aggregated forest growing stock can thus be considered as modelling of single dimensional time series data, while modelling growing stock distributed to DBH classes can be considered as modelling multi-dimensional time series (MDTS) data, where the multiple dimensions represent the growing stock values of DBH growing classes A, B and C as defined above.

The developments described in this paper were motivated by the needs of data analysis of forest growing stock data aggregated per tree class. There are several methods that analyse and cluster time series data from the domain of environmental sciences [4]. To model time series data, these approaches typically use hidden Markov models, neural networks, genetic programming, regression-based approaches (e.g., autoregressive integrated moving average) or analyse the data in the frequency domain. Such methods are often used to forecast weather conditions (e.g., rainfall), predict river water levels (or flood protection), analyse temporal remotely sensed data about land use and land cover. These approaches have a number of limitations: They limit the type of variables that can be used, they make assumptions about prior distributions or missing values, and they offer limited interpretability of the learned models. To overcome these limitations, we propose to use predictive clustering trees [5] that do not make such prior assumptions and are readily interpretable.

The task of modelling forest growing stock data aggregated per tree class is a task of MDTS modelling. To address this task, we extend the approach of learning predictive clustering trees (PCTs) to handle MDTS. Predictive clustering produces both clusters (of MDTS) and explanations of the clusters simultaneously. We apply the proposed method of PCTs for MDTS to Slovenian forest inventory data, in order to find explanations for structural changes in Slovenian forests over the period from 1970 to 2010. This is a new methodological approach for explaining time dynamics of growing stock at the level of DBH classes. It is much more informative then modelling time dynamics of overall growing stock (at an aggregated level), and as such complements the classic mechanistic modelling methodologies and single dimensional modelling of time series data.

2   Predictive Clustering for Multi-dimensional Time Series

Predictive modelling aims at constructing models that can predict a target property of an object from a description of the object. Predictive models are learned from sets of examples, where each example has the form (D, T), with D being an object description and T a target property value. Clustering [6], on the other hand, is concerned with grouping objects into subsets of objects (called clusters) that are similar w.r.t. their description D. There is no target property defined in clustering tasks. In conventional clustering, the notion of a distance (or conversely, similarity) is crucial: examples are considered to be points in a metric space and clusters are constructed such that examples in the same cluster are close according to a particular distance metric. A centroid (or prototypical example) may be used as a representative for a cluster. The centroid is the point with the lowest average (squared) distance to all the examples in the cluster, i.e., the mean or medoid of the examples. Hierarchical clustering and k-means clustering are the most commonly used algorithms for this type of clustering.

Predictive clustering [7] combines elements of both predictive modelling and clustering. As in clustering, we seek clusters of examples that are similar to each other, but in general taking both the descriptive part and the target property into account (the distance measure can be defined on $D \cup T$ or any subset thereof). In addition, a predictive model must be associated to each cluster. The predictive model assigns new instances to clusters based on their description D and provides a prediction for the target property T. A well-known type of model that can be used to this end is a decision tree. A decision tree that is used for predictive clustering is called a predictive clustering tree (PCT). Each node of a PCT represents a cluster. The conjunction of conditions on the path from the root to that node gives a description of the cluster.

Džeroski et al. [8] have proposed to use predictive clustering trees (PCTs) for clustering time series data. The main advantage of using PCTs over other clustering algorithms, such as hierarchical agglomerative clustering and k-means, is that PCTs cluster the time series and provide a description of the clusters at the same time. This allows one to relate various heterogeneous data types and to draw conclusions about their relations.

Table 1 presents the generic induction algorithm for PCTs [7]. It is a variant of the standard greedy recursive top-down decision tree induction algorithm used, e.g., in C4.5 [9]. It takes as input a set of instances I (in our case forest compartments described by phyto-geographical properties and their associated overall forest growing stock time series). The procedure *BestTest* (Table 1, right) searches for the best acceptable test (on a phtyogeographic property) that can be put in a node. If such a test t^* can be found, then the algorithm creates a new internal node labeled t^* and calls itself recursively to construct a subtree for each cluster in the partition $\mathcal{P}^*$ induced by t^* on the instances. If no acceptable test can be found, then the algorithm creates a leaf, and the recursion terminates. (The procedure *Acceptable* defines the stopping criterion of the algorithm, e.g., specifying maximum tree depth or a minimum number of instances in each leaf).

Up to this point, the algorithm is identical to a standard decision tree learner. The main difference is in the heuristic that is used for selecting the tests and the prototype function. For PCTs, this heuristic is the reduction in variance (weighted by cluster size, see line 4 of *BestTest*). Maximizing variance reduction maximizes cluster homogeneity. The variance and prototype function for performing the clustering of the instances need to be instantiated depending on the prediction task at hand.

Table 1. The generic PCT induction algorithm Clus.

procedure $\mathrm{PCT}(I)$ **returns** tree	**procedure** $\mathrm{BestTest}(I)$
1: $(t^*, h^*, \mathcal{P}^*) = \mathrm{BestTest}(I)$	1: $(t^*, h^*, \mathcal{P}^*) = (none, 0, \emptyset)$
2: **if** $t^* \neq none$ **then**	2: **for each** possible test t **do**
3:　　**for each** $I_k \in \mathcal{P}^*$ **do**	3:　　$\mathcal{P} =$ partition induced by t on I
4:　　　$tree_k = \mathrm{PCT}(I_k)$	4:　　$h = Var(I) - \sum_{I_k \in \mathcal{P}} \frac{|I_k|}{|I|} Var(I_k)$
5:　　**return** $\mathrm{node}(t^*, \bigcup_k \{tree_k\})$	5:　　**if** $(h > h^*) \wedge \mathrm{Acceptable}(t, \mathcal{P})$ **then**
6: **else**	6:　　　$(t^*, h^*, \mathcal{P}^*) = (t, h, \mathcal{P})$
7:　　**return** $\mathrm{leaf}(\mathrm{centroid}(I))$	7: **return** $(t^*, h^*, \mathcal{P}^*)$

In this work, we focus on learning PCTs for the prediction and clustering of time series. Learning PCTs for time series data is non-trivial because for many distance measures (the correlation-based, dynamic time warping, and qualitative distances), no closed algebraic form for the centroid is known. Therefore, Džeroski et al. [8] propose to compute cluster variance based on the sum of squared pairwise distances (SSPD) between the cluster elements. Also, two possible representations for c are considered: (a) the centroid is an arbitrary time series, and (b) the centroid is one of the time series from the cluster (the cluster prototype). We use the representation (b), where the centroid can be computed with $|C|^2$ distance computations by substituting q with each time series in the cluster.

The cluster variance can be defined by using the sum of the squared pairwise distances (SSPD) between the cluster elements, i.e.,

$$Var(C) = \frac{1}{2|C|^2} \sum_{X \in C} \sum_{Y \in C} d^2(X, Y) . \tag{1}$$

The factor 2 in the denominator of Eq. 1 ensures that it is identical to the equation for the Euclidean distance. The advantage of this approach is that no centroid needs to be computed to calculate the variance. This approach also requires $|C|^2$ distance computations to calculate the variance, just as the approach with the centroid in representation (b). Hence, using the definition based on a centroid is only more efficient if the centroid can be computed in time linear in the cluster size. This is the case for the Euclidean distance in combination with using the time-point-wise average of the time series as centroid. For the other distance measures, no such centroids are known. Therefore, we choose to estimate cluster variance using the SSPD. The PCT induction algorithm places cluster centroids in its leaves, which can be inspected by the domain expert and used both as predictions and as cluster prototypes. For these centroids, we use representation (b) as discussed above.

The algorithm for the induction of PCTs for modelling MDTS use variance and prototype functions instantiated for the task of modelling MDTS. More specifically, we first define distances for each time seies and then aggregate the distances per time series into a single overall distance.

Different distances on time series can be used for clustering MDTS (i.e., tuples of time series). We can either use one distance measure for all time series/dimensions, or even different distance measure for different dimensions. When choosing which distance to use, we should take into account the application at hand and the properties of the distances. Here, we use the qualitative distance (QD) [10].

QD primarily focuses on the shape of the time series, i.e., the qualitative changes (increase/decrease/steady) between pairs of time points. It can capture non-linear dependencies, as well as linear ones. However, it ignores almost completely the magnitudes of the observed values of the time series. Consider two time series X and Y. Then choose a pair of time points i and j and observe the qualitative change of the value of X and Y at these points. There are three possibilities: increase ($X_i > X_j$), no-change ($X_i \approx X_j$), and decrease ($X_i < X_j$). d_{qual} is obtained by summing the difference in qualitative change observed for X and Y for all pairs of time points, i.e.,

$$d_{\text{qual}}(X, Y) = \sum_{i=1}^{n-1} \sum_{j=i+1}^{n} \frac{2 \cdot Diff(q(X_i, X_j), q(Y_i, Y_j))}{N \cdot (N - 1)} , \tag{2}$$

where the $Diff(q_1, q_2)$ function (the difference between pairs of qualitative changes can be 0 for having the same trend, 0.5 if only one changes and 1 for having opposite trends). Roughly speaking, d_{qual} counts the number of disagreements in change of X and Y. d_{qual} does not have the drawbacks of the correlation based measure. First, it can be computed for very short time series, without decreasing the quality of the estimate. Second, it captures the similarity in shape of the time series, regardless of whether their dependence is linear or non-linear.

3   Data Analysis Task and Study Design

The data used here come from the spatial information database Silva-SI [11]. The data have been gathered from various sources, mostly forest inventories within Forest Management Plans (FMP), which are typically performed every 10 years (each year approximately 1/10 of FMP are revised). The database currently comprises digitalised data for 21052 forest compartments covering 7452 km^2 or 64% of the Slovenian forests for the period from the year 2010 back to 1970. In our study, we use only compartments that are 100% owned by the state, which amounts to 5237 compartments.

Forest compartments are permanent and their size and borders have not changed since the first forest inventories. All compartments are described with the same suite of 47 environmental and stand variables, which indicate the state of forest stands. The variables describing the geographic characteristics of compartments were acquired from a digital elevation model with a spatial resolution of 25 m × 25 m. From these variables, based on a previous study [12], we selected a subset of the 6 most important variables: mean inclination (INC), mean elevation (ELV), the type of bedrock (BEDR), prevailing aspect of the compartment (ASP), phyto-geographical region (PHYTOREG) and site productivity index(PI).

The dynamics of growing stock is described by the time course profiles of growing stock values, recorded each 10 years, from 1970 to 2010. At the same 5 time points, we also have the growing stock values per DBH class (A,B,C), representing the growing stock dynamics in more detail. This more detailed view constitutes a multi-dimensional (three-dimensional) time series.

The data analysis task that we address is to group/cluster forest compartments with similar growing stock dynamics as described by the growing stock DBH classes. In addition, we want to describe the groups/clusters of forest compartments in terms of their geo-physical properties. The goal is thus to relate the dynamics of growing stock for DBH classes A, B and C and the variables describing forest compartments (INC, ELV, BEDR, ASP, PHYTOREG, PI).

We use PCTs for predicting/clustering MDTS. After building a tree (and a PCT), it is typical to prune it, in order to deal with noise and other types of imperfection in the data. We employ two pruning algorithms: *F-test* pruning and *MaxSize* pruning. In our experiments, we set the significance level for the *F-test* pruning to 0.01 and the maximal tree size to 6 leaves (i.e., clusters).

4   Results

We first discuss the results of the quantitative scenario and give the obtained predictive clustering tree (Fig. 1) and a map of Slovenia illustrating the distribution of the locations that share similar forest growing stock dynamics (Fig. 3(left)). Next, in a similar way, we present the results for the qualitative scenario (Fig. 2 gives the predictive clustering tree and Fig. 3(right) gives the distribution of locations with similar dynamics).

The structure of the model describing the quantitative aspect of temporal multi-dimensional dynamics of growing stock shows that the most important variable (in state owned forests) related to the dynamics is the phyto-geographic region ($PHYTOREG$), followed by the site productivity index (PI), altitude (ELV), inclination (INC) and bedrock ($BEDR$) (Fig. 1). While all forest compartments (all clusters) show permanent increase of the total growing stock (e.g., due to the impact of forest policies, the increasing importance of nature conservation, protection of large trees ...), the dynamics of DBH classes A, B and C of forest stands is still different. In most cases, DBH class C shows the largest accumulation of growing stock, which has the lowest increase at the most extreme growing productive site due to slower forest growth. However, some groups (e.g., group four) show relatively even increase of all three DBH classes, while in other groups (e.g., group six and partly one, two and three) the changes are uneven.

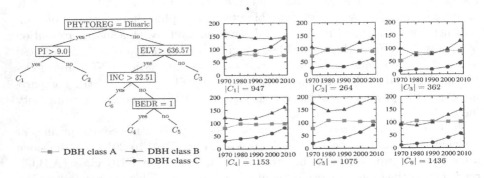

Fig. 1. The predictive clustering tree (top) for modelling MDTS of growing stock DBH classes A, B and C constructed by using the Euclidean distance measure between time series. In the graphs representing the centroids/prototypes of the clusters (bottom), the x-axis denotes the time in years and the y-axis the forest growing stock of DBH classes A, B, and C, in m^3 per ha.

A more detailed examination of the model shows that forests from the Dinaric phyto-geographical region located at high productivity sites (C_1) have the largest change of growing stock in DBH class C. For these forests, the growing stock in DBH class B declines slowly over the entire period, while the growing stock in Class A remains more or less the at the same level. The different types of change of growing stock in DBH classes B and C show that the growing stock moves from DBH class B to DBH class C, which means that growing stock in these forests continuously increases, while the stability of the growing stock in DBH class A does not indicate aging of these forests.

Such a pattern of change is present also in all other groups, but with a lower intensity of accumulation of growing stock in DBH class C. However, growing

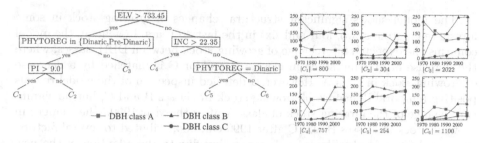

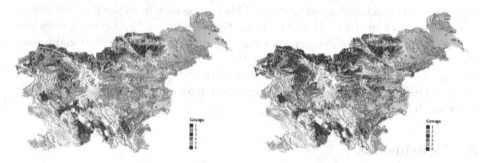

Fig. 2. The predictive clustering tree (top) for modelling MDTS of growing stock DBH classes A, B, and C, constructed by using the qualitative distance measure between time series. In the graphs representing the centroids/prototypes of the clusters (bottom), the x-axis denotes the time in years and the y-axis the forest growing stock in of DBH classes A, B and C in m^3 per *ha*.

Fig. 3. Maps of Slovenia that illustrates the distribution of the forest compartments (private and state-owned forests) belonging to the different clusters identified with the predictive clustering tree from Fig. 1 (left) and Fig. 2 (right).

stock in DBH class C is continually increasing, which indicates the accumulation of growing stock in larger trees. In contrast to group one (C_1), growing stock in DBH class B in increasing from year 1980, while DBH class A stays stable. If we compare the absolute values of growing stock for DBH classes A and B between the six groups, we can notice that forests with lower growing stock are located either in the areas with non-favourable conditions for forest growth (e.g., low site productivity index, high inclination, less suitable carbonate bedrock) or at lower altitudes, which were in the past more exposed to human exploitation, due to their vicinity to more densely populated regions.

The model constructed in the second scenario, where we use the qualitative distance between the growing stock DBH classes A, B and C time series is given in Fig. 2. It shows that the most important variable distinguishing between different patterns of growing stock dynamics is the elevation (ELV), followed by phyto-geographical region (PHYTOREG), inclination (INC), site productivity index (PI) and phyto-geographical region (PHYTOREG) again.

The model shows significant structural changes of growing stock in some forests (e.g., groups two, four and six) in the last 40 years. In general, the model confirms the expected interchange of growing stock between DBH classes, where the increase of growing stock in classes B and/or C is matched by a decrease of growing stock in class A. However, a detailed inspection of the model reveals positive temporal dynamics of growing stock in classes B and C, but no significant change of the growing stock of class A after the year 1990. The changes in growing stock in classes B and C after 1990 can be attributed to several factors, such as changes in the Slovenian forest policy due to the adoption of the new Law on Forests, lower logging intensity because of demographic and economic reasons (e.g., decreasing importance of the economic functions of forests and increasing importance of nature conservation).

The continuous change of forest stock also indicates that most of the forests are not yet in a steady state balance with the growing conditions despite the very high and well structured growing stock. This can be seen from the positive trends of growing stock in classes B and C, while the growing stock of class A does not change significantly. Besides organizational changes in 1990, the National Forest Program was adopted in 1996 which set the allowable cut limit to a maximum of 60% of the total increment. This is reflected in the increasing trend of growing stock in DBH classes B and C and has resulted in an increase of the total growing stock in the period 1990–2000.

5   Conclusions

This paper describes an extension of predictive clustering trees for the analysis of multi-dimensional time series and its successful application to a complex forestry data set. The analysis of MDTS has been thus far treated by using standard clustering algorithms to detect the potential groups/clusters of examples. The obtained clusters have been then typically described by post facto search for cluster descriptions. The proposed algorithm provides both the clusters of MDTS and their symbolic descriptions simultaneously, and to the best of our knowledge, it is the first algorithm of this kind.

We used the proposed algorithm to model Slovenian forest inventory data consisting of phyto-geographical properties and forest growing stock of forest compartments. The inventory data spans over four decades, from 1970 until 2010. The task concerns the modelling of the dynamics of the forest growing stock distributed into DBH classes. The forest growing stock is the basic attribute for describing the forest ecosystem response to natural and anthropogenic impacts and allows us to follow the structural, functional and compositional changes of forest ecosystems.

Most often, the changes of growing stock given with DBH classes are considered at stand level and smaller changes are expected at landscape/regional level. However, change of growing stock of forest stands can occur at landscape/regional level too, and our study has identified quite large changes. Since most of the analyzed data are from nature based/uneven-aged forest stands, we expected that the growing stock would remain relatively stable but our results have revealed that this was not the case.

Based on the results, we can make two important conclusions concerning the changes of growing stock structure and its temporal dynamics. First, the quantitative model shows that growing stock in Slovenian forests has continuously increased in the last 40 years. At the same time, the growing stock of DBH classes show intensive dynamics, which indicates that despite the large total growing stock, Slovenian forests have not yet reached their steady state of structural development. This is confirmed also with the second conclusion based on the qualitative model, which shows steady and progressive temporal dynamics of growing stock in classes B and C, while the dynamics of class A does not yet show any degressive dynamics. Thus, Slovenian state-owned forests are increasing their growing stock while their structure is balanced, which gives them large structural stability.

Acknowledgement. We would like to acknowledge the support of the European Commission through the project MAESTRA - Learning from Massive, Incompletely annotated, and Structured Data (Grant number ICT-2013-612944).

References

1. McElhinny, C., Gibbons, P., Brack, C., Bauhus, J.: Forest and woodland stand structural complexity: its definition and measurement. For. Ecol. Manage. **218** (1–3), 1–24 (2005)
2. Jørgensen, S.E., Bendoricchio, G.: Fundamentals of Ecological Modelling. Elsevier, Padova (2001)
3. Rules on the forest management and silviculture plans. Official Gazette of the Republic of Slovenia, no. 5/1998 of 23-1-1998 (1998)
4. Liao, T.W.: Clustering of time series data-a survey. Pattern Recogn. **38**(11), 1857–1874 (2005)
5. Blockeel, H.: Top-down induction of first order logical decision trees. Ph.D thesis, Katholieke Universiteit Leuven, Leuven, Belgium (1998)
6. Kaufman, L., Rousseeuw, P. (eds.): Finding Groups in Data: An Introduction to Cluster Analysis. Wiley, Hoboken (1990)
7. Blockeel, H., Raedt, L.D., Ramon, J.: Top-down induction of clustering trees. In: Proceedings of the 15th International Conference on Machine Learning, pp. 55–63. Morgan Kaufmann (1998)
8. Džeroski, S., Gjorgjioski, V., Slavkov, I., Struyf, J.: Analysis of time series data with predictive clustering trees. In: Džeroski, S., Struyf, J. (eds.) KDID 2006. LNCS, vol. 4747, pp. 63–80. Springer, Heidelberg (2007). doi:10.1007/978-3-540-75549-4_5
9. Quinlan, J.: C4.5: Programs for Machine Learning. Morgan Kaufmann Series in Machine Learning. Morgan Kaufmann, San Francisco (1993)
10. Todorovski, L., Cestnik, B., Kline, M., Lavrač, N., Džeroski, S.: Qualitative clustering of short time-series: a case study of firms reputation data. In: ECML/PKDD 2002 Workshop on Integration and Collaboration Aspects of Data Mining, Decision Support and Meta-Learning, pp. 141–149 (2002)
11. Poljanec, A.: Changes in forest stand structure in Slovenia in period 1970–2005. Ph.D thesis, Biotechnical Faculty, University of Ljubljana, Slovenia (2008)
12. Debeljak, M., Poljanec, A., Ženko, B.: Modelling forest growing stock from inventory data: a data mining approach. Ecol. Ind. **41**, 30–39 (2014)

New Decoding Algorithm for Cryptcodes Based on Quasigroups for Transmission Through a Low Noise Channel

Aleksandra Popovska-Mitrovikj$^{(\boxtimes)}$, Verica Bakeva, and Daniela Mechkaroska

Faculty of Computer Science and Engineering,
Ss. Cyril and Methodius University, Skopje, Macedonia
{aleksandra.popovska.mitrovikj,verica.bakeva}@finki.ukim.mk,
daniela-mec@hotmail.com

Abstract. Random Codes Based on Quasigroups (RCBQ) are crypt-codes, so they provide (in one algorithm) a correction of certain amount of errors in the input data and an information security. Cut-Decoding and 4-Sets-Cut-Decoding algorithms are proposed elsewhere and they improve decoding of these codes.

In the decoding process of these codes, three types of errors appear: *more-candidate-error*, *null-error* and *undetected-error*. *More-candidate-errors* can occur even all bits in the message are correctly transmitted. So, the packet-error (and bit-error) probability can be positive for very small bit-error probability in the noise channel. In order to eliminate this problem, here we define new decoding algorithms (called Fast-Cut-Decoding and Fast-4-Sets-Cut-Decoding algorithms) that enable more efficient and faster decoding, especially for transmission through a low noise channel. We present several experimental results obtained with these new algorithms. Also, we analyze the results for bit-error and packet-error probabilities and decoding speed when messages are transmitted through Gaussian channel with different values of signal-to-noise ratio (SNR).

Keywords: Cryptcoding · Low noise channel · Bit-error probability · Packet-error probability · Quasigroup

1 Introduction

The need for secure data transmission requires continuous improvement of existing and developing new algorithms that will provide correct and secure transmission of data. Due to necessity of obtaining efficient and secure transmission of data at the same time, the concept of cryptcoding begins to develop. Cryptcoding merges processes of encoding and encryption. An usual way to obtain codes resistant to an intruder attack consists of application of some of the known ciphers on the codewords, before sending them through an insecure channel [4,10,11]. Then two algorithms are used, one for correction of errors and another for obtaining information security.

© Springer International Publishing AG 2017
D. Trajanov and V. Bakeva (Eds.): ICT Innovations 2017, CCIS 778, pp. 196–204, 2017.
DOI: 10.1007/978-3-319-67597-8_19

Random Codes Based on Quasigroups (RCBQ) are defined (in [1]) by using a cryptographic algorithm during the encoding/decoding process, i.e., they are cryptcodes. Therefore, they allow not only correction of certain amount of errors in the input data, but they also provide an information security, all built in one algorithm. For improving the performances of these codes, Cut-Decoding and 4-Sets-Cut-Decoding algorithms are proposed in [5,6]. In the decoding process of these codes, three types of errors appear: *more-candidate-error*, *null-error* and *undetected-error*. *More-candidate-errors* can occur even all bits in the message are correctly transmitted. Therefore, the packet-error and bit-error probabilities can be positive for very small bit-error probability in the noise channel. In order to eliminate this problem, in this paper we define new decoding algorithms called Fast-Cut-Decoding and Fast-4-Sets-Cut-Decoding algorithms. These algorithms enable more efficient and faster decoding, especially for transmission through a low noise channel.

The paper is organized as follows. In Sect. 2, we briefly describe Cut-Decoding and 4-Sets-Cut-Decoding algorithms for RCBQ. The description of decoding with new algorithms (Fast-Cut-Decoding and Fast-4-Sets-Cut-Decoding) is given in Sect. 3. In Sect. 4, we present several experimental results obtained with these new algorithms. We analyze the results for packet-error, bit-error probabilities and decoding speed when messages are transmitted through Gaussian channel with different values of signal-to-noise ratio (SNR). Also, we compare these results with the results obtained with the old algorithms. At the end, we give some conclusions for presented results.

2 Description of Cut-Decoding and 4-Sets-Cut-Decoding Algorithms

RCBQs are designed using algorithms for encryption and decryption from the implementation of TASC (Totally Asynchronous Stream Ciphers) by quasigroup string transformation [2]. These cryptographic algorithms use the alphabet Q and a quasigroup operation $*$ on Q together with its parastrophe $\backslash$.

The notions of quasigroups and quasigroup string transformations are given in the previous papers for these codes [5,7–9]. Here, we use the same terminology and notations as there.

2.1 Description of Coding

At first, let describe Standard coding algorithm for RCBQs proposed in [1]. The message $M = m_1 m_2 \ldots m_l$ (of $N_{block} = 4l$ bits where $m_i \in Q$ and Q is an alphabet of 4-bit symbols (nibbles)) is extended to message $L = L^{(1)} L^{(2)} \ldots L^{(s)} = L_1 L_2 \ldots L_m$ by adding redundant zero symbols. The produced message L has $N = 4m$ bits $(m = rs)$, where $L_i \in Q$ and $L^{(i)}$ are sub-blocks of r symbols from Q. In this way we obtain (N_{block}, N) code with rate $R = N_{block}/N$. The codeword is produced after applying the encryption algorithm of TASC

Encryption	Decryption
Input: Key $k = k_1 k_2 \ldots k_n$ and	**Input**: The pair
$L = L_1 L_2 \ldots L_m$	$(a_1 a_2 \ldots a_r, k_1 k_2 \ldots k_n)$
Output: codeword	**Output**: The pair
$C = C_1 C_2 \ldots C_m$	$(c_1 c_2 \ldots c_r, K_1 K_2 \ldots K_n)$
For $j = 1$ to m	For $i = 1$ to n
$\quad X \leftarrow L_j;$	$\quad K_i \leftarrow k_i;$
$\quad T \leftarrow 0;$	For $j = 0$ to $r - 1$
$\quad$ For $i = 1$ to n	$\quad X, T \leftarrow a_{j+1};$
$\quad\quad X \leftarrow k_i * X;$	$\quad temp \leftarrow K_n;$
$\quad\quad T \leftarrow T \oplus X;$	$\quad$ For $i = n$ to 2
$\quad\quad k_i \leftarrow X;$	$\quad\quad X \leftarrow temp \setminus X;$
$\quad k_n \leftarrow T$	$\quad\quad T \leftarrow T \oplus X;$
$\quad$ **Output**: $C_j \leftarrow X$	$\quad\quad temp \leftarrow K_{i-1};$
	$\quad\quad K_{i-1} \leftarrow X;$
	$\quad X \leftarrow temp \setminus X;$
	$\quad K_n \leftarrow T;$
	$\quad c_{j+1} \leftarrow X;$
	Output: $(c_1 c_2 \ldots c_r, K_1 K_2 \ldots K_n)$

Fig. 1. Algorithms for encryption and decryption

(given in Fig. 1) on the message L. For this aim, a key $k = k_1 k_2 \ldots k_n \in Q^n$ should be chosen. The obtained codeword of M is $C = C_1 C_2 \ldots C_m$, where $C_i \in Q$.

In Cut-Decoding algorithm, instead of using (N_{block}, N) code with rate R, we use together two $(N_{block}, N/2)$ codes with rate $2R$ for coding/decoding the same message of N_{block} bits. Namely, for coding we apply the encryption algorithm (given in Fig. 1) two times, on the same redundant message L using different parameters (different keys or quasigroups). In this way we obtain the codeword of the message as concatenation of two codewords of $N/2$ bits. In 4-Sets-Cut-Decoding algorithm we use four $(N_{block}, N/4)$ codes with rate $4R$, on the same way as in coding with Cut-Decoding algorithm and the codeword of the message is a concatenation of four codewords of $N/4$ bits.

2.2   Description of Decoding

The decoding in all three algorithms is actually a list decoding and it is described below.

In Standard decoding algorithm for RCBQs, after transmission through a noise channel (for our experiments we use Gaussian channel), the codeword C will be received as message $D = D^{(1)} D^{(2)} \ldots D^{(s)} = D_1 D_2 \ldots D_m$ where $D^{(i)}$ are blocks of r symbols from Q and $D_i \in Q$. The decoding process consists of four steps: (i) procedure for generating the sets with predefined Hamming distance, (ii) inverse coding algorithm, (iii) procedure for generating decoding candidate sets and (iv) decoding rule.

Let B_{max} be a given integer which denotes the assumed maximum number of bit errors that occur in a block during transmission. We generate the sets $H_i = \{\alpha | \alpha \in Q^r, \quad H(D^{(i)}, \alpha) \leq B_{max}\}$, for $i = 1, 2, \ldots, s$, where $H(D^{(i)}, \alpha)$ is Hamming distance between $D^{(i)}$ and α.

The decoding candidate sets S_0, S_1, $S_2, \ldots, S_s$ are defined iteratively. Let $S_0 = (k_1 \ldots k_n; \lambda)$, where λ is the empty sequence. Let S_{i-1} be defined for $i \geq 1$. Then S_i is the set of all pairs $(\delta, w_1 w_2 \ldots w_{4ri})$ obtained by using the sets S_{i-1} and H_i as follows (w_j are bits). For each element $\alpha \in H_i$ and each $(\beta, w_1 w_2 \ldots w_{4r(i-1)}) \in S_{i-1}$, we apply the inverse coding algorithm (i.e., algorithm for decryption given in Fig. 1) with input (α, β). If the output is the pair (γ, δ) and if both sequences γ and $L^{(i)}$ have the redundant zeros in the same positions, then the pair $(\delta, w_1 w_2 \ldots w_{4r(i-1)} c_1 c_2 \ldots c_r) \equiv (\delta, w_1 w_2 \ldots w_{4ri})$ $(c_i \in Q)$ is an element of S_i.

In Cut-Decoding algorithm, after transmission through a noisy channel, we divide the outgoing message $D = D^{(1)} D^{(2)} \ldots D^{(s)}$ in two messages $D_1 = D^{(1)} D^{(2)} \ldots D^{(s/2)}$ and $D_2 = D^{(s/2+1)} D^{(s/2+2)} \ldots D^{(s)}$ with equal lengths and we decode them parallel with the corresponding parameters. In this decoding algorithm we make modification in the procedure for generating decoding candidate sets. Let $S_i^{(1)}$ and $S_i^{(2)}$ be the decoding candidate sets obtained in the ith iteration of both parallel decoding processes, $i = 1, \ldots, s/2$. Then, before the next iteration we eliminate from $S_i^{(1)}$ all elements whose second part does not match with the second part of an element in $S_i^{(2)}$, and vice versa. In the $(i+1)^{th}$ iteration the both processes use the corresponding reduced sets $S_i^{(1)}$ and $S_i^{(2)}$.

In [6] authors proposed 4 different versions of decoding with 4-Sets-Cut-Decoding algorithm. The best results are obtained using 4-Sets-Cut-Decoding algorithm#3. In our experiments we use only this version and further on we briefly describe it. After transmitting through a noisy channel, we divide the outgoing message $D = D^{(1)} D^{(2)} \ldots D^{(s)}$ in four messages $D^1 = D^{(1)} D^{(2)} \ldots D^{(s/4)}$, $D^2 = D^{(s/4+1)} D^{(s/4+2)} \ldots D^{(s/2)}$, $D^3 = D^{(s/2+1)} D^{(s/2+2)} \ldots D^{(3s/4)}$ and $D^4 = D^{(3s/4+1)} D^{(3s/4+2)} \ldots D^{(s)}$ with equal lengths and we decode them parallelly with the corresponding parameters. Similarly, as in Cut-Decoding algorithm, in each iteration of the decoding process we reduce the decoding candidate sets obtained in the four decoding processes, as follows. Let $S_i^{(1)}$, $S_i^{(2)}$, $S_i^{(3)}$ and $S_i^{(4)}$ be the decoding candidate sets obtained in the ith iteration of four parallel decoding processes, $i = 1, \ldots, s/4$. Let $V_1 = \{w_1 w_2 \ldots w_{r \cdot a \cdot i} | (\delta, w_1 w_2 \ldots w_{r \cdot a \cdot i}) \in S_i^{(1)}\}$, $\ldots$, $V_4 = \{w_1 w_2 \ldots w_{r \cdot a \cdot i} | (\delta, w_1 w_2 \ldots w_{r \cdot a \cdot i}) \in S_i^{(4)}\}$ and $V = V_1 \cap V_2 \cap V_3 \cap V_4$. If $V = \emptyset$ then $V = (V_1 \cap V_2 \cap V_3) \cup (V_1 \cap V_2 \cap V_4) \cup (V_1 \cap V_3 \cap V_4) \cup (V_2 \cap V_3 \cap V_4)$. Before the next iteration we eliminate from $S_i^{(j)}$ all elements whose second part is not in V, $j = 1, 2, 3, 4$.

After the last iteration, if all reduced sets $S_{s/2}^{(1)}$, $S_{s/2}^{(2)}$ in Cut-Decoding algorithm (or $S_{s/4}^{(1)}$, $S_{s/4}^{(2)}$, $S_{s/4}^{(3)}$, $S_{s/4}^{(4)}$ in 4-Sets-Cut-Decoding) have only one element with a same second component then this component is the decoded message L. In this case, we say that we have a *successful decoding*. If the decoded message is not the correct one then we have an *undetected-error*. If the reduced sets obtained in the last iteration have more than one element then we have a *more-candidate-error*. If we obtain $S_i^{(1)} = S_i^{(2)} = \emptyset$ in some iteration of Cut-Decoding or $S_i^{(1)} = S_i^{(2)} = S_i^{(3)} = S_i^{(4)} = \emptyset$ in some iteration of 4-Sets-Cut-Decoding algorithm, then the process will finish (a *null-error* appears). But, if we obtain at least one nonempty decoding candidate set in an iteration then the decoding continues with the nonempty sets (the reduced sets are obtained by intersection of the non-empty sets only).

3   Description of Fast-Cut-Decoding and Fast-4-Sets-Cut-Decoding Algorithms

As we mentioned previously, decoding with Cut-Decoding and 4-Sets-Cut-Decoding algorithms is actually list decoding. Therefore, the speed of decoding process depends on the list size (a shorter list gives faster decoding). In both algorithms, the list size depends on B_{max} (the maximal assumed number of bit errors in a block). For smaller values of B_{max}, we obtain shorter lists. But, we do not know in advance how many errors appear during transmission of a block. If this number of errors is larger than assumed number of bit errors B_{max} in a block, the errors will not be corrected. On the other side, if B_{max} is too large, we have long lists and the process of decoding is too slow. Also, larger value of B_{max} can lead to ending of the decoding process with *more-candidate-error* (the correct message will be in the list of the last iteration, if there are no more than B_{max} errors during transmission). Therefore, with all decoding algorithms for RCBQ, *more-candidate-errors* can be obtained, although the bit-error probability of the channel is too small and the number of bit errors in a block is not greater than B_{max} (or no errors during transmission).

In order to solve this problem, we propose the following modification of Cut-Decoding and 4-Sets-Cut-Decoding algorithms, called Fast-Cut-Decoding and Fast-4-Sets-Cut-Decoding algorithms.

Here, instead of fixed value B_{max} in the both algorithms, we start with $B_{max} = 1$. If we have successful decoding, the procedure is done. If not, we increase the value of B_{max} with 1 and repeat the decoding process with new value of B_{max}, etc. The decoding finishes with $B_{max} = 4$ (for rate 1/4) or with $B_{max} = 5$ (for rate 1/8).

The new algorithms try to decode the message using the shorter lists and in the case of successful decoding with small value of B_{max} ($B_{max} < 4$), we avoid long lists and slower decoding. Also, we decrease the number of *more-candidate-errors*.

4   Experimental Results

In this section, we give the experimental results obtained with new Fast-Cut-Decoding and Fast-4-Sets-Cut-Decoding algorithms for rate $R = 1/4$ and $R = 1/8$, for different values of SNR in Gaussian channel. We compare these results with corresponding results obtained with Cut-Decoding and 4-Sets-Cut-Decoding algorithms. In order to show the efficiency of new algorithms we present percentages of messages which decoding finished with $B_{max} = 1, 2, 3, 4$ or 5.

First, we present results for code $(72, 288)$ with rate $R = 1/4$. For this code, we made experiments with Cut-Decoding algorithm and Fast-Cut-Decoding algorithm using the following parameters:

- Redundancy pattern: 1100 1100 1000 0000 1100 1000 1000 0000 1100 1100 1000 0000 1100 1000 1000 0000 0000 0000, for rate $1/2$,
- Two different keys of 10 nibbles, and
- Quasigroup $(Q, *)$ on $Q = \{0, 1, 2, ..., 9, a, b, c, d, e, f\}$ given in Table 1 and the corresponding parastrophe.

In experiments with Cut-Decoding algorithm we use $B_{max} = 4$ and for Fast-Cut-Decoding algorithm the maximum value of B_{max} is 4.

Table 1. Quasigroup of order 16 used in the experiments

*	0	1	2	3	4	5	6	7	8	9	a	b	c	d	e	f
0	3	c	2	5	f	7	6	1	0	b	d	e	8	4	9	a
1	0	3	9	d	8	1	7	b	6	5	2	a	c	f	e	4
2	1	0	e	c	4	5	f	9	d	3	6	7	a	8	b	2
3	6	b	f	1	9	4	e	a	3	7	8	0	2	c	d	5
4	4	5	0	7	6	b	9	3	f	2	a	8	d	e	c	1
5	f	a	1	0	e	2	4	c	7	d	3	b	5	9	8	6
6	2	f	a	3	c	8	d	0	b	e	9	4	6	1	5	7
7	e	9	c	a	1	d	8	6	5	f	b	2	4	0	7	3
8	c	7	6	2	a	f	b	5	1	0	4	9	e	d	3	8
9	b	e	4	9	d	3	1	f	8	c	5	6	7	a	2	0
a	9	4	d	8	0	6	5	7	e	1	f	3	b	2	a	c
b	7	8	5	e	2	a	3	4	c	6	0	d	f	b	1	9
c	5	2	b	6	7	9	0	e	a	8	c	f	1	3	4	d
d	a	6	8	4	3	e	c	d	2	9	1	5	0	7	f	b
e	d	1	3	f	b	0	2	8	4	a	7	c	9	5	6	e
f	8	d	7	b	5	c	a	2	9	4	e	1	3	6	0	f

In Table 2, we give experimental results for bit-error probabilities BER_{cut} and packet-error probabilities PER_{cut} (obtained with Cut-Decoding algorithm) and the corresponding probabilities BER_{f-cut} and PER_{f-cut} (obtained with Fast-Cut-Decoding algorithm). The results for BER_{cut} and PER_{cut} for Cut-Decoding algorithm are published in [3]. In this paper, the authors concluded that for values of SNR smaller than 0, the coding does not have sense since the bit-error probability obtained with Cut-Decoding algorithm is larger than the bit-error probability in the channel (without coding).

Analyzing the results in Table 2, we can conclude that for all values of SNR, results for BER_{f-cut} are better than the corresponding results of BER_{cut}. For

Table 2. Experimental results with R=1/4

SNR	BER_{cut}	BER_{f-cut}	PER_{cut}	PER_{f-cut}
0	0.07153	0.06122	0.10001	0.08993
1	0.01830	0.01493	0.02722	0.02275
2	0.00249	0.00155	0.00410	0.00274
3	0.00073	0.00006	0.00230	0.00014
4	0.00052	0.00001	0.00252	0.00007

$SNR = 4$, BER_{f-cut} is even 50 times smaller than BER_{cut}. The same conclusions can be derived for comparison of PER_{f-cut} and PER_{cut}.

In Table 3, we present the percentage of messages which decoding ended with $B_{max} = 1$, $B_{max} = 2$, $B_{max} = 3$ or $B_{max} = 4$. From the results given there, we can see that for smaller values of SNR (0 or 1), we have larger percentage of messages which decoding needed $B_{max} = 3$ or 4. On the other side, for $SNR = 4$, the decoding of more than 90% of the messages successfully finished with $B_{max} = 1$. From these results, we can conclude that for greater values of SNR (low noise in the channel) decoding with the new proposed algorithm is much faster than the old one.

Table 3. Percentage of messages decoded with different values of B_{max}

SNR	$B_{max} = 1$	$B_{max} = 2$	$B_{max} = 3$	$B_{max} = 4$
0	9.88%	26.21%	37.62%	26.29%
1	16.61%	44.79%	29.39%	9.21%
2	44.36%	41.71%	12.10%	1.83%
3	76.40%	20.76%	2.64%	0.20%
4	93.37%	6.17%	0.45%	0.01%

Further on, we present the results for code $(72, 576)$ with rate $R = 1/8$. We compare experimental results obtain with Cut-Decoding, Fast-Cut-Decoding, 4-Sets-Cut-Decoding and Fast-4-Sets-Cut-Decoding algorithms. In the experiments we used the following parameters:

– In Cut-Decoding and Fast-Cut-Decoding algorithms - redundancy pattern: 1100 1100 1000 0000 1100 1000 1000 0000 1100 1100 1000 0000 1100 1000 1000 0000 0000 0000, for rate 1/4 and two different keys of 10 nibbles.
– In 4-Sets-Cut-Decoding and Fast-4-Sets-Cut-Decoding algorithms - redundancy pattern: 1100 1110 1100 1100 1110 1100 1100 1100 0000 for rate 1/2 and four different keys of 10 nibbles.
– In all experiments we used the same quasigroup on Q given in Table 1.

Here, in the experiments with old algorithms we use $B_{max} = 5$ and for Fast-Cut-Decoding and Fast-4-Sets-Cut-Decoding algorithms the maximum value of B_{max} is 5.

In Table 4, we present experimental results for bit-error probabilities BER_{cut}, BER_{f-cut}, BER_{4sets}, $BER_{f-4sets}$ obtained with Cut-Decoding algorithm, Fast-Cut-Decoding algorithm, 4-Sets-Cut-Decoding algorithm and Fast-4-Sets-Cut-Decoding algorithm, respectively, and the corresponding packet-error probabilities PER_{cut}, PER_{f-cut}, PER_{4sets}, $PER_{f-4sets}$. For SNR smaller than -1, using of Cut-Decoding algorithm does not have sense since the values of bit-error probabilities are larger than the bit-error probability in the channel.

In Table 5, we give the percentage of messages which decoding ended with $B_{max} = 1$, $B_{max} = 2$, $B_{max} = 3$, $B_{max} = 4$ or $B_{max} = 5$ for both new algorithms.

Table 4. Experimental results with R=1/8

SNR	BER_{cut}	BER_{f-cut}	BER_{4sets}	$BER_{f-4sets}$	PER_{cut}	PER_{f-cut}	PER_{4sets}	$PER_{f-4sets}$
−2	/	/	0.06548	0.04905	/	/	0.11283	0.09086
−1	0.05591	0.03920	0.02225	0.00795	0.10491	0.07171	0.03831	0.01656
0	0.01232	0.00872	0.00449	0.00074	0.02376	0.01598	0.00835	0.00151
1	0.00224	0.00069	0.00066	0.00013	0.00425	0.00187	0.00136	0.00021
2	0.00037	0	0.00008	0	0.00058	0	0.00014	0

Table 5. Percentage of messages decoded with different values of B_{max}

SNR	Fast-cut-decoding					Fast-4-sets-cut-decoding				
	B_{max}									
	1	2	3	4	5	1	2	3	4	5
−2	/	/	/	/	/	8.29%	1.06%	7.50%	51.42%	31.73%
−1	6.15%	1.53%	25.44%	44.43%	22.46%	1.74%	7.18%	32.27%	49.60%	9.20%
0	1.44%	13.59%	49.07%	28.33%	7.57%	3.59%	27.58%	48.83%	18.77%	1.22%
1	1.97%	47.47%	39.68%	9.42%	1.47%	20.36%	52.04%	25.06%	2.42%	0.12%
2	19.40%	65.50%	13.31%	1.60%	0.19%	58.03%	37.30%	4.54%	0.12%	0.01%

From the results given in Tables 4 and 5, we can derived similar conclusions for rate 1/8 as for rate 1/4. Namely, we can conclude that for all values of SNR, results for BER and PER obtained with the new algorithms are better than the corresponding results obtained with the old versions of the algorithms. Again, this improvement is more significant for larger values of SNR, i.e., for low noise in the channel. Even more, for $SNR \geq 2$, the values of BER and PER obtained with new algorithms are equal to 0. From Table 5, we can see that with new Fast-4-Sets-Cut-Decoding algorithm if values of SNR are smaller then we have larger percentage of messages which decoding needed $B_{max} = 4$ or 5. On the other side, for $SNR = 2$, the decoding of more than 80% of the messages successfully finished with $B_{max} = 1$ or $B_{max} = 2$. Therefore, we can conclude that for greater values of SNR (low noise in the channel) decoding with the new proposed algorithms is much faster than the old one.

5 Conclusion

In this paper we define new modification of decoding algorithms for RCBQ called Fast-Cut-Decoding and Fast-4-Sets-Cut-Decoding algorithms. With these algorithms, the number of *more-candidate-errors* that appeared during decoding process decrease. Even more, for larger values of SNR, this type of errors are completely eliminated. In this way, we obtain better results for packet-errors and bit-error probabilities than with previous decoding algorithms. Also, we present the percentage of messages which decoding ends with different values of B_{max}. From these results we can conclude that new proposed algorithms provide more efficient and faster decoding, especially for transmission through a low noise Gaussian channel.

Acknowledgment. This research was partially supported by Faculty of Computer Science and Engineering at "Ss Cyril and Methodius" University in Skopje.

References

1. Gligoroski, D., Markovski, S., Kocarev, L.: Error-correcting codes based on quasigroups. In: Proceedings of 16th International Conference on Computer Communications and Networks, pp. 165–172 (2007)
2. Gligoroski, D., Markovski, S., Kocarev, L.: Totally asynchronous stream ciphers + Redundancy = Cryptcoding. In: Aissi, S., Arabnia, H.R. (eds.) Proceedings of the International Conference on Security and management, SAM 2007, pp. 446–451. CSREA Press, Las Vegas (2007)
3. Mechkaroska, D., Popovska-Mitrovikj, A., Bakeva, V.: Cryptcodes based on quasigroups in gaussian channel. Quasigroups Relat. Syst. **24**(2), 249–268 (2016)
4. Mathur, C.N., Narayan, K., Subbalakshmi, K.P.: High diffusion cipher: encryption and error correction in a single cryptographic primitive. In: Zhou, J., Yung, M., Bao, F. (eds.) ACNS 2006. LNCS, vol. 3989, pp. 309–324. Springer, Heidelberg (2006). doi:10.1007/11767480_21
5. Popovska-Mitrovikj, A., Markovski, S., Bakeva, V.: Increasing the decoding speed of random codes based on quasigroups. In: Markovski, S., Gusev, M. (eds.) ICT Innovations 2012, Web proceedings, pp. 93–102 (2012). ISSN 1857-7288
6. Popovska-Mitrovikj, A., Markovski, S., Bakeva, V.: 4-Sets-Cut-Decoding algorithms for random codes based on quasigroups. Int. J. Electron. Commun. (AEU) **69**(10), 1417–1428 (2015). Elsevier
7. Popovska-Mitrovikj, A., Bakeva, V., Markovski, S.: On random error correcting codes based on quasigroups. Quasigroups Relat. Syst. **19**(2), 301–316 (2011)
8. Popovska-Mitrovikj, A., Markovski, S., Bakeva, V.: Performances of error-correcting codes based on quasigroups. In: Davcev, D., Gomez, J.M. (eds.) ICT-Innovations 2009, pp. 377–389. Springer, Heidelberg (2009)
9. Popovska-Mitrovikj, A., Markovski, S., Bakeva, V.: Some new results for random codes based on quasigroups. In: 10th Conference on Informatics and Information Technology with International Participants, Bitola, Macedonia, pp. 178–181 (2013)
10. Hwang, T., Rao, T.R.N.: Secret error-correcting codes (SECC). In: Goldwasser, S. (ed.) CRYPTO 1988. LNCS, vol. 403, pp. 540–563. Springer, New York (1990). doi:10.1007/0-387-34799-2_39
11. Zivic, N., Ruland, C.: Parallel joint channel coding and cryptography. Int. J. Electr. Electron. Eng. **4**(2), 140–144 (2010)

Authorization Proxy for SPARQL Endpoints

Riste Stojanov(✉) and Milos Jovanovik

Faculty of Computer Science and Engineering,
Ss. Cyril and Methodius University in Skopje, Skopje, Macedonia
{riste.stojanov,milos.jovanovik}@finki.ukim.mk

Abstract. A large number of emerging services expose their data using various Application Programming Interfaces (APIs). Consuming and fusing data form various providers is a challenging task, since separate client implementation is usually required for each API. The Semantic Web provides a set of standards and mechanisms for unifying data representation on the Web, as well as means of uniform access via its query language – SPARQL. However, the lack of data protection mechanisms for the SPARQL query language and its HTTP-based data access protocol might be the main reason why it is not widely accepted as a data exchange and linking mechanism. This paper presents an authorization proxy that solves this problem using query interception and rewriting. For a given client, it solely returns the permitted data for the requested query, defined via a flexible policy language that combines the RDF and SPARQL standards for policy definition.

1 Introduction

Services intended for consumption over HTTP usually expose their data through an Application Programming Interface (API). The APIs provide means for data access by using different methods, along with various types of authentication and authorization, which provide their clients with the intended service. However, this diversity represents a large hurdle for the process of fusing data from different APIs in a sustainable manner, since it requires multiple clients or modules, and since API specifications evolve over time. The former impacts scalability, while the latter requires continuous development and maintenance. An additional challenge is the diversity of the data which the APIs provide – they are, in general, differently modeled and therefore cannot be seamlessly integrated and consolidated. However, the principles of Linked Data and the Semantic Web provide means to solve the data interoperability problems, by annotation and aligning data from different sources, represented in different models [8].

The SPARQL protocol[1] can overcome the challenges regarding API consumption and data integration. It provides REST-based data access using the SPARQL 1.1 query language[2]. It provides a data layer abstraction and is able to perform all standard database actions, via the SPARQL 1.1 query. Since it

---

[1] https://www.w3.org/TR/sparql11-protocol/.
[2] https://www.w3.org/TR/sparql11-query/.

© Springer International Publishing AG 2017
D. Trajanov and V. Bakeva (Eds.): ICT Innovations 2017, CCIS 778, pp. 205–218, 2017.
DOI: 10.1007/978-3-319-67597-8_20

can be perceived as a single API, usable for all RDF/Linked Data datasets, the integration of data coming from APIs based on the SPARQL protocol can be executed by a single client. Since the retrieved data conforms to the RDF framework, the consolidation task gets significantly simpler as well. Additionally, it provides SPARQL federated query[3] support – a portion of the query sent to the SPARQL API can be further sent to another SPARQL API. The retrieved results will then be used by the initial SPARQL API, to compute the response for the client. This is possible due to the RDF model of the data and the usage of the same SPARQL protocol for the API.

However, a significant drawback of this approach is that the main parameter for such API is a SPARQL query. Generally, with a REST-based service, the data access protection mechanisms are usually based on authentication (OAuth[4], WebID[5], etc.), which allows certain parts of the API to be accessible or not. This can be solved on the HTTP level, by allowing/blocking access to certain URL paths. But, with the SPARQL-based APIs, all requests are sent to a fixed URL, which can only be allowed for the authorized users. This fixed URL receives a SPARQL query as a parameter, which determines the actual action to be performed. In other words, the actual 'meaning' of the SPARQL query provided in the API request cannot be evaluated on HTTP level. Therefore, the usual HTTP-level protection of APIs in not available for SPARQL APIs and more granular protection should be provided. The data which can be accessed via SPARQL queries[6] is modeled using RDF – the Resource Description Framework[7]. RDF organized data in *triples*, which consist of a *subject*, a *predicate* and an *object*. One RDF triple indicates that the subject is in a relation defined by the predicate with the object. These RDF triples are organized in RDF graphs – an unordered set of RDF triples. Just like subjects, predicated and objects, RDF graph have their unique identifier. Therefore, the usual physical organization of RDF graph is a database entry of an RDF quad: a subject, a predicate, an object, and a graph the triples they form belongs to. So, one RDF dataset can contain one or more RDF graphs.

The protection of any data always starts with a requirement given in natural language. The security policies should provide flexibility to cover the requirement for data protection. Additionally, these requirement can link the requester with some of the protected data, and can use some contextual conditions. In order to incorporate and link these information, and the authorization system should be flexible enough to define the rules that should be meet the requester and its context, and which data should be protected with those conditions. The acceptance of such policy language depends on its maintainability, and it should be approximately one-to-one correspondence between the requirements and the policies. The natural language requirements often combine rules that allow or

---

[3] https://www.w3.org/TR/sparql11-federated-query/.
[4] https://oauth.net/2/.
[5] https://www.w3.org/wiki/WebID.
[6] https://www.w3.org/TR/rdf-sparql-query/.
[7] https://www.w3.org/RDF/.

deny access to some data, and the policy language should support them, as well should be able to resolve the possible conflicts that may arise from them.

In this paper we present an authorization framework that behaves as a proxy for the SPARQL protocol. The proxy provides context aware fine-grained protection of the underlaying dataset with a policy language that combines the W3C's RDF and SPARQL standards. The authorization proxy is able to combine the policies based on their priorities for conflict resolution. The query rewriting algorithm employed by the authorization proxy modifies each incoming query such that only the allowed data for the requester is returned.

In Sect. 2 we describe the state of the art authorization approaches for the Semantic Web that use query rewriting. In Sect. 3 we describe our framework with its policy language and query rewriting algorithm. The discussion in Sect. 4 shows a use-case that shows the flexibility of the policy language and confirms the correctness of the query rewriting algorithm, after which the conclusion is presented in Sect. 5.

2 Related Work

It is hard to provide flexible data protection for SPARQL protocol since it is able to select every particular resource using various path and triple expressions, combined with built-in filtering predicates. Unlike protecting standard HTTP resources, where each URL pattern can be protected in combination with the accessing method [9], the graph structure that is queried with SPARQL is more difficult to be defined by the policy. Additionally, the requester may be represented as some of the protected resources, and the available data may be connected with it. When the context is included in the whole story, the access control becomes even harder.

One of the possible approaches for protecting semantic data is to build a temporal model that contains only the available data for the requester [3,4,12] and than to execute each of the incoming queries against that model. This approach allows using a flexible policy languages, such as in [3], where the protected data is selected using SPARQL. Another similar option is annotating the data for each requester [11], and than rewrite the query to be executed including the requester annotation. However, this kind of enforcement is not scalable, since they require preprocessing of the dataset on each request, which is often time consuming and imposes significant overall query processing time increase.

Other enforcement type is by query rewriting, where each incoming query is modified so that it will return only the data available for the requester. Due to complexity of the query transformation process, most of the available approaches use fairly simple policy languages, suitable for their enforcement algorithm [2,5, 10,13]. In the fQuery system [13] the policies are in a form of a triple pattern ?s ?p ?o where each element can be constrained with multiple filters, and the matching algorithm appends the filters for each of the query's matched patterns in an *OPTIONAL* group. In [10] a tuple is used to constrain which requesters can access which quad patterns, and those policies are enforced using *FILTER (NOT) EXISTS* query elements. The both approaches [10,13] have fairly simple policies with a limited protection possibility, which makes them hard to use them for protection in production. The work in [5] defines a protection framework which is based on a policy that are able to protect queries that starts with an instance of a given class. The policies are composed of multiple triples that forms a chain, and only the last triple's object can have a constrained value using filter expression. Their rewriting algorithm appends the policy triple patterns to the query for each of the variables used as a subject. The rigid policy language is hard to manage, since allow-only policies are used, and there is no possibility to link the requester with the data that is being protected. In [2] the XACML [6] policy language is used to protect a semantic resources based on their class or instance, and their query rewriting algorithm appends *OPTIONAL* and *FILTER* element for protection. Since XACML is not designed for semantic resource protection, only limited number of protections scenarios are supported.

In [1], a flexible policy language is used, which is composed of a set of triple patterns (referred to as policy body) that imply which triple pattern is protected (referred to as policy head). The rewriting algorithm appends the policy body to the query *WHERE* element using conjunction operator, and uses filters to align the different variables used in the query and in the policy. The denying policies are appended using *MINUS* element. Policy language with similar expressiveness is used in [14], but without possibility to deny access. The rewriting algorithm replaces each of the matched query's triple patterns with the policy body (referred to as precondition it [14]). The disadvantage of [1,14] is that their policies will eliminate all solutions that fail to bind all patterns from the policy body (or precondition), although they should remain, since they do not match the policy's preconditions.

3 Authorization Framework

Our authorization proxy operates by intercepting a HTTP requests toward the SPARQL endpoints. It extracts the query attributes from the request, together with the additional request parameters and headers, and uses this data to modify the query such that it will return only the information available for the given request. Thus, the request can be observed as a representation of its requester intention. The authorization proxy first transforms the request into a semantic graph that contains the description of the requester, its context and the query it tries to execute. This semantic graph is referred to as Intent in this paper. The semantic format of the intent enables linking the requester's info and context with the data being protected. The authorization proxy described in this paper is not responsible for requester authentication. Instead it depends on a *IntentProvider* implementation to obtain the description about the requester and its current contextual environment. The proxy is implemented in the Spring Boot framework and leverages its auto-configuration ability for *IntentProvider* discovery and registration [7]. The simple integration of this framework with Spring security [15] and other authentication libraries enables the intent provider to integrate with various authentication mechanisms and then to extract the requesters' info. The *IntentProvider* implementation is notified for each new request intercepted by the proxy, so that it can update the intent for the corresponding request.

After the intent is obtained from the request, the authorization proxy locates the query based on its configuration[8] and parses it for better manipulation. The parsed query is then transformed, and as such is executed against the dataset. During the transformation process, an explanation is constructed and returned in the *auth-explain*[9] header of the response. The authorization proxy can protect multiple datasets, and the policy language enables definition of policies per dataset.

The authorization proxy currently does not have any user interface and relies fully on its configuration. Among the configurable *query* parameter and *auth-explanation* header, the proxy requires configuration of policy directory and model directory. The policy directory is monitored and each *.rdf* file is processed for policy registration. The policies are stored in memory in a list sorted by their priority in ascending direction. The *PolicyRegistry* component is responsible for registering and obtaining them whenever needed. The model directory is observed for *.owl* ontology definitions, and they are parsed and stored for constraining the property domains and ranges. The models are not required for proper functioning of the system, but their usage can significantly improve the rewritten query processing time.

This policy language used by the authorization proxy leverages the W3C's RDF and SPARQL standards for policy definition. Each policy is an instance of the *p:Policy* class. This policy language assumes that same variable names among the policy's properties mean a same thing, i.e. the variables has scope defined by the policy. This is because the authorization proxy combines together the values of the properties *p:intent_binding*, *p:protected_data* and *p:head_quad* in the query transformation process. The following properties may be configured for the policies:

- *p:intent_binding* contains a SPARQL SELECT query that is used to activate the policy, through extracting the corresponding variables from the intent that satisfy the query's WHERE constrain. The policy is not applicable for a particular intent when this query does not return any results.
- *p:protected_data* defines the data that is being protected. It accepts a SPARQL CONSTRUCT queries as value. Only the WHERE element of this query is used in the enforcement process, since SPARQL does not provide a way to select quads.
- *p:head_quad* is a quad pattern that defines which quads will be protected.
- *p:permission* defines whether the policy *allow* or *deny* the protected data.
- *p:protected_datasets* defines the datasets for which this policy will be applied.
- *p:priority* is used for conflict resolution, by or ordering the policies in the enforcement process. It has a *xsd:double* value.

When the *PolicyRegistry* detects a new policy, it first replaces all existing variables with new, generated ones, so that the conflicts among variables with

---

[8] The query can be obtained from a request parameter or header, which is configurable in this system.

[9] The explanation can be turned on/off, and header name can also changed using the proxy's configuration.

same name can be avoided. During this process, the policy's *p:head_quad* is rewritten with *?v1 ?v2 ?v3 ?vg*, and the corresponding variables replacement is applied to the *p:intent_binding* and *p:protected_data* queries. When the head quad pattern contains IRIs or literals, they are also replaced with a variable. In this case, additional *FILTER (?var=term)* is added to the policy's *protected_data* element, where *?var* is the new variable name that replaces the *term* IRI or literal in the policy's *head_quad*. In this way, all policies are aligned to have same *head_quad* element and they can be easily combined. The solution to the rewritten *head_quad* pattern *?v1 ?v2 ?v3 ?vg* denotes the data that is protected. Next, the *PolicyRegistry* uses the registered models to infer the possible classes for each of the policy's variables, and stores this mapping in the computed *varConstr* policy's property. This process only uses the property domains and ranges for type inference, together with the *rdf:type* property that is used to infer the class of the triple's subject variable to its object's value.

3.1   Authorized Query Transformation

The *PermissionTransformer* component is responsible for the query rewriting process. This component tries to induce the authorization restrictions for each of the queries' quads, using the *PermissionVisitor* implementation. In this process, the query is first transformed using the *AlgebraQuad*[10], which replaces each of the query's triple patterns with a quad pattern, while the triple blocks are transformed in quad blocks. Then, the *PermissionVisitor* processes each of these quad blocks, which are composed of a set of quads. Listing 3.1 presents the *PermissionVisitor.visit* method using a functional programming notation. The *visit* method is invoked for each of the query's *Quad Blocks*. The visitor initializes Q_{filter} to an empty algebra operation, and it is modified through the visiting process and represents the restriction that will be added to the query after all quad blocks are visited.

Line 4 from Listing 3.1 transforms the stream with policies into a stream with pairs $\langle p, m \rangle$, where m is the mapping that contains the result of the policy's *p:intent_binding* query execution against the intent. A pair is returned since the both elements are necessary in the next processing steps. In line 5 the pairs that contain empty mappings will be removed, and only the activated policies for the intent remain for further processing. Then, in line 6, the policy - mapping pair is transformed into another pair that replaces the mapping with a rewritten version of the policy's *p:protected_data* query. The *apply* procedure is used to extend the *PermissionVisitor* internal Q_{filter} element for each activated policy and quad block. During this process, each of the quad block's quad patterns is aligned against the policy's *p:head_quad* pattern using the *mapping* function. This mapping function first check if the two quad pattern arguments match with each other[11], and when they do, it check if their corresponding variables are not

---

[10] Jena ARQ algebra transformer http://bit.ly/2rgvvLw.

[11] Two quad pattern are matched if all of their element match, which is the case when at least one of those elements is a variable, or when they are same.

from a disjoint types. When this is the case, the variables from the first quad are mapped to the corresponding elements of the second quad, while the variables from the second quad are mapped only to a concrete values from the first one.

Listing 3.1. PermissionVisitor implementation

```
1   visit(QuadBlock) {
2     qbConstr:=induceVariableClasses(QuadBlock,model)
3     policies.stream()
4       .map(p → ⟨p, m := executeSelect(p.intent_binding, intent)⟩)
5       .filter(⟨p, m⟩ → m ≠ ∅)
6       .map(⟨p, m⟩ → p_rw := rewrite(p, m))
7       .foreach(p_rw → apply(p_rw, QuadBlock, qbConstr))
8   }
9
10  apply(p, QuadBlock, qbConstr) {
11    QuadBlock.stream()
12      .map(qp → ⟨qp, m := mapping(p.head_quad, qp, qbConstr, p.varConstr)⟩)
13      .filter(⟨qp, m⟩ → m ≠ ∅)
14      .foreach(⟨qp, m⟩ → {
15        Q_filter := appendFilter(Q_filter, m)
16        dir := (dir ≠ null ? dir : p.permission)
17        Q_filter := (p.permission = dir ?
18          Q_filter ∪ p.permitted_data :
19          Q_filter \ p.permitted_data)
20      });
21  }
```

After the filtering in line 13, only the mappings from the quad patterns relevant for the policy remain and then they are combined to form the Q_{filter} element. The mapping m is then appended in the Q_{filter} element in the form of *FILTER (?v1=hq[0] && ?v2=hq[1] && ?v3=hq[2] && ?vg=hq[3])*[12]. The *appendFilter* function combines the different quad pattern's mapping with *OR* (‖) operator. The *dir* field used in line 16 is an internal property of the *PermissionVisitor*, and it is initialized with the lowest applicable priority policy's permission value. All the rewritten policies are combined together into the Q_{filter} operation with SPARQL UNION or MINUS element, depending on the policy's permission and the value of the *dir* field.

Once all query's quad blocks are processed, the Q_{filter} element represents a combination of the activated and relevant policies' permitted data restrictions, with a variables that correspond to the one from the original query. This operation actually represents the data that is available or denied for the query, depending on the *PermissionVisitor.dir* field. The lowest priority policy is generally the default behavior, and this allows denying or allowing the whole dataset by default. Thus, if the lowest priority policy denies access, then Q_{filter} holds the data that is denied for the query, or the allowed data in the opposite case.

---

[12] Here *hq[i]* denotes the i-th element of the policy's *p:head_quad*.

Once the *Permission Visitor* has finished visiting all query's quad blocks, its Q_{filter} and *dir* fields hold the results of this process. Then, the *Permission-Transformer* appends Q_{filter} element inside a *FILTER EXISTS* clause when *dir* is *ALLOW*, and in *FILTER NOT EXISTS* otherwise. In the first case, only the results from the original query that intersects with the results of Q_{filter} will be returned. The *FILTER EXISTS* expression is chosen because it uses the result bindings from the original query and filters them, and does not reduce the result that failure to bind the additional constrains in its patterns. Similarly, the *FILTER NOT EXISTS* removes only the results that are found by its element from the original query, without requiring additional bindings.

4   Discussion

The Listing 4.1 shows an example data from the university ontology[13], where for each study program there is a separate graph that logically organizes the students with their grades that are enrolled at that study program.

Listing 4.1. Example data

```
:f a u:Faculty; u:network_address '10.10.0.0/16';.
:cs a u:StudyProgram; u:faculty :f.
:john a u:User; u:works_at :f; u:phone '070111222'.
:ben a u:User; u:works_at :f; u:phone '075333444'.
:sw_17 a u:Course; u:has_professor :ben; u:year 2017.
:sec_17 a u:Course; u:has_professor :john; u:year 2017.
:cs {
    :bob a u:User; u:enrolled_at :cs; u:phone '077123456'.
    :alice a u:User; u:enrolled_at :cs; u:phone '071654321'.
    :g1 a u:Grade; u:for_student :alice; u:for_course :sw_17; u:grade_value 'B'.
    :g2 a u:Grade; u:for_student :bob; u:for_course :sec_17.
}
```

Listing 4.2. Example query

```
SELECT * WHERE {
    ?s ?p ?o.
    ?g u:for_student ?s.
    OPTIONAL {
        ?g u:grade_value ?v
    }
}
```

The Listings 4.3 and 4.4 show two pair of conflicting policies. The policy *:publicUser* allows all user's data, *:protectedPhone* protects the phone number for everyone but the user, *otherGrades* denies the grades for everyone but its student, and *:profGrades* allows the professors to access the grade from their faculty network. The priorities used in this policies define their processing order in the rewriting

---

[13] https://github.com/ristes/univ-datasets/ont/univ.owl.

algorithm, and the higher order policies will override the results from the lower order one. All of these policies, except *:publicUser*, require that the intent should contain a requester in it, and they link the requester with the data using the its variable in *p:protected_data* query. This way the security administrator has a flexibility to link the requester's properties with the underlaying data, which significantly simplifies the policy management process and reduces the number of policies required to protect the system. The policy *:publicUser* does not contain the *p:intent_binding* property, which is again simplification of the policy maintenance process, and it represents a policy that does not depend on the incoming intent. The authorization proxy will automatically activate this policy and will try to apply it for every query that have a quad with a *u:User* subject. The policy *:profGrades* leverages the intent flexibility, and if the *IntentProvider* injects the requester's agent IP address and its network, this policy will be activated when the professor is accessing the system from its faculty network. The intent providing mechanism, together with the rewrite algorithm that combines the intent mappings with the protected data query provide a flexible context aware polices, where the context can be used for user activation and for data filtering.

Listing 4.3. Deny policies

```
:otherGrades a p:Policy;
  p:intent_binding
   'SELECT ?r WHERE {
   ?r a int:Requester }';
  p:protected_data ' WHERE {
   ?g u:for_student ?s .
   ?g ?p ?o.
   FILTER (?s != ?r) }';
  p:head_quad '?g ?p ?o ?x';
  p:permission 'DENY';
  p:priority 10.

:protectedPhone a p:Policy;
  p:intent_binding
   'SELECT ?r WHERE {
   ?r a int:Requester }';
  p:protected_data ' WHERE {
   ?s u:phone ?o .
   FILTER (?s != ?r) }';
  p:head_quad
   '?s u:phone ?o ?x';
  p:permission 'DENY';
  p:priority 30.
```

Listing 4.4. Allow policies

```
:publicUser a p:Policy;
  p:protected_data ' WHERE {
   ?u a u:User.
   ?u ?p ?o }';
  p:head_quad '?u ?p ?o ?x';
  p:permission 'ALLOW';
  p:priority 20.

:profGrades a p:Policy;
  p:intent_binding
   'SELECT ?r, ?net WHERE {
   ?r a int:Requester .
   ?ag a int:Agent .
   ?ag int:ip_address ?ip.
   ?ip int:network ?net }';
  p:protected_data ' WHERE {
   ?c u:has_professor ?r.
   ?r u:works_at ?f.
   ?f u:network_address ?net.
   ?g u:for_course ?c.
   ?g ?p ?o
   }';
  p:head_quad '?g ?p ?o ?x';
  p:permission 'ALLOW';
  p:priority 50.
```

Listing 4.5 shows the rewritten version of the query from Listing 4.2, using the policy defined in Listings 4.3 and 4.4 , where the requester is :john. For this query all configured policies are activated, and their activated versions are shown in Listing 4.6. Here, the variables are replaced in the policy initialization phase such that the heads are always in form ?v1 ?v2 ?v3 ?vg, and the other variables are sequentially replaced as processed. Additionally, the :protectedPhone policy's head is modified and the u:phone property is replaced with ?v2 in the p:head_quad property, while a FILTER (?v2=u:phone) is appended in the p:protected_data element. These activated policies are linked with the intent through replacement of the variables ?r and ?net with :john and 10.10.0.0/16, correspondingly.

Listing 4.5. Example protected query

```
SELECT * WHERE {
?s ?p ?o.
?g u:for_student ?s.
OPTIONAL {?g u:grade_value ?v}
FILTER NOT EXISTS {
  FILTER (
  (?v1=?g && ?v3=?s &&
      ?v2=u:for_student)||
  (?v1=?g && ?v3=?v &&
      ?v2=u:grade_value)||
  (?v1=?s && ?v2=?p && ?v3=?o)
  )
  {
  { Q_otherGrades
    MINUS Q_publicUser
  } UNION Q_protectedPhone
  } MINUS Q_profGrades
}
}
```

Listing 4.6. Activated policies

```
Q_otherGrades='{ ?v1 ?v2 ?v3.
  ?v1 u:for_student ?v4.
  FILTER (?v4 != :john)
}'
Q_publicUser='{
  ?v1 a u:User; ?v2 ?v3
}'
Q_protectedPhone='{ ?v1 ?v2 ?v3.
  FILTER (?v1!=:john
      && ?v2=u:phone)
}'
Q_profGrades='{
  ?v5 u:has_professor :john.
  :john u:works_at ?v6.
  ?v6 u:network_address
    "10.10.0.0/16".
  ?v1 u:for_course ?v5.
  ?v1 ?v2 ?v3
}'
```

The FILTER (NOT) EXISTS element is chosen because it does not include additional bindings in the query. It returns a variable bindings that are selected/excluded from the query's results. In this way, if the policy requires presence of the u:network_address property for some faculty, this will not affect the original query's result when this property is absent. The FILTER (NOT) EXISTS query element in Listing 4.5 is composed of two main parts: FILTER expression, and Quad block that is composed of multiple UNION/MINUS elements. The quad block combines the activated policies for the query and it may return variable bindings that do not correspond to the variables in the originally requested query. The FILTER expressions are used to bind the policy's variables with the one in the query. In this way, the FILTER (NOT) EXISTS element will output the variable bindings present in the original query, therefore constraining the results.

When the query from Listing 4.2 is intercepted, the authorization proxy infers that the variable $?g$ is a $u{:}Grade$, $?s$ is a $u{:}User$, and $?v$ is a *double* from the ontology model definitions of the properties $u{:}for\_student$ and $u{:}grade\_value$. The triples $?g\ u{:}for\_student\ ?s$ and $?g\ u{:}grade\_value\ ?v$ are matching the quad $?v1$ $?v2\ ?v3\ ?vg$ bounded with $?v1 \rightarrow u{:}Grade$ from the policies $:otherGrades$ and $:profGrades$. Therefore, these policies are activated and the *FILTER* expressions ($?v1{=}?g$ && $?v3{=}?s$ && $?v2{=}u{:}for\_student$) and ($?v1{=}?g$ && $?v3{=}?v$ && $?v2{=}u{:}grade\_value$) are appended in the rewritten query from Listing 4.5. Similarly, the triple $?s\ ?p\ ?o$ activates the policies $:protectedPhone$ and $:publicUser$, appending the *FILTER* ($?v1{=}?s$ && $?v3{=}?o$ && $?v2{=}?p$) variable expression. Listing 4.7 shows the data protected by each activated policy and Listing 4.8 shows the forbidden quads[14] Q_{deny} for the example query from Listing 4.2. Q_{deny} shows that the grade $:g1$ is forbidden since it is not assigned to the student $:john$, nor it is for a course that is held by the professor $:john$. Additionally, the phone numbers for every user except for the logged one ($:john$) are denied for access.

Listing 4.7. Activated policy's data

```
:otherGrades {
    :g2 rdf:type u:Grade; u:for_student :bob; u:for_course :sec_17.
    :g1 rdf:type u:Grade; u:for_student :alice; u:for_course :sw_17; u:grade_value "B".
}
:publicUser {
  :john rdf:type u:User; u:works_at :f; u:phone "070111222".
  :ben rdf:type u:User; u:works_at :f; u:phone "075333444".
  :bob rdf:type u:User; u:phone "077123456"; u:enrolled_at :cs.
  :alice rdf:type u:User; u:phone "071654321"; :enrolled_at :cs.
}
:protectedPhone {
  :ben u:phone "075333444". :bob u:phone "077123456".
  :alice u:phone "071654321"
}
:profGrades {
    :g2 rdf:type u:Grade; u:for_student :bob; u:for_course :sec_17.
}
```

Listing 4.8. Restricted data

```
Q_deny = {{Q_otherGrades MINUS Q_publicUser} UNION Q_protectedPhone}
MINUS Q_profGrades
Q_deny = {
  :g1 rdf:type u:Grade; u:for_student :alice; u:for_course :sw_17; u:grade_value "B".
  :ben u:phone "075333444".
  :bob u:phone "077123456".
  :alice u:phone "071654321".
}
```

---

[14] The Listings 4.8 and 4.7 omit the quad's graph element for simplicity.

Listing 4.9. Original query results

g	s	p	o	v
: g2	: bob	rdf : type	u : User	
: g2	: bob	u : phone	"077123456"	
: g2	: bob	u : enrolled_at	: cs	
: g1	: alice	rdf : type	u : User	"B"
: g1	: alice	u : phone	"071654321"	"B"
: g1	: alice	u : enrolled_at	: cs	"B"

Listing 4.10. Rewritten query results

g	s	p	o	v
: g2	: bob	rdf : type	u : User	
: g2	: bob	u : enrolled_at	: cs	

Listing 4.9 shows the results that the example query from Listing 4.2 returns. After the rewriting process, the Q_{deny} from Listing 4.8 data is binded to its variables in the *FILTER NOT EXISTS* expression and the result obtained from its execution is shown in Listing 4.10, where the forbidden data for the grade *:g1* is removed, together with *:ben*'s phone number. This way, the results are the one that are available for the requester *:john*, when he access the data from its faculty network.

The variable constrains leveraged by the *mapping* function can reduce the number of the appended elements in the *FILTER (NOT) EXISTS* expression, which optimizes the query processing time. However, if there is no ontology that describes the domains and the ranges of the properties, the rewriting algorithm will still produce a valid query, but its execution time can be significantly longer. The example query from Listing 4.2 activates all of the defined policies, but if the query is *SELECT * WHERE { ?s a u:User. ?s ?p ?o}*, only the policies *:publicUser* and *:protectedPhone* would have been activated, and again the user will access only the allowed data, but this time for shorter time, since the *FILTER EXISTS* query will have only two quad blocks. A *FILTER EXISTS* expression will be appended because *:publicUser* is the lowest priority policy of the applicable ones, and it has *ALLOW* permission.

In [1] is presented a query rewriting algorithm that is flexible enough to protect the triples and enables context dependent policies. It is the most flexible approach from the reviewed ones. However, the policies are not able to protect data stored in graphs and in different datasets, and the authors do not consider the conflicts and their resolution. The authorization proxy is a step in this direction, providing context dependent policies that are able to protect data from different graphs and datasets. The policy priorities are incorporated in the query rewriting algorithm and provide flexible conflict resolution mechanism. The *IntentProvider* enables dynamic context definition and the configurable parameters make the authorization suitable for different implementations of the SPARQL protocol.

5   Conclusion

This paper describes a complete platform for authorization of SPARQL endpoints. It uses a context dependent policies, which can link the requester with the underlaying data through the Intent provisioning mechanism, and the query rewriting algorithm guaranties that only permitted data will be returned.

The policy language provides flexibility equivalent to the SPARQL language for protected data selection and policy activation conditions specification. The policies enable interlinking of the protected data with the requester's properties and its context. The *ALLOW* and *DENY* policy permissions simplify their maintenance, making the requirement transformation strait forward translation from natural language into SPARQL. The conflicts that arise from the opposite permission policies are resolved using priorities, which complies to the way they are overcame in the requirement definition process. The policies are intended to follow and promote the interoperability of the semantic web, and multiple datasets may be protected by a shared set of policies. There is also ability to define a policy for a specific dataset. The injected *IntentProvider* enables support for various authentication mechanisms, enabling various environmental evidences to be dynamically included in the intents. This enables defining context aware policies that will be activated based on the data provided in the *Intent* semantic graph.

The authentication proxy intercepts each query and transforms it based on the execution environment provided through the intent. The query rewriting transformation ensures that only the permitted data that is configured through the policies will be returned. The algorithm is designed to add as little overhead as possible, and uses the ontology model for this purpose, but its correctness is not constrained from it. The policies priorities are used for conflict resolution, where the higher priority policies override the lower order ones.

The authorization proxy is intended to protect the data exposed using the SPARQL protocol, but also provides ability to configure each of its parameters in order to be able to protect any arbitrary endpoint and to leverage any desired authentication mechanism. Even though the policy language can support protection of semantic data modification and graph management operations, the authorization proxy in its current implementation does not offer support for this kind of protection. Its extension in this direction is planned as a part of the future work.

References

1. Abel, F., De Coi, J.L., Henze, N., Koesling, A.W., Krause, D., Olmedilla, D.: Enabling advanced and context-dependent access control in RDF stores. In: Aberer, K., et al. (eds.) ASWC/ISWC -2007. LNCS, vol. 4825, pp. 1–14. Springer, Heidelberg (2007). doi:10.1007/978-3-540-76298-0_1
2. Chen, W., Stuckenschmidt, H.: A model-driven approach to enable access control for ontologies. Wirtschaftsinformatik **1**, 663–672 (2009)
3. Dietzold, S., Auer, S.: Access control on RDF triple stores from a semantic wiki perspective. In: ESWC Workshop on Scripting for the Semantic Web, Citeseer (2006)

4. Flouris, G., Fundulaki, I., Michou, M., Antoniou, G.: Controlling access to RDF graphs. In: Berre, A.J., Gómez-Pérez, A., Tutschku, K., Fensel, D. (eds.) FIS 2010. LNCS, vol. 6369, pp. 107–117. Springer, Heidelberg (2010). doi:10.1007/978-3-642-15877-3_12
5. Franzoni, S., Mazzoleni, P., Valtolina, S., Bertino, E.: Towards a fine-grained access control model and mechanisms for semantic databases. In: IEEE International Conference on Web Services (ICWS 2007), pp. 993–1000. IEEE (2007)
6. Godik, S., Anderson, A., Parducci, B., Humenn, P., Vajjhala, S.: Oasis extensible access control 2 markup language (xacml), vol. 3. Technical report, OASIS (2002)
7. Gutierrez, F.: Pro Spring Boot. Springer, Heidelberg (2016)
8. Heath, T., Bizer, C.: Linked data: evolving the web into a global data space. Synth. Lect. Semant. Web Theor. Technol. 1(1), 1–136 (2011)
9. Hollenbach, J., Presbrey, J., Berners-Lee, T.: Using rdf metadata to enable access control on the social semantic web. In: Proceedings of the Workshop on Collaborative Construction, Management and Linking of Structured Knowledge (CK 2009), vol. 514 (2009)
10. Kirrane, S.: Linked data with access control. Ph.D. thesis (2015)
11. Lopes, N., Kirrane, S., Zimmermann, A., Polleres, A., Mileo, A.: A logic programming approach for acess control over RDF. Ph.D. thesis (2012)
12. Muhleisen, H., Kost, M., Freytag, J.-C.: SWRL-based access policies for linked data. In: Procs of SPOT, vol. 80 (2010)
13. Oulmakhzoune, S., Cuppens-Boulahia, N., Cuppens, F., Morucci, S.: fQuery: SPARQL query rewriting to enforce data confidentiality. In: Foresti, S., Jajodia, S. (eds.) DBSec 2010. LNCS, vol. 6166, pp. 146–161. Springer, Heidelberg (2010). doi:10.1007/978-3-642-13739-6_10
14. Padia, A., Finin, T., Joshi, A.: Attribute-based fine grained access control for triple stores. In: 14th International Semantic Web Conference (2015)
15. Scarioni, C.: Pro Spring Security. Apress, Berkeley (2013)

Number of Errors that the Error-Detecting Code Surely Detects

Nataša Ilievska[✉]

Faculty of Computer Science and Engineering,
Ss. Cyril and Methodius University, Skopje, Republic of Macedonia
natasa.ilievska@finki.ukim.mk

Abstract. In this paper we consider an error-detecting code based on linear quasigroups. We give a proof that the code is linear. Also, we obtain the generator and the parity-check matrices of the code, from where we obtain the Hamming distance of the code when a linear quasigroup of order 4 from the best class of quasigroups of order 4 for coding, i.e., the class of quasigroups of order 4 that gives smallest probability of undetected errors is used for coding. With this we determine the number of errors that the code will detect for sure.

Keywords: Linear code · Hamming distance · Generator matrix · Parity-check matrix · Error-detecting code · Linear quasigroup · Probability of undetected errors

1 Introduction

In this paper we consider an error-detecting code based on linear quasigroups. For every error-detecting code two parameters are important: the probability of undetected errors and the number of errors that the code detects for sure. Until now, the considered code is completely analyzed from the aspect of the probability of undetected errors. Namely, for the analysed code the probability of undetected errors depends on the quasigroup used for coding. The formula for the probability of undetected errors is derived in [1]. Best for coding are the quasigroups that provide smallest probability of undetected errors. The best class of quasigroups of order 4 for the considered codes, i.e., the class of quasigroups of order 4 that gives smallest probability of undetected errors and the corresponding probability of undetected errors are found in [2]. Now, in this paper we are focused on the second important parameter, i.e., the number of errors that the code detects for sure. First, we will show that when linear quasigroup of arbitrary order is used for coding, the code is linear. Then, we will obtain the generator matrix, the parity-check matrix and the Hamming distance of the code which determines the number of errors that the code detects for sure in the case when linear quasigroup of order 4 from the best class of quasigroups of order 4 is used for coding.

© Springer International Publishing AG 2017
D. Trajanov and V. Bakeva (Eds.): ICT Innovations 2017, CCIS 778, pp. 219–228, 2017.
DOI: 10.1007/978-3-319-67597-8_21

2   Mathematical Preliminaries

Definition 1. *Quasigroup is algebraic structure* $(Q, *)$ *such that*

$$(\forall u, v \in Q)(\exists! x, y \in Q) \ (x * u = v \ \& \ u * y = v) \tag{1}$$

Definition 2. *The quasigroup* $(Q, *)$ *of order* 2^q *is linear if there are non-singular binary matrices* A *and* B *of order* $q \times q$ *and a binary matrix* C *of order* $1 \times q$, *such that*

$$(\forall x, y \in Q) \ x * y = z \Leftrightarrow z = xA + yB + C \tag{2}$$

where **x**, **y** *and* **z** *are binary representations of* x, y *and* z *as vectors of order* $1 \times q$ *and* $+$ *is binary addition.*

In what follows when we say that $(Q, *)$ is a quasigroup of order 2^q, then we take $Q = \{0, 1, ..., 2^q - 1\}$. Also, with bolded symbol we will represent a binary representation of that element, i.e., if x is element from a quasigroup Q of order 2^q, then **x** is a binary representation of x as $1 \times q$ vector. Additionally, for shorter record with **x** * **y** will be denoted the binary representation of the quasigroup product $x * y$.

In this paper we will also use the following well known definitions and a theorem ([3]).

With $V_k(F)$ we will denote the set of all k-tuples over a field F.

Definition 3. *A linear* (n, k) *code over* F *is* k-*dimensional subspace of* $V_n(F)$

Definition 4. *A generator matrix* G *for* (n, k) *code* C_W *is any* $k \times n$ *matrix whose rows are a vector space basis for* C_W.

Since every vector space can have more different basis, the generator matrix for a given linear code is not unique. If a generator matrix have a form $G = [I_k \ D]$, where I_k is $k \times k$ identity matrix and D is $k \times (n - k)$ matrix, then a matrix G is said that is in standard form.

Definition 5. *If* $G = [I_k \ D]$ *is a generator matrix for* (n, k) *code* C_W, *then* $H = [-D^T \ I_{n-k}]$ *is a parity-check matrix for* C_W.

Theorem 1. *Let* H *be a parity-check matrix for an* (n, k) *code* C_W *over* F. *Then every set of* $s - 1$ *columns of* H *are linearly independent if and only if* C_W *has Hamming distance at least* s.

The previous theorem is used to determine the Hamming distance of a linear code, given by its parity-check matrix. Namely, as direct consequence of Theorem 1 follows:

Corollary 1. *A linear code* C_W *with parity-check matrix* H *has Hamming distance* d *if and only if every set of* $d - 1$ *columns of* H *are linearly independent, and some set of* d *columns are linearly dependent.*

3 Linear Error-Detecting Code

Let $(Q, *)$ be a linear quasigroup of order 2^q and let $a_0 a_1 a_2 \ldots a_{n-1}$ be an input block of length n. The redundant characters are defined in the following way:

$$d_i = a_i * a_{i+1}, \quad i \in \{0, 1, \ldots, n-1\} \tag{3}$$

where all operations in indexes are per modulo n. This means that $d_0 = a_0 * a_1$, $d_1 = a_1 * a_2$, ..., $d_{n-2} = a_{n-2} * a_{n-1}$, $d_{n-1} = a_{n-1} * a_0$. Now, the extended message $a_0 a_1 a_2 \ldots a_{n-1} d_0 d_1 d_2 \ldots d_{n-1}$, previously turned into binary form, is transmitted through the binary symmetric channel. A block of length n is extended to a block of length $2n$, therefore the rate of the code is $1/2$.

Under the influence of the noises in the channel, some of the characters may not be correctly transmitted. After receiving the block, the receiver checks if all Eq. (3) are satisfied. If there is some $i \in \{0, 1, \ldots, n-1\}$ for which the equation is not satisfied, the receiver concludes that there are errors in transmission and it asks the sender to send the block once again. But, since the redundant characters are transmitted through the binary symmetric channel, it is possible that they are incorrectly transmitted too, in a way that all Eq. (3) are satisfied, although some of the information characters $a_0, a_1, \ldots, a_{n-1}$ are incorrectly transmitted. For this reason, it is possible to have undetected errors in transmission.

In what follows, we will represent the messages as n tuples, i.e., the input block $a_0 a_1 \ldots a_{n-1}$ will be represent as $(a_0, a_1, \ldots, a_{n-1})$, the coded block $a_0 a_1 \ldots a_{n-1} d_0 d_1 \ldots d_{n-1}$ will be represent as $(a_0, a_1, \ldots, a_{n-1}, d_0, d_1, \ldots d_{n-1})$, while their binary representations will be represented as $(\boldsymbol{a_0}, \boldsymbol{a_1}, \ldots, \boldsymbol{a_{n-1}})$ and $(\boldsymbol{a_0}, \boldsymbol{a_1}, \ldots, \boldsymbol{a_{n-1}}, \boldsymbol{d_0}, \boldsymbol{d_1}, \ldots, \boldsymbol{d_{n-1}})$, respectively.

Example 1. Let for coding be used the following quasigroup of order 4:

*	0	1	2	3
0	0	1	3	2
1	3	2	0	1
2	2	3	1	0
3	1	0	2	3

For this quasigroup, the matrices A, B and C satisfying (2) are:

$$A = \begin{bmatrix} 1 & 0 \\ 1 & 1 \end{bmatrix}, \quad B = \begin{bmatrix} 1 & 1 \\ 0 & 1 \end{bmatrix} \text{ and } C = [0 \ 0].$$

Let the input block be 32021, i.e., $a_0 = 3, a_1 = 2, a_2 = 0, a_3 = 2$ and $a_4 = 1$. Then, this input block can be represented as $(3, 2, 0, 2, 1)$. Using (3), we obtain the redundant characters: $d_0 = 2, d_1 = 2, d_2 = 3, d_3 = 3$, and $d_4 = 1$, i.e., the coded message is $(3, 2, 0, 2, 1, 2, 2, 3, 3, 1)$. The binary representations are: $\boldsymbol{a_0} = (1, 1), \boldsymbol{a_1} = (1, 0), \boldsymbol{a_2} = (0, 0), \boldsymbol{a_3} = (1, 0), \boldsymbol{a_4} = (0, 1), \boldsymbol{d_0} = (1, 0), \boldsymbol{d_1} = (1, 0), \boldsymbol{d_2} = (1, 1), \boldsymbol{d_3} = (1, 1)$ and $\boldsymbol{d_4} = (0, 1)$, therefore the coded block turned into binary form is $(\boldsymbol{a_0}, \boldsymbol{a_1}, \boldsymbol{a_2}, \boldsymbol{a_3}, \boldsymbol{a_4}, \boldsymbol{d_0}, \boldsymbol{d_1}, \boldsymbol{d_2}, \boldsymbol{d_3}, \boldsymbol{d_4}) = (1, 1, 1, 0, 0, 0, 1, 0, 0, 1, 1, 0, 1, 0, 1, 1, 1, 1, 0, 1)$.

Theorem 2. *If a linear quasigroup Q of order 2^q for which the constant term C in the linear representation (2) is zero matrix is used for coding and input blocks of length n are coded, then the code defined with (3) is linear $(2nq, nq)$ code.*

Proof. If a linear quasigroup of order 2^q is used for coding, then the binary representation of every element of the quasigroup Q has q bits. If the input block is $a_0 a_1 \ldots a_{n-1}$, where $a_i \in Q$, then every character a_i can be represented in a binary form as $a_i = (x_{qi}, x_{qi+1}, \ldots, x_{qi+q-1})$. From here it follows that the binary representation of the input block is $(a_0, a_1, \ldots, a_{n-1}) = (x_0, x_1, \ldots, x_{nq-1})$. Also, every binary nq - tuple is a binary representation of a block of length n with elements that belong to a quasigroup of order 2^q. Follows, the input message space is

$$V_{nq}(F) = \{(x_0, x_1, \ldots, x_{nq-1}) | x_i \in \{0,1\}, i = 0, 1, \ldots nq - 1\}$$

where $F = \{0,1\}$. The space of code words is

$$C_W = \{(a_0, a_1, \ldots, a_{n-1}, a_0 * a_1, a_1 * a_2, \ldots, a_{n-1} * a_0) | a_0 \in Q,$$
$$a_1 \in Q, \ldots, a_{n-1} \in Q\} \qquad (4)$$

Recall that in (4) a_i ($i \in \{0, 1, \ldots, n-1\}$) is a binary representation of $a_i \in Q$ with length q, while $a_i * a_{i+1}$ ($i \in \{0, 1, \ldots, n-1\}$) is a binary representation of the quasigroup product $a_i * a_{i+1}$, also with length q. This means that C_W contains $2nq$ - tupels over the set $\{0,1\}$, i.e., $C_W \subseteq V_{2nq}(F)$.

We will show that C_W is a vector subspace of $V_{2nq}(F)$, where $F = \{0,1\}$. It is obvious that the zero vector belongs to C_W. Namely, from the definition of the code (3), from (2) and $C = O$ follows that the block in which all characters are zeros will be coded into the zero vector.

Let $a = (a_0, a_1, \ldots, a_{n-1}, a_0 * a_1, a_1 * a_2, \ldots, a_{n-1} * a_0) \in C_W$ and $b = (b_0, b_1, \ldots, b_{n-1}, b_0 * b_1, b_1 * b_2, \ldots, b_{n-1} * b_0) \in C_W$. We should show that

(i) $a + b \in C_W$, i.e. $a + b$ is a code word
(ii) $\lambda a \in C_W$, for all $\lambda \in F$.

The proof of (ii) is trivial. Namely,

– if $\lambda = 1$, then $\lambda a = 1 \cdot a = a \in C_W$.
– if $\lambda = 0$, then $\lambda a = 0 \cdot a = 0 \in C_W$.

Now, we will prove (i).

$a + b = (a_0 + b_0, a_1 + b_1, \ldots, a_{n-1} + b_{n-1}, a_0 * a_1 + b_0 * b_1, a_1 * a_2 + b_1 * b_2, \ldots, a_{n-1} * a_0 + b_{n-1} * b_0)$. In order to prove that $a + b \in C_W$, we should show that $a_i * a_{i+1} + b_i * b_{i+1}$, $i = 0, 1, \ldots, n-1$ are redundant characters for the input message $(a_0 + b_0, a_1 + b_1, \ldots, a_{n-1} + b_{n-1})$, i.e.

$$a_i * a_{i+1} + b_i * b_{i+1} = (a_i + b_i) * (a_{i+1} + b_{i+1}), \qquad (5)$$

$i = 0, 1, \ldots, n - 1$.

From (2) and $C = O$, we obtain:

$$a_i * a_{i+1} + b_i * b_{i+1} = a_i A + a_{i+1} B + b_i A + b_{i+1} B$$
$$= (a_i + b_i)A + (a_{i+1} + b_{i+1})B$$
$$= (a_i + b_i) * (a_{i+1} + b_{i+1})$$

therefore (5) is satisfied.

So far we have shown that C_W is vector subspace from $V_{2nq}(F)$. Now, we should show that the dimension of C_W is nq. We will show that one basis of C_W is $B = \{(e_i, d_0^i, d_1^i, \ldots, d_{n-1}^i) | i = 1, 2, \ldots, nq\}$, where e_i is a binary $1 \times nq$ vector such that on the i-th position is 1, and on all other positions is 0, while d_j^i $(j = 0, 1, \ldots, n-1)$, are the binary representations of the redundant characters calculated using (3) applied to the input message e_i previously turned into the word over the alphabet Q. Namely, $1 \times nq$ binary vector e_i is divided into n parts of length q bits and then each part is turned into symbol from the quasigroup Q with what e_i is turned into a word e_i over the alphabet Q. Then this word is coded using Eq. (3) and each of the obtained redundant characters $d_0^i, d_1^i, \ldots, d_{n-1}^i$ are turned into a binary form of length q, with what are obtained $d_0^i, d_1^i, \ldots, d_{n-1}^i$.

Clearly, the set B is linearly independent set. Now we will show that it is generating set for the space C_W. For that purpose, we define a map $f : V_{nq}(F) \to C_W$, in a way that every element of $V_{nq}(F)$ is maped into the word in which it is coded, i.e.

$$f(a_0, a_1, \ldots, a_{n-1}) = (a_0, a_1, \ldots, a_{n-1}, a_0 * a_1, a_1 * a_2, \ldots, a_{n-1} * a_0) \quad (6)$$

Then, it is obvious that $B = \{f(e_i) | i = 1, 2, \ldots, nq\}$.

It can be easily shown that f is linear map, i.e. that

(iii) $f(a + b) = f(a) + f(b)$, for all $a, b \in V_{nq}(F)$
(iv) $f(\lambda a) = \lambda f(a)$, for all $\lambda \in F$ and $a \in V_{nq}(F)$

The correctness of (iii) follows directly from the proof of (i), while the correctness of (iv) can be checked with direct substitution of the values of $\lambda \in \{0, 1\}$ into the equation.

Let $y \in C_W$. This means that y is a code word, i.e. there is an element $x \in V_{nq}(F)$ (i.e. $x \in Q^n$), such that $f(x) = y$. Since $\{e_1, e_2, \ldots, e_{nq}\}$ is a base of $V_{nq}(F)$, follows that there are some $\lambda_i \in \{0, 1\}$, $i = 1, 2, \ldots, nq$, such that $x = \sum_{i=1}^{nq} \lambda_i e_i$. Since f is a linear map and $f(x) = y$, follows that

$$y = f(x) = f(\sum_{i=1}^{nq} \lambda_i e_i) = \sum_{i=1}^{nq} \lambda_i f(e_i)$$

which implies that $B = \{f(e_i) | i = 1, 2, \ldots, nq\}$ is a base of C_W. From this follows that the dimension of the vector space C_W is nq.

In this way we proved that the code defined with (3) is a linear $(2nq, nq)$ code.

The proof of the above theorem gives us procedure for constructing a generator matrix in a standard form. Namely, the generator matrix is

$$
G = \begin{bmatrix}
e_1 & d_0^1 & d_1^1 & \cdots & d_{n-1}^1 \\
e_2 & d_0^2 & d_1^2 & \cdots & d_{n-1}^2 \\
\vdots & \vdots & \vdots & \vdots & \vdots \\
e_{nq} & d_0^{nq} & d_1^{nq} & \cdots & d_{n-1}^{nq}
\end{bmatrix}_{nq \times 2nq}
\tag{7}
$$

where e_i and d_j^i, $i = 1, 2, \ldots, nq$, $j = 0, 1, \ldots, n-1$, are as defined in the proof of the previous theorem.

Since for every $i \in \{1, 2, \ldots, nq\}$, e_i has length nq bits and each of d_j^i, $j = 0, 1, \ldots, n-1$ has length q bits, the matrix G has $2nq$ columns and nq rows, i.e., G is a matrix of order $nq \times 2nq$.

Example 2. Let obtain the generator and the parity-check matrices for the quasigroup given in Example 1 when the input block has length $n = 6$ characters. Since the quasigroup is of order 4, follows that $q = 2$. The vectors e_i, $i \in \{1, 2, \ldots, 12\}$ have length 12 bits. First, $e_1 = (1, 0, 0, 0, 0, 0, 0, 0, 0, 0, 0, 0)$. This binary vector turned into a vector over the quasigroup of order 4 is $e_1 = (2, 0, 0, 0, 0, 0)$. Now, we compute the redundant characters for this vector: $d_0^1 = 2 * 0 = 2, d_1^1 = d_2^1 = d_3^1 = d_4^1 = 0 * 0 = 0, d_5^1 = 0 * 2 = 3$, therefore $d_0^1 = (1, 0), d_1^1 = d_2^1 = d_3^1 = d_4^1 = (0, 0), d_5^1 = (1, 1)$.

$e_2 = (0, 1, 0, 0, 0, 0, 0, 0, 0, 0, 0, 0)$. This binary vector turned into a vector over the quasigroup of order 4 is $e_2 = (1, 0, 0, 0, 0, 0)$. Now, we compute the redundant characters for this vector: $d_0^2 = 1 * 0 = 3, d_1^2 = d_2^2 = d_3^2 = d_4^2 = 0 * 0 = 0, d_5^2 = 0 * 1 = 1$, therefore $d_0^2 = (1, 1), d_1^2 = d_2^2 = d_3^2 = d_4^2 = (0, 0), d_5^2 = (0, 1)$.

$e_3 = (0, 0, 1, 0, 0, 0, 0, 0, 0, 0, 0, 0)$. This binary vector turned into a vector over the quasigroup of order 4 is $e_3 = (0, 2, 0, 0, 0, 0)$. The redundant characters for this vector $d_0^3 = 0 * 2 = 3, d_1^3 = 2 * 0 = 2, d_2^3 = d_3^3 = d_4^3 = d_5^3 = 0 * 0 = 0$, therefore $d_0^3 = (1, 1), d_1^3 = (1, 0), d_2^3 = d_3^3 = d_4^3 = d_5^3 = (0, 0)$.

In the same manner we compute all values d_j^i for all $i \in \{1, 2, \ldots, 12\}$ and $j \in \{0, 1, 2, 3, 4, 5\}$, with what we obtain the generator matrix G:

$$
G = \begin{bmatrix}
1 & 0 & 0 & 0 & 0 & 0 & 0 & 0 & 0 & 0 & 0 & 0 & 1 & 0 & 0 & 0 & 0 & 0 & 0 & 0 & 0 & 0 & 1 & 1 \\
0 & 1 & 0 & 0 & 0 & 0 & 0 & 0 & 0 & 0 & 0 & 0 & 1 & 1 & 0 & 0 & 0 & 0 & 0 & 0 & 0 & 0 & 0 & 1 \\
0 & 0 & 1 & 0 & 0 & 0 & 0 & 0 & 0 & 0 & 0 & 0 & 1 & 1 & 1 & 0 & 0 & 0 & 0 & 0 & 0 & 0 & 0 & 0 \\
0 & 0 & 0 & 1 & 0 & 0 & 0 & 0 & 0 & 0 & 0 & 0 & 0 & 1 & 1 & 1 & 0 & 0 & 0 & 0 & 0 & 0 & 0 & 0 \\
0 & 0 & 0 & 0 & 1 & 0 & 0 & 0 & 0 & 0 & 0 & 0 & 0 & 0 & 1 & 1 & 1 & 0 & 0 & 0 & 0 & 0 & 0 & 0 \\
0 & 0 & 0 & 0 & 0 & 1 & 0 & 0 & 0 & 0 & 0 & 0 & 0 & 0 & 0 & 1 & 1 & 1 & 0 & 0 & 0 & 0 & 0 & 0 \\
0 & 0 & 0 & 0 & 0 & 0 & 1 & 0 & 0 & 0 & 0 & 0 & 0 & 0 & 0 & 0 & 1 & 1 & 1 & 0 & 0 & 0 & 0 & 0 \\
0 & 0 & 0 & 0 & 0 & 0 & 0 & 1 & 0 & 0 & 0 & 0 & 0 & 0 & 0 & 0 & 0 & 1 & 1 & 1 & 0 & 0 & 0 & 0 \\
0 & 0 & 0 & 0 & 0 & 0 & 0 & 0 & 1 & 0 & 0 & 0 & 0 & 0 & 0 & 0 & 0 & 0 & 1 & 1 & 1 & 0 & 0 & 0 \\
0 & 0 & 0 & 0 & 0 & 0 & 0 & 0 & 0 & 1 & 0 & 0 & 0 & 0 & 0 & 0 & 0 & 0 & 0 & 1 & 1 & 1 & 0 & 0 \\
0 & 0 & 0 & 0 & 0 & 0 & 0 & 0 & 0 & 0 & 1 & 0 & 0 & 0 & 0 & 0 & 0 & 0 & 0 & 0 & 1 & 1 & 1 & 0 \\
0 & 0 & 0 & 0 & 0 & 0 & 0 & 0 & 0 & 0 & 0 & 1 & 0 & 0 & 0 & 0 & 0 & 0 & 0 & 0 & 0 & 1 & 1 & 1
\end{bmatrix}_{12 \times 24}
\tag{8}
$$

Since, the generator matrix is $G = [I_{12}\ D]$, the parity-check matrix is $H = [-D^T\ I_{12}] = [D^T\ I_{12}]$, i.e.,

$$H = \begin{bmatrix} 1 & 1 & 1 & 0 & 0 & 0 & 0 & 0 & 0 & 0 & 0 & 0 & 1 & 0 & 0 & 0 & 0 & 0 & 0 & 0 & 0 & 0 & 0 & 0 \\ 0 & 1 & 1 & 1 & 0 & 0 & 0 & 0 & 0 & 0 & 0 & 0 & 0 & 1 & 0 & 0 & 0 & 0 & 0 & 0 & 0 & 0 & 0 & 0 \\ 0 & 0 & 1 & 1 & 1 & 0 & 0 & 0 & 0 & 0 & 0 & 0 & 0 & 0 & 1 & 0 & 0 & 0 & 0 & 0 & 0 & 0 & 0 & 0 \\ 0 & 0 & 0 & 1 & 1 & 1 & 0 & 0 & 0 & 0 & 0 & 0 & 0 & 0 & 0 & 1 & 0 & 0 & 0 & 0 & 0 & 0 & 0 & 0 \\ 0 & 0 & 0 & 0 & 1 & 1 & 1 & 0 & 0 & 0 & 0 & 0 & 0 & 0 & 0 & 0 & 1 & 0 & 0 & 0 & 0 & 0 & 0 & 0 \\ 0 & 0 & 0 & 0 & 0 & 1 & 1 & 1 & 0 & 0 & 0 & 0 & 0 & 0 & 0 & 0 & 0 & 1 & 0 & 0 & 0 & 0 & 0 & 0 \\ 0 & 0 & 0 & 0 & 0 & 0 & 1 & 1 & 1 & 0 & 0 & 0 & 0 & 0 & 0 & 0 & 0 & 0 & 1 & 0 & 0 & 0 & 0 & 0 \\ 0 & 0 & 0 & 0 & 0 & 0 & 0 & 1 & 1 & 1 & 0 & 0 & 0 & 0 & 0 & 0 & 0 & 0 & 0 & 1 & 0 & 0 & 0 & 0 \\ 0 & 0 & 0 & 0 & 0 & 0 & 0 & 0 & 1 & 1 & 1 & 0 & 0 & 0 & 0 & 0 & 0 & 0 & 0 & 0 & 1 & 0 & 0 & 0 \\ 0 & 0 & 0 & 0 & 0 & 0 & 0 & 0 & 0 & 1 & 1 & 1 & 0 & 0 & 0 & 0 & 0 & 0 & 0 & 0 & 0 & 1 & 0 & 0 \\ 1 & 0 & 0 & 0 & 0 & 0 & 0 & 0 & 0 & 0 & 1 & 1 & 0 & 0 & 0 & 0 & 0 & 0 & 0 & 0 & 0 & 0 & 1 & 0 \\ 1 & 1 & 0 & 0 & 0 & 0 & 0 & 0 & 0 & 0 & 0 & 1 & 0 & 0 & 0 & 0 & 0 & 0 & 0 & 0 & 0 & 0 & 0 & 1 \end{bmatrix}_{12 \times 24} \qquad (9)$$

4   Number of Errors that the Linear Error-Detecting Code Surely Detects When Linear Quasigroup of Order 4 Is Used for Coding

In this section we will construct the generator and the parity-check matrices for the code when a linear quasigroup from the best class of quasigroups of order 4 is used for coding. The best class of quasigroups of order 4 is defined to be the class of quasigroups that contains exactly those quasigroups of order 4 that give smallest probability of undetected errors. This class of quasigroups is obtained in [2]. The best class of quasigroups of order 4 contains 4 pairs of matrices A and B, while the matrix C can be any of the four matrices of order 1×2. The best class of quasigroups of order 4, represented by the pair of non-singular binary matrices A and B is the following.

$$A_1 = \begin{bmatrix} 1 & 0 \\ 1 & 1 \end{bmatrix}, B_1 = \begin{bmatrix} 1 & 1 \\ 0 & 1 \end{bmatrix}$$

$$A_2 = \begin{bmatrix} 0 & 1 \\ 1 & 1 \end{bmatrix}, B_2 = \begin{bmatrix} 1 & 1 \\ 1 & 0 \end{bmatrix}$$

$$A_3 = \begin{bmatrix} 1 & 1 \\ 1 & 0 \end{bmatrix}, B_3 = \begin{bmatrix} 0 & 1 \\ 1 & 1 \end{bmatrix}$$

$$A_4 = \begin{bmatrix} 1 & 1 \\ 0 & 1 \end{bmatrix}, B_4 = \begin{bmatrix} 1 & 0 \\ 1 & 1 \end{bmatrix}$$

Fig. 1. The best class of linear quasigroups of order 4.

Theorem 3. *When a linear quasigroup Q from the best class of quasigroups of order 4 for which the constant term C in the linear representation (2) is zero matrix is used for coding, the code defined with (3) has Hamming distance 4 bits.*

Proof. We will obtain the Hamming distance of the code using the Corollary 1. For that purpose we need to know the generator matrix of the code, which will be obtained using the proof of the Theorem 2, i.e., we will obtain the matrix (7).

Let the quasigroup used for coding is represented with the binary non-singular 2×2 matrices:

$$A = \begin{bmatrix} a_1 \\ a_2 \end{bmatrix} \text{ and } B = \begin{bmatrix} b_1 \\ b_2 \end{bmatrix}$$

where a_1 is the first row and a_2 is the second row of the matrix A, while b_1 is the first row and b_2 is the second row of the matrix B. This means that a_1, a_2, b_1 and b_2 are 1×2 binary vectors.

Let the input block has length n and e_i, $i \in \{1, 2, \ldots, 2n\}$ are binary $1 \times 2n$ vectors such that on the i-th position is 1 and on all other positions is 0. In order to obtain the generator matrix, we should apply (3) on each of the vectors e_i, $i \in \{1, 2, \ldots, 2n\}$, previously turned into a string e_i over the alphabet $Q = \{0, 1, 2, 3\}$. In that way, we obtain the redundant characters $d_0^i, d_1^i, \ldots, d_{n-1}^i$. By replacing their binary representations into (7), we obtain the generator matrix G:

$$G = \begin{bmatrix} e_1 & a_1 & 0 & 0 & 0 & \ldots & 0 & b_1 \\ e_2 & a_2 & 0 & 0 & 0 & \ldots & 0 & b_2 \\ e_3 & b_1 & a_1 & 0 & 0 & \ldots & 0 & 0 \\ e_4 & b_2 & a_2 & 0 & 0 & \ldots & 0 & 0 \\ e_5 & 0 & b_1 & a_1 & 0 & \ldots & 0 & 0 \\ e_6 & 0 & b_2 & a_2 & 0 & \ldots & 0 & 0 \\ e_7 & 0 & 0 & b_1 & a_1 & \ldots & 0 & 0 \\ e_8 & 0 & 0 & b_2 & a_2 & \ldots & 0 & 0 \\ \vdots & \vdots & \vdots & \vdots & \vdots & \vdots & \vdots & \vdots \\ e_{2n-1} & 0 & 0 & 0 & 0 & \ldots & b_1 & a_1 \\ e_{2n} & 0 & 0 & 0 & 0 & \ldots & b_2 & a_2 \end{bmatrix} \quad (10)$$

where 0 is 1×2 zero-vector. With this, we obtained the generator matrix in a standard form $G = [I_{2n} \ D]$, from where the parity-check matrix is obtained as $H = [-D^T \ I_{2n}]$. Since D is a binary matrix, $-D^T = D^T$, from where follows that the parity-check matrix of the code is $H = [D^T \ I_{2n}]$, where

$$D^T = \begin{bmatrix} a_1^T & a_2^T & b_1^T & b_2^T & 0 & 0 & 0 & 0 & \ldots & 0 & 0 \\ 0 & 0 & a_1^T & a_2^T & b_1^T & b_2^T & 0 & 0 & \ldots & 0 & 0 \\ 0 & 0 & 0 & 0 & a_1^T & a_2^T & b_1^T & b_2^T & \ldots & 0 & 0 \\ 0 & 0 & 0 & 0 & 0 & 0 & a_1^T & a_2^T & \ldots & 0 & 0 \\ \vdots & \vdots & \vdots & \vdots & \vdots & \vdots & \vdots & \vdots & \vdots & \vdots & \vdots \\ 0 & 0 & 0 & 0 & 0 & 0 & 0 & 0 & \ldots & b_1^T & b_2^T \\ b_1^T & b_2^T & 0 & 0 & 0 & 0 & 0 & 0 & \ldots & a_1^T & a_2^T \end{bmatrix} \quad (11)$$

where a_1^T is transposition of the first row and a_2^T is transposition of the second row of the matrix A, while b_1^T is transposition of the first row and b_2^T is transposition of the second row of the matrix B. This means that $a_1{}^T, a_2{}^T, b_1{}^T$ and $b_2{}^T$ are 2×1 binary vectors. Also, $\mathbf{0}$ in the matrix (11) is 2×1 zero-vector.

Note that in Example 2 are obtained the generator and the parity-check matrices for the first quasigroup in Fig. 1, i.e., the quasigroup of order 4 from the best class of quasigroups of order 4 for coding determined with A_1 and B_1 when the length of the input block is $n = 6$ characters from the quasigroup.

From the quasigroups given in Fig. 1 can be seen that regardless which quasigroup of order 4 from the best class of quasigroups of order 4 is used for coding, each column of D^T contains exactly three 1's. Also, since matrices A and B do not contain zero-row and neither A nor B have equal rows (i.e. $a_i \neq \mathbf{0}$, $b_i \neq \mathbf{0}$ for $i \in \{1, 2\}$, $a_1 \neq a_2$ and $b_1 \neq b_2$), the matrix H does not have two equal columns.

Now, using Corollary 1 we will obtain the Hamming distance of the code. First, since there is no null column in H, follows that every set of one column of H is linearly independent. Two binary vectors can be linearly dependent if and only if they are equal. Since there are no two equal columns in H, every set of two columns of H is linearly independent. Next, we will show that every set of three columns of H is linearly independent. The following cases are possible:

(i) All three columns are from the matrix I_{2n}. It is obvious that they are linearly independent.

(ii) Two columns are from the matrix I_{2n} and one column is from the matrix D^T. If these three columns are linearly dependent, then the column from D^T can be represented as a sum of the other two columns. But, the sum of the two columns from the matrix I_{2n} contains two 1's, while the column from the matrix D^T contains three 1's. This means that the column from D^T can not be represented as a sum of the other two columns, which means that these three columns are linearly independent.

(iii) One column is from I_{2n} and two columns are from D^T. If these three columns are linearly dependent, then the one column of D^T can be represented as a sum of the other column of D^T and the column from I_{2n}. But, the sum of these two columns is a column that has four 1's (if the column of D^T has 0 at the position at which the column from I_{2n} has 1) or two 1's (if the column of D^T has 1 at the position at which the column from I_{2n} has 1). Since the column from D^T has three 1's, follows that it is impossible by adding of one column of D^T with a column of I_{2n} to obtain a column of D^T. Follows, the three columns are linearly independent.

(iv) All three columns are from the matrix D^T. By summarizing of any two of them is obtained a column that has six 1's (if there is no common position at which the two columns have 1), four 1's (if there is exactly one common position at which the two columns have 1) or two 1's (if there are exactly two common positions at which the two columns have 1). Since the third column of D^T has also three 1's, it is impossible to represent one of the columns as a sum of the other two. Follows, these three columns are linearly independent.

Until now, we proved that every set of at most 3 columns from the parity-check matrix H is linearly independent. It is obvious that there is a set of four columns of the parity-check matrix H that is linearly dependent. Namely, let d be a column from D^T. Since every column from D^T has exactly three 1's, follows that d has exactly three 1's. Let the 1's are at positions i, j, and k. Then, the sum of the i-th, j-th and k-th column of I_{2n} gives the column d, which means that these four columns from H are linearly dependent.

From all above and Corollary 1, follows that when a linear quasigroup of order 4 from the best class of quasigroups of order 4 for which the constant term C in the linear representation (2) is zero matrix is used for coding, the Hamming distance of the code is 4 bits.

Direct consequence of the Theorem 3 is the following theorem.

Theorem 4. *When a linear quasigroup from the best class of quasigroups of order 4 for which the constant term C in the linear representation (2) is zero matrix is used for coding, the code defined with (3) detects for sure up to 3 incorrectly transmitted bits.*

5   Conclusion

In this paper we considered an error-detecting code based on linear quasigroups. We proved that when linear quasigroup of arbitrary order is used for coding, but for which the constant term in the linear representation is zero matrix, then the code is linear. Also, we showed that when such linear quasigroup of order 4 from the best class of quasigrous of order 4, i.e., the class of quasigroups of order 4 that have smallest probability of undetected errors is used for coding, the code has Hamming distance 4 bits, therefore the code detects for sure up to 3 incorrectly transmitted bits.

Acknowledgments. This work was partially financed by the Faculty of Computer Science and Engineering at the "Ss.Cyril and Methodius" University.

References

1. Ilievska, N., Bakeva, V.: A model of error-detecting codes based on quasigroups of order 4. In: Proceedings of 6th International Conference for Informatics and Information Technology, Bitola, Republic of Macedonia, pp. 7–11 (2008)
2. Bakeva, V., Ilievska, N.: A probabilistic model of error-detecting codes based on quasigroups. Quasigroups Relat. Syst. **17**(2), 135–148 (2009)
3. Vanstone, S., Oorschot, P.: An Introduction to Error Correcting Codes with Applications. Kluwer academic publishers, Boston/Dordrecht/London (1989)

Representation of Algebraic Structures by Boolean Functions and Its Applications

Smile Markovski, Verica Bakeva, Vesna Dimitrova[(✉)],
and Aleksandra Popovska-Mitrovikj

Faculty of Computer Science and Engineering,
Ss. Cyril and Methodius University, Skopje, Macedonia
smile.markovski@gmail.com, {verica.bakeva,vesna.dimitrova,
aleksandra.popovska.mitrovikj}@finki.ukim.mk

Abstract. Boolean functions are mappings $\{0,1\}^n \to \{0,1\}$, where n is a nonnegative integer. It is well known that each Boolean function $f(x_1, \ldots, x_n)$ with n variables can be presented by its Algebraic Normal Form (ANF). If (G, F) is an algebra of order $|G|$, $2^{n-1} \leq |G| < 2^n$, where F is a set of finite operations on G, then any operation $f \in F$ of arity k can be interpreted as a partial vector valued Boolean function $f_{v.v.} : \{0,1\}^{kn} \to \{0,1\}^n$. By using the function $f_{v.v.}$ and ANF of Boolean functions, we can characterize different properties of the finite algebras, and here we mention several applications. We consider especially the case of groupoids, i.e., the case when $F = \{f\}$ consists of one binary operation and we classify groupoids of order 3 according to the degrees of their Boolean functions. Further on, we give another classification of linear groupoids of order 3 using graphical representation. At the end, we consider an application of Boolean representation for solving a system of equations in an algebra.

Keywords: Boolean function · ANF · Finite algebraic structures · Quasigroups · Representations by Boolean functions

1 Introduction

The Boolean functions are applied for solving many problems in computer sciences, cryptology, coding theory, electronic circuits, etc. In this paper we intend to emphasis their application for solving some problems in algebraic structures. The idea for representing algebras by Boolean functions can be found in several papers, for example [2]. In that paper, the authors also give a classification of quasigroups of order 4 by the degree of their Boolean functions. In [1], a classification of quasigroups of order 4 by image patterns is given. These classifications are very useful for choosing the quasigroup in designing cryptographic primitives and error-correcting or error-detecting codes. In [6], for a given quasigroup $(Q, *)$, a transformation $e : Q^+ \to Q^+$ is defined and few properties of it (useful in cryptography) are proved.

© Springer International Publishing AG 2017
D. Trajanov and V. Bakeva (Eds.): ICT Innovations 2017, CCIS 778, pp. 229–239, 2017.
DOI: 10.1007/978-3-319-67597-8_22

In this paper, we generalize the previous ideas to arbitrary algebraic structures. As illustration, we consider groupoids of order 3 (since the number of groupoids of higher order is too large), but the same analyses can be done for every other finite algebraic structure.

The rest of this paper is organized as follows. For the aim of representation, ANF (algebraic normal form) of a Boolean function is needed, and in Sect. 2 an algorithm for finding ANF is given. How an algebra can be represented by Boolean functions is discussed in Sect. 3. In Sect. 4, using degrees of the polynomials in the Boolean representation of the groupoids of order 3, a classification is given. Also, a relationship between the degree of quasigroups of order 4 and their classes of isomorphism is found. We make another classification of linear groupoids of order 3 by graphical representation in Sect. 5. An application for solving equations in algebras is presented in Sect. 6. In Sect. 7, some conclusions are given.

2   ANF of Boolean Functions

Let $B = \{0, 1\}$ be a set of two elements. Any mapping $B^n \rightarrow B$, where n is a nonnegative integer, is said to be an n-ary Boolean function. The Boolean functions are also known as logical functions, when 0 is interpreted as False and 1 as True. The logical function conjunction is the Boolean function $\wedge : B^2 \rightarrow B$ defined by $\wedge(x, y) = 1 \Leftrightarrow x = y = 1$, and the logical function exclusive disjunction, also known as XOR function, is the Boolean function $\veebar : B^2 \rightarrow B$ defined by $\veebar(x, y) = 1 \Leftrightarrow x \neq y$. Further on, instead of $\wedge$ and $\veebar$, we will use the symbols $\cdot$ and $+$, respectively.

A set $\{f_1, f_2, \ldots, f_k\}$ of Boolean functions is said to be complete if any other Boolean function can be represented only with them. An example of a complete set of Boolean functions is $\{\cdot, +\}$, as we will show below. Namely, each Boolean function $f : B^n \rightarrow B$ can be represented as polynomial in so called Algebraic Normal Form (ANF) as follows:

$$f(x_1, x_2, \ldots, x_n) = a_0 + a_1 x_1 + \cdots + a_n x_n + a_{12} x_1 x_2 + \cdots + a_{n-1n} x_{n-1} x_n$$
$$+ a_{123} x_1 x_2 x_3 + \cdots + a_{12 \ldots n} x_1 x_2 \ldots x_n,$$

where $a_\lambda \in B$, or shortly, by

$$f(x_1, \ldots, x_n) = \sum_{I \subseteq \{1, 2, \ldots, n\}} a_I x^I,$$

where $a_I \in \{0, 1\}$, $x^{\{k_1, k_2, \ldots, k_p\}} = x_{k_1} x_{k_2} \ldots x_{k_p}$. Let stress that for $I = \emptyset$, $a_I x^I = a_0$.

The ANF of a Boolean function $f(x_1, x_2, \ldots, x_n)$ can be obtained by the following algorithm [3].

ANF Algorithm

Step 0. Present each element $(b_1, b_2, \ldots, b_n) \in B^n$ as integer
$$k = b_1 + b_2 2 + b_3 2^2 + \cdots + b_n 2^{n-1}.$$

Step 1. Set $g(x_1, x_2, \ldots, x_n) = f(0, 0, \ldots, 0)$.

Step 2. For $k = 1$ to $2^n - 1$, do

Step 2.1. Take the binary representation of the integer k,
$$k = b_1 + b_2 2 + b_3 2^2 + \cdots + b_n 2^{n-1};$$

Step 2.2. If $g(b_1, b_2, \ldots, b_n) \neq f(b_1, b_2, \ldots, b_n)$ then set
$$g(x_1, x_2, \ldots, x_n) = g(x_1, x_2, \ldots, x_n) + \prod_{i=1}^n x_i^{b_i}.$$

Step 3. ANF$(f) = g(x_1, x_2, \ldots, x_n)$.

Here, x^0 denotes the empty symbol and $x^1 \equiv x$.

In the sequel, if $k = b_1 + b_2 2 + \cdots + b_n 2^{n-1}$ is in binary representation, then we denote $\bar{k} = (b_1, b_2, \ldots, b_n) \in B^n$.

Example 1. The function f and the computation of its ANF g are given in Table 1. Thus, ANF$(f) = x_2 + x_3 + x_1 x_2$.

Table 1. A computation of ANF

k	x_1	x_2	x_3	$f(x_1, x_2, x_3)$	$g(x_1, x_2, x_3)$
0	0	0	0	0	0
1	1	0	0	0	0
2	0	1	0	1	x_2
3	1	1	0	0	$x_2 + x_1 x_2$
4	0	0	1	1	$x_2 + x_3 + x_1 x_2$
5	1	0	1	1	$x_2 + x_3 + x_1 x_2$
6	0	1	1	0	$x_2 + x_3 + x_1 x_2$
7	1	1	1	1	$x_2 + x_3 + x_1 x_2$

A partial (n-ary) Boolean function is a mapping $f : D \to B$, where $D \subseteq B^n$. For obtaining ANF for the partial functions **Step 2.2** in the previous algorithm should be replaced by

Step 2.2′. If $((b_1, b_2, \ldots, b_n) \in D$ and $g(b_1, b_2, \ldots, b_n) \neq f(b_1, b_2, \ldots, b_n))$
then set $g(x_1, x_2, \ldots, x_n) = g(x_1, x_2, \ldots, x_n) + \prod_{i=1}^n x_i^{b_i}$.

Note that if $D \subset B^n$ then the ANF for a partial Boolean function is not unique one. Namely, for any $(b_1, \ldots, b_n) \in B^n \setminus D$, we can define values on different ways ($|B|^{|B^n \setminus D|}$ ways). For each selection of values for $(b_1, \ldots, b_n) \in$

$B^n \setminus D$, we obtain different ANF. All of them are ANF for the considered partial Boolean function. With proposed algorithm, for a given partial Boolean function, we obtain the ANF with the smaller number of terms (and of course, smallest degree). Further on, we will consider only this kind of ANF.

A mapping $f : B^n \to B^m$ is called a vector valued (or vectorial) Boolean function. If $f(x_1, x_2, \ldots, x_n) = (y_1, y_2, \ldots, y_m)$ then each y_i can be considered as a Boolean function $y_i = h_i(x_1, x_2, \ldots, x_n)$. So, each vectorial Boolean function $f : B^n \to B^m$ can be represented with an m-tuple of Boolean functions $(h_1, h_2, \ldots, h_m)$. As previously, for $D \subseteq B^n$, a partial vectorial Boolean function $f : D \to B^m$ can be defined.

3   Representation of Algebras by Boolean Functions

Here we consider finite algebras only, like groupoids, groups, quasigroups, rings. Recall that a groupoid $(Q, *)$ is a quasigroup iff the equations $a * x = b$ and $y * a = b$ have unique solutions x, $y \in Q$ for each a, $b \in Q$. When Q is finite, the main body of the multiplication table of a quasigroup is a Latin square.

Given a finite set $A = \{a_1, a_2, \ldots, a_k\}$ and a function $f : A^n \to A$, we can represent the function f by a partial vectorial Boolean function as follows. Let $2^{p-1} \le k < 2^p$, $p \ge 1$. We represent each element a_i by the p-tuple $\bar{i} \in \{0, 1\}^p$. We represent the function f by the partial vectorial Boolean function $f_{v.v.} : D \to B^p$ where $D \subseteq B^{pn}$ and if $f(a_{i_1}, a_{i_2}, \ldots a_{i_n}) = a_j$ then $f_{v.v.}(\bar{i}_1, \bar{i}_2, \ldots, \bar{i}_n) = \bar{j}$. Consequently, the function f can be represented by p partial Boolean functions $(h_1, h_2, \ldots, h_p)$, corresponding to $f_{v.v.}$.

Example 2. Let $A = \{a, b, c\}$ and consider the groupoid

$$
\begin{array}{c|ccc}
f & a & b & c \\
\hline
a & a & b & a \\
b & a & a & a \\
c & b & a & c
\end{array}
$$

If we use the correspondence $a \to (0, 0)$, $b \to (0, 1)$, $c \to (1, 0)$, then the partial vectorial Boolean function $f_{v.v.}$ is defined by $f_{v.v.}(0, 0, 0, 0) = (0, 0)$, $f_{v.v.}(0, 0, 0, 1) = (0, 1)$, $f_{v.v.}(0, 0, 1, 0) = (0, 0)$, $f_{v.v.}(0, 1, 0, 0) = (0, 0)$, $f_{v.v.}(0, 1, 0, 1) = (0, 0)$, $f_{v.v.}(0, 1, 1, 0) = (0, 0)$, $f_{v.v.}(1, 0, 0, 0) = (0, 1)$, $f_{v.v.}(1, 0, 0, 1) = (0, 0)$, $f_{v.v.}(1, 0, 1, 0) = (1, 0)$. Now, the function f is represented by the pair of Boolean functions

$$
h_1(x_1, x_2, x_3, x_4) = x_1 x_3,
$$
$$
h_2(x_1, x_2, x_3, x_4) = x_1 + x_4 + x_1 x_3 + x_2 x_4.
$$

Example 3. The ring $(\mathbb{Z}_4, +, \cdot)$ has vectorial Boolean representation with the corresponding functions $\bar{+}, \bar{\cdot}$ as usual, given in Table 2. It can be represented by the following four Boolean functions:

$$h_1^+(x_1, x_2, x_3, x_4) = x_1 + x_3 + x_2 x_4, \qquad h_2^+(x_1, x_2, x_3, x_4) = x_2 + x_4,$$
$$h_1^-(x_1, x_2, x_3, x_4) = x_1 x_4 + x_2 x_3, \qquad h_2^-(x_1, x_2, x_3, x_4) = x_2 x_4.$$

Note that the functions h_1^+ and h_2^+ are Boolean representations of the cyclic group C_4.

Table 2. The ring $(\mathbb{Z}_4, +, \cdot)$

$\dot{+}$	$\bar{0}$	$\bar{1}$	$\bar{2}$	$\bar{3}$
$\bar{0}$	$\bar{0}$	$\bar{1}$	$\bar{2}$	$\bar{3}$
$\bar{1}$	$\bar{1}$	$\bar{2}$	$\bar{3}$	$\bar{0}$
$\bar{2}$	$\bar{2}$	$\bar{3}$	$\bar{0}$	$\bar{1}$
$\bar{3}$	$\bar{3}$	$\bar{0}$	$\bar{1}$	$\bar{2}$

$\dot{\cdot}$	$\bar{0}$	$\bar{1}$	$\bar{2}$	$\bar{3}$
$\bar{0}$	$\bar{0}$	$\bar{0}$	$\bar{0}$	$\bar{0}$
$\bar{1}$	$\bar{0}$	$\bar{1}$	$\bar{2}$	$\bar{3}$
$\bar{2}$	$\bar{0}$	$\bar{2}$	$\bar{0}$	$\bar{2}$
$\bar{3}$	$\bar{0}$	$\bar{3}$	$\bar{2}$	$\bar{1}$

Example 4. The quasigroup has Boolean representation with the functions

$*$	0	1	2	3
0	1	0	2	3
1	3	2	1	0
2	2	3	0	1
3	0	1	3	2

$$h_1(x_1, x_2, x_3, x_4) = x_1 + x_2 + x_3,$$
$$h_2(x_1, x_2, x_3, x_4) = 1 + x_1 + x_3 + x_4 + x_1 x_3 + x_2 x_3.$$

Table 3. A quasigroup of order 16

$*$	0	1	2	3	4	5	6	7	8	9	a	b	c	d	e	f
0	a	4	5	9	6	0	e	1	2	c	d	f	3	8	b	7
1	5	b	c	8	4	e	0	7	3	2	f	a	1	9	d	6
2	c	5	2	d	f	8	a	e	1	3	6	7	b	0	9	4
3	7	d	3	e	2	1	b	c	5	9	4	8	0	f	6	a
4	1	2	4	a	b	7	8	9	0	d	3	e	6	c	5	f
5	4	a	8	b	d	2	c	6	e	f	5	9	7	3	1	0
6	0	e	d	2	8	3	6	5	c	b	7	4	9	a	f	1
7	b	6	0	5	9	d	4	8	7	a	2	3	f	1	e	c
8	d	8	6	1	c	a	f	0	b	5	9	2	4	7	3	e
9	2	f	1	0	7	c	5	b	9	6	8	d	a	e	4	3
a	6	c	b	7	a	f	1	3	4	8	e	0	d	5	2	9
b	8	1	f	6	3	9	7	4	a	e	c	5	2	d	0	b
c	f	3	9	4	e	6	2	d	8	7	0	1	c	b	a	5
d	e	9	7	3	1	b	d	f	6	0	a	c	5	4	8	2
e	3	0	e	c	5	4	9	a	f	1	b	6	8	2	7	d
f	9	7	a	f	0	5	3	2	d	4	1	b	e	6	c	8

We say that an algebra is of degree d if d is the maximum degree of all Boolean functions in its Boolean representation. Each algebra in Examples 1 – 4 has degree 2. On the other hand, the quasigroup of order 16 given on Table 3 is of degree 6.

One of its Boolean functions looks like follows:

$h_1(x_1, \ldots, x_8) = 1 + x_2 + x_6 + x_7 + x_8 + x_1x_2 + x_1x_3 + x_2x_4 + x_1x_5 + x_2x_5 + x_4x_5 + \cdots + x_1x_2x_5 + x_2x_3x_5 + x_2x_5x_8 + x_1x_2x_5x_6 + x_2x_3x_4x_5 + \cdots + x_2x_3x_6x_7 + x_3x_5x_7x_8 + \cdots + x_1x_2x_3x_5x_8 + x_2x_4x_5x_7x_8 + \cdots + x_1x_2x_4x_6x_7x_8 + x_2x_3x_4x_6x_7x_8.$

It consists of 121 members.

4   Classification of Set of Algebras of Same Type According to Their Boolean Representation

The Boolean representation of algebras allows a classification of the set of all algebras of the same type according to their Boolean representations. This can be done using the degree of the ANF polynomials of the Boolean functions. The classification has sense when the set of all algebras of the same type is enough large.

The number of quasigroups of order k increases more than exponentially. There are 576 quasigroups of order 4, 161280 of order 5, 812851200 of order 6, and so on. Also, for many applications, the classification of the set of all quasigroups (or some other types of groupoids) of order k is important. In the paper [2], the authors classified the quasigroups of order 4 according to their Boolean functions h_1, h_2 representation. They are classified in three classes: the class of linear quasigroups (h_1 and h_2 are both linear), the class of semi-linear quasigroups (one of the functions h_1 or h_2 is linear and the another one is quadratic) and the class of quadratic quasigroups (both of the functions h_1 and h_2 are quadratic). The number of linear quasigroups is 144, the number of semi-linear is 288, and the number of quadratic quasigroups is 144. According to definition of degree of an algebraic structure (given before), the quasigroups in the first class have a degree 1, and the quasigroups in the other two classes have a degree 2.

Here, we made analyses for the relationship between degrees of the quasigroups of order 4 and the classes of isomorphism of them. We obtain that all quasigroups of one class of isomorphism have the same degree. There are 35 classes of isomorphism of quasigroups of order 4, and 15 of them contain only quasigroups of degree 1. The other 20 classes of isomorphism consist of quasigroups of degree 2.

4.1   Classification of the Set of Groupoids of Order 3 According to Their Boolean Representation

In this paper we made a classification of all groupoids of order 3, and there are 19683 of them. Their classification is given in Table 4.

Table 4. Classes of qroupoids of order 3 by Boolean representation

Classes	No. of qroupoids	Degree
Linear	369	1
Semi-linear	4500	2
Quadratic	14814	2

We use the same terminology as in the case of quasigroups given before. Namely, the class of linear qroupoids contains qroupoids where h_1 and h_2 are both linear, the class of semi-linear qroupoids contains qroupoids where one of the functions h_1 or h_2 is linear and the another one is quadratic, and the class of quadratic qroupoids where both of the functions h_1 and h_2 are quadratic. The qroupoids in the first class have a degree 1, and the qroupoids in the other two classes have a degree 2.

Also, as for quasigroups, we investigate the relationship between degrees of the qroupoids of order 3 and the classes of isomorphism of them. But, in this case, we did not find any connection between these two classifications. There are classes of isomorphism which contain qroupoids of degree 1 and degree 2. But, we must stress again that for a groupoid of order 3, ANF is not unique. This is probably the reason why this connection cannot be done.

5   Transformations Defined by Groupoids

Consider an alphabet (i.e., a finite set) G, and denote by G^+ the set of all finite nonempty strings of elements of G. Let $*$ be a binary operation on the set G, i.e. consider a groupoid $(G, *)$. For each $l \in G$ we define functions $e_{l,*}, e'_{l,*} : G^+ \longrightarrow G^+$ as follows. Let $a_i \in G$, $i = 1, \ldots, n$ and $\alpha = a_1 a_2 \ldots a_n$. Then

$$e_{l,*}(\alpha) = b_1 b_2 \ldots b_n \iff b_1 = l * a_1, \ b_2 = b_1 * a_2, \ldots, \ b_n = b_{n-1} * a_n, \quad (1)$$

$$e'_{l,*}(\alpha) = b_1 b_2 \ldots b_n \iff b_1 = a_1 * l, \ b_2 = a_2 * b_1, \ldots, \ b_n = a_n * b_{n-1}. \quad (2)$$

The function $e_{l,*}$ is called e-transformation of G^+ based on the operation $*$ with leader l, and its graphical representation is shown on Fig. 1.

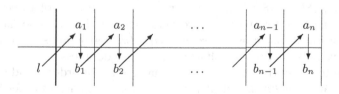

Fig. 1. Graphical representation of $e_{l,*}$ function

Example 5. Consider the groupoid $(\{a, b, c\}, *)$ given in Example 2 and take a leader c. The string $\alpha = b\,a\,c\,c\,a\,b\,b\,a\,c\,b\,a\,c\,c\,a\,b\,b\,c\,a\,a\,b$ is transformed into the string $e_{c,*}(\alpha) = a\,a\,a\,a\,a\,b\,a\,a\,a\,b\,a\,a\,a\,a\,b\,a\,a\,a\,a\,b$.

Theorem 1 [5]. *The transformations $e_{l,*}$ and $e'_{l,*}$, defined on a quasigroup $(G, *)$, are permutations of G^+.*

Theorem 2. *If there are transformations $e_{l,*}$ and $e'_{r,*}$ defined on a finite groupoid $(G, *)$ that are permutations of G^+, then $(G, *)$ is a quasigroup.*

Proof. Let $e_{l,*}$ be a permutation on G^+. Then for all $n \in \mathbb{N}$, the restrictions $\varphi_n : G^n \to G^n$ of $e_{l,*}$ on the subset $G^n = \{(x_1, \ldots, x_n) | x_i \in G\}$ of G^+ are permutations too. Now, let $b_1 * a_2 = b_1 * a'_2$ for some $b_1, a_2, a'_2 \in G$. Since φ_1 is a permutation, we can find an $a_1 \in G$ such that $e_{l,*}(a_1) = b_1$. By definition of $e_{l,*}$, we have $b_2 = b_1 * a_2 = b_1 * a'_2$. This means that $\varphi_2(a_1, a_2) = \varphi_2(a_1, a'_2) = (b_1, b_2)$, which implies $(a_1, a_2) = (a_1, a'_2)$, i.e., $a_2 = a'_2$. Hence, the groupoid $(G, *)$ is with left cancellation. By using the permutation $e'_{r,*}$ we can show that $(G, *)$ is with right cancellation as well. Since $(G, *)$ is a finite groupoid with cancellation, it is a quasigroup.

5.1   Classification of Linear Groupoids of Order 3 by Graphical Presentation

In [1], using image pattern, a classification of quasigroups of order 4 as fractal and non-fractal is given. Here, we make a similar classification for linear qroupoids of order 3 since the number of all groupoids of order 3 is too large. This classification is made on the following way. Let start with a periodical sequence (for example, 012012012...) with length 90 and apply 90 times the e-transformation given in (1) with given leaders. We present all transformed sequences visually using different color for each symbol 0, 1, and 2. On this way, we obtain an image pattern for each groupoid and we analyze the structure of this patterns. For different initial sequence and different leader, the different patterns can be obtained. Some of the patterns have fractal or symmetric structure, some of them have no structure. If all obtained patterns for a groupoid have no structure, this groupoid is called *non-structured*. In opposite case, the groupoid is called *structured*. The number of *non-structured* linear groupoids is 14. Their lexicographic numbers are the following: 2630, 3326, 7726, 8950, 10734, 11059, 11140, 11958, 12929, 16358, 16569, 17054, 18537, 18753. In Fig. 2, we present patterns of three *structured* groupoids. The first two patterns are periodic and the last one has a fractal structure. In Fig. 3, patterns of two *non-structured* groupoids are given. These patterns have no structure and show a kind of randomness.

We have to stress that in the case of quasigroups of order 3 and 4, all linear quasigroups are *structured*. But, it is not case for groupoids of order 3. Namely, the groupoids presented in Fig. 3 are linear, but they are *non-structured*.

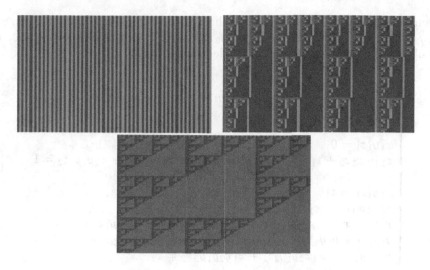

Fig. 2. Pattern presentation of *structured* groupoids

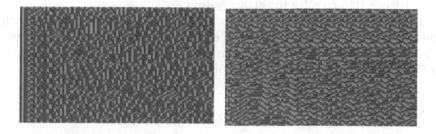

Fig. 3. Pattern presentation of *non-structured* groupoids

6   Solving Equations in Groupoids

In this section, we show how the Boolean representations of algebras can be used for solving equations in algebras. It will be illustrated by solving equations in groupoids, and we consider the groupoid (A, f) from the Example 2. Let we have the following system of three equations with three variables x, y, z.

$$\begin{cases} x * (z * ((y * x) * z)) & = y \\ (x * (z * y)) * z & = b \\ (y * (c * z)) * (x * (y * x)) = z, \end{cases} \tag{3}$$

where $x, y, z \in A = \{0, 1, 2\}$. The system (3) in the groupoid (A, f) can be solved only by exhaustive search, since neither associative nor commutative law hold in the groupoid. Hence, we have to take all of the $3^3 = 27$ possibilities and to check if some of them is a solution.

By using the functions $h_1(x_1, x_2, x_3, x_4) = x_1 x_3$ and $h_2(x_1, x_2, x_3, x_4) = x_1 + x_4 + x_1 x_3 + x_2 x_4$ of the Boolean representation of the groupoid (A, f)

(given in Example 2), if we represent the variables x, y, z as $x = (x_1, x_2)$, $y = (y_1, y_2)$, $z = (z_1, z_2)$, where $x_1, x_2, y_1, y_2, z_1, z_2 \in B$ are Boolean variables, we will obtain the following system of Boolean equations:

$$
\begin{cases}
x_1 y_1 z_1 = y_1 \\
x_1 + z_1 + x_1 y_1 + x_2 z_1 + x_1 x_2 y_1 + x_1 y_1 z_2 + x_1 y_1 z_1 + \\
x_1 y_1 z_1 z_2 + x_1 x_2 y_1 z_2 + x_1 x_2 y_1 z_1 z_2 = y_2 \\
\\
x_1 y_1 z_1 = 0 \\
z_2 + x_1 z_2 + z_1 z_2 + y_1 z_1 z_2 + x_2 z_1 z_2 + x_1 y_1 z_1 z_2 + x_2 y_1 z_1 z_2 = 1 \\
\\
x_1 y_1 z_1 = z_1 \\
y_1 + x_1 y_1 + x_1 z_1 + x_1 z_2 + x_1 y_2 + y_1 y_2 + y_1 z_1 + y_1 z_2 + \\
x_2 y_1 + x_1 x_2 y_1 + x_1 y_2 z_1 + x_1 y_2 z_2 + x_2 y_1 y_2 + x_2 y_1 z_2 + \\
y_1 y_2 z_1 + y_1 y_2 z_2 + x_1 x_2 y_1 y_2 + x_1 x_2 y_1 z_2 + x_2 y_1 y_2 z_1 + \\
x_2 y_1 y_2 z_2 + x_1 x_2 y_1 y_2 z_1 + x_1 x_2 y_1 y_2 z_2 = z_2
\end{cases}
\tag{4}
$$

There are several tools for solving Boolean equations, for example we can use Gröbner bases ([4]) to solve the above system.

This example is a small one, but for a little more complex groupoid with 8 element underlying set and a system of equations with 10 variables we need to make $8^{10} \approx 10^9$ exhaustive search for finding the solutions, while by computer and Gröbner bases the system can be sold in seconds.

7   Conclusion

In this paper, we propose a representation of any algebraic structure by Boolean functions. Using this representation, some classifications of groupoids of order 3 are given. On the same way, these classifications can be made for any other algebraic structures, although in many cases their number is too large. This representation can be used for solving many problems in many areas, especially in cryptology and coding theory.

Acknowledgment. This research was partially supported by Faculty of Computer Science and Engineering at "Ss Cyril and Methodius" University in Skopje.

References

1. Dimitrova, V., Markovski, S.: Classification of quasigroups by image patterns. In: Proceedings of the 5-th Conference CIIT 2007, Bitola, Macedonia, pp. 152–159 (2007)
2. Gligoroski, D., Dimitrova, V., Markovski, S.: Quasigroups as Boolean functions, their equation systems and Gröbner Bases. In: Sala, M., Mora, T., Perret, L., Sakata, S., Traverso, C. (eds.) Gröbner Bases, Coding, and Cryptography, pp. 415–420. Springer, Heidelberg (2009). doi:10.1007/978-3-540-93806-4_31

3. Joux, A.: Algorithmic Cryptanalyses, Cryptography and Network Security. Chapman&Hall/CRC (2009)
4. Lazard, D.: Gröbner bases, Gaussian elimination and resolution of systems of algebraic equations. In: van Hulzen, J.A. (ed.) EUROCAL 1983. LNCS, vol. 162, pp. 146–156. Springer, Heidelberg (1983). doi:10.1007/3-540-12868-9_99
5. Markovski, S.: Quasigroup string processing and applications in cryptography. In: Proceedings of the 1st Conference MII, Thessaloniki, pp. 278–290 (2003)
6. Markovski, S., Gligoroski, D., Bakeva, V.: Quasigroup string processing: Part 1, Prilozi, Mat.-Tehn. Nauki, MANU Skopje, XX 1–2, pp. 13–28 (1999)

Software Quality Metrics While Using Different Development Methodologies

Simona Tudjarova, Ivan Chorbev[✉], and Boban Joksimoski

Faculty of Computer Science and Engineering, University of Ss. Cyril and Methodius,
Rugjer Boshkovikj 16, P.O. Box 393 1000 Skopje, Macedonia
simona.tudjarova@gmail.com,
{ivan.chorbev,boban.joksimoski}@finki.ukim.mk

Abstract. The initial idea for this research is to study the testing metrics throughout different software development methodologies, their importance and the way they influence the software quality. For that purpose, the research is conducted within an international IT company, where the teams are formed on geographically different locations. The main problem investigated is whether the team is committed to flawless and impeccable test management, regardless of the methodology used and how the clients' requirements can affect it. The conclusion is distributed among the case studies and it is based on how the testing metrics can help monitor the development and testing process.

Keywords: Software testing · Testing metrics · Methodologies for software development · Scrum · Kanban · Planned iterative model

1 Introduction

The development of software technologies and the advancement of the IT industry have raised software testing as a dominant engineering practice that helps controll expenses, improving quality and smartly managing time and risks. Software testing is an essential part of the software development process. "Metrics are a powerful and even mandatory tool for both effective test management and extracting the test results for decision making. Metrics provide objective and exact information which is mandatory for having control over quality. There are two phrases that are often quoted as justifications for using metrics as tools of control. The first one is: "You can't control what you can't measure" by Tom DeMarco." [1] Software quality needs constant monitoring and measurement. Testing metrics act as indicators or references for acceptance of the functional and non-functional software features, the speed of accomplishing the requirements, the predictability in future estimations, etc. Testing metrics represent quantitative and visualization values that make it possible to easily notice the problems, but also to emphasize the positive sides of the software. Measurements are a key element of the effective and efficient process of software testing, as they grade the quality and the efficiency of the process.

© Springer International Publishing AG 2017
D. Trajanov and V. Bakeva (Eds.): ICT Innovations 2017, CCIS 778, pp. 240–250, 2017.
DOI: 10.1007/978-3-319-67597-8_23

The success in the software industry relies on development of quality software with no defects, within the defined time frame and planned budget. The defects can also be defined as undesired product characteristics. They can appear in any phase, including requirements analysis, design or implementation, but also due to incorrect testing [2].

2   Related Work

Software testing is the most critical phase of the software development lifecycle. The software subjected to testing, goes through different phases: analysis, planning, environment preparation, test case design, test execution, reporting defects, deployment and maintenance. Much research has been done to optimize the whole process, with intention to improve the software quality in shortest time possible. After numerous evaluations of the different testing processes, it turns out that different software development models are used for different application types and different testing techniques are adopted in order to test them [3, 9]. According to the research conducted by Itti Hooda and Rajender Singh Chhillar, every company modifies the testing process according to its own needs or its clients' needs and conducts testing depending on the importance of the application.

Many software projects start being developed with small number of not sufficiently detailed requirements. Those requirements are sensitive and fragile. The stakeholders and the end users have no clear vision for the functionalities and the characteristics of the product at the beginning of the development. That is why the agile approach and its scrum implementation fit best in projects where the requirements are constantly changing. [7] Every software project has its uniqueness and differs from the others. The differences range from the essential nature of the requirements, the capability of the team members, the approach to work and development, up to the organizational and cultural factors [4]. The teams that use scrum implementation of the agile development method demand for their own specialized set of metrics, which is a consequence of the specifications that scrum brings [8].

Sprint Burndown chart
Burndown chart is a commonly used chart in the scrum implementation which allows tracking down the progress of the project. It helps identifying the current status and deviations from ideal planning. In addition, it can also anticipate the last standing of the project from any point of the chart [5]. When using scrum the software development is organized in time frames, so called sprints. At the beginning of every sprint, all the team members agree on how many tasks and requirements can be completed in the current sprint and that quantity represents a goal that the team has to accomplish during the sprint. The chart called **Sprint Burndown** Chart is updated with every complete requirement - user story. The X-axis represents the time, while the Y-axis represents the quantity of remaining work intended to be finished in the current sprint. The remaining work can be measured in hours or in **Story Points**. The team that continually accomplishes the goals it sets can serve as a reference of a well-organized scrum team. However, every mistake and wrong estimation of the team capability, leads to gaining experience and improvement.

Control chart

Kanban provides a visual working environment which helps recognizing the bottlenecks and highlights the issues, so that the team can easily track and resolve them as quickly as possible (Morris). When using a kanban implementation, various charts and computations are used. One characteristic chart is the Control chart.

This chart is one of the charts well suited for kanban. Its main characteristic is the **cycle time** of individual tasks. That is the total time it takes for a task to change from "In work" to "Done". Using this chart, performances and future behaviors can be identified. Measuring the cycle time represents efficient and flexible way of obtaining information about the team progress, but also shows how it can be improved, as the result data are visible instantly. Using this chart, we come across the **rolling average** value, which is based on defects and is not time based. For every defect represented on the chart, the rolling average at that moment is computed as an average of the cycle time of the observed defect, X defects before and X defects after. The gray shaded surface on the chart represents **standard deviation**, which gives the deviation from the real data to the rolling average, Fig. 3. The goal of every team is to gain consistent and short cycle time and to maintain it no matter what kind of task it has to accomplish.

The problem with the well-known traditional waterfall model, which is not suitable to adapt to changes and development, and the great dynamics of the agile methods, created need of establishing a new method for developing software products. In the iterative development, every project is divided to subsections. This enables the development team to present the results earlier in the process and to ensure early feedback from the stakeholders. Usually, every iteration represents a mini waterfall model. (Govardhan, 2010) [1]

The main difference between the agile methodologies and the planned iterative model is that planned iterative phases, so called iterations, last longer than agile increments (sprints) and not all of them have the same duration.

No matter which methodology is used, there are numerous test management tools that offer the possibility to track testing metrics, such as: reported and fixed defects per iteration, reported and not fixed defects per iteration, high priority defects, number of successfully executed test scenarios, number of failed test scenarios. One representative tool, which is quite common nowadays, is Jira and it is used by the projects in the case studies below.

3   Case Studies

Case Study 1: Scrum implementation

Project: Software for an insurance company. End users are brokers, underwriters, risk managers of insurance policies and they are all distributed worldwide. An implementation of different functionalities is needed, depending on the countries where the end users are located, the markets where they operate and their requirements.

Methodology: In this case, an entirely new project is being developed with many of the requirements not yet defined. Therefore, continuous communication with the clients is needed and changes during the development process are possible. So, the scrum implementation is being used. Each sprint lasts for two weeks, but effort estimations are calculated for 8 days, as 1 day is reserved for planning and 1 day is reserved for presenting the results to the client.

Team: The team consists of 4 software developers, 2 testers, a business analyst, project manager and product owner. The team is geographically distributed.

Results: The analysis given in Fig. 1 refers to duration of 2 weeks or 1 sprint. All the data are gathered at the end of the sprint. Figure 1 shows a Burndown chart where the values refer to the testing team only.

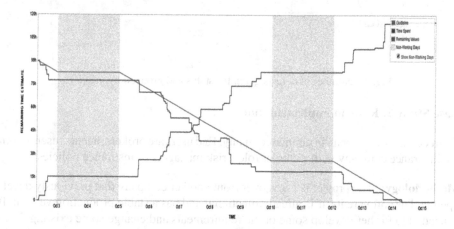

Fig. 1. Burndown chart (Color figure online)

- X-axis - time; Y-axis - remaining time to execute the tasks in hours - remaining estimated time; gray shaded parts - holidays or weekends
- Red line - the remaining time until the end of the sprint
- Green line - the time that the team of testers actually spent
- Gray line - line that needs to be followed in order to ideally complete the tasks within the estimated time

The testing team estimated that the anticipated workload can be finished in 90 working hours. At the end of the sprint, the green line reaches value of around 115 h. That means that the team spent 27% more time than the estimated, which means that some tasks were underestimated. From the testing perspective, it is considered that the team successfully fulfilled the goal as the red line meets the gray line close before the sprint end. This means that the team managed to accomplish the requirements within the sprint.

Summary: It is essential that the business analysts, in accordance with the product owner define the criteria for a completed task (Definition of done), the Entry/Exit Criteria for testing, as well as the criteria for a stable version. So, the success of the sprints is evaluated according to the realization of the planned requirements, but also according to the fulfillment of the criteria - all the reported defects with a "Major" priority to be closed, Fig. 2. This sprint, according to the above mentioned, from testing perspective, is successful.

Two Dimensional Filter Statistics:	Bugs			
Scrum Team	⚐ Major	→ Normal	⚐ Minor	T:
OPEN	0	5	0	5
IN PROGRESS	0	1	0	1
RESOLVED	2	5	0	7
BACKLOG	0	0	3	3
Total Unique Issues:	2	11	3	16

Grouped by: Priority   Showing **4** of **4** statistics.

Fig. 2. Number of defects given for status and grouped by priority

Case Study 2: Kanban implementation

Project: Software for an insurance company. End users are brokers, agents (users) from the insurance company with different roles, risk managers of insurance policies.

Methodology: The project is taken over from another company that previously developed it. The software itself is already functional and has a stable production version. It is needed to further develop some of the requirements and change some existing functionalities upon the end users' requests, etc. There is no need for frequent and dynamic changes during the development. That is why the clients suggested kanban implementation for development.

Team: The team consists of 3 software developers, 2 testers and a project manager. The testers work as part of the "Testing as a Service" program, and the rest of the team is formed on the client's side.

Results: One of the charts that shows work progress overview when using Kanban implementation is the Control chart. The project has been ongoing for one month when the analysis has been performed.

The Control chart shown on the Fig. 3 gives an overview of the progress of the team of testers. The graph reveals the following: The average time a defect remains in status "Selected for development" is about 9 days (1w, 1d, 20 h), the standard deviation is still varying, the range is wide and thus it is not possible to easily predict the behavior and the performances of the team. The last 2 defects shown on the chart (green circles with no fill), remained in the development process more than 16 and more than 20 days accordingly. In total, there are 26 defects shown on the chart, meaning that 26 defects have completed the phase "Selected for development". Some of them are now in the testing phase and some of them are successfully closed.

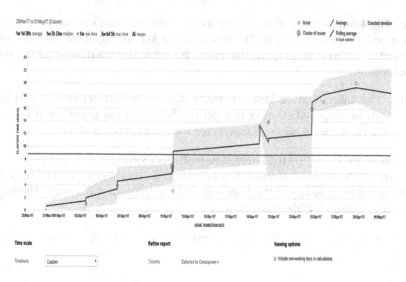

Fig. 3. Control chart of a project where the defects are presented according to the status "Selected for development" (Color figure online)

Summary: The rolling average is a metric for the team's efficiency while fixing defects. The chart in the Fig. 3 leads to a conclusion that the team's progress cannot be easily predicted because the rolling average is not consistent and standard deviation varies. This kind of beginning for a project is expected, as the team needs some time to get to know the environment and the technology which is used. If the trend of narrow range of the standard deviation continues, as in the case of the last week shown on the chart, then there would be a greater confidence while doing estimations for the team, as the cycle time would be close to the rolling average.

Case study 3: Planned iterative development

Project: Software for insurance company. In this case, there are already established systems which are currently in use. Those systems are developed by some legacy technologies that are unable to support the new requirements and cannot follow the latest trends. This project, supported by new technologies, aims to develop software from scratch, having almost the same purpose as the old one, but enriched with even more features. The biggest challenge is to integrate this software with systems which are already developed.

Methodology: Even though the requirements are based on an existing system, the team avoids the traditional methodology – waterfall and the planned iterative model is adopted. That enables the team to be aware of the project's global picture, although there are some changed and still undefined requirements. These systems support users worldwide. Therefore, there are different legislations which lead to many dependencies in the development. The planned iterative development is chosen to be the most suitable methodology that provides control over such systems.

Team: The team consists of 8 developers, 6 testers, 2 business analysts, a project manager and a product manager. The development and management team is working distributed on different geographical locations, but the testing team is completely formed on one location.

Results: The analysis given below is made during the first iteration, which lasts 6 months (still in progress). During that period, based on the client's and manager's demand, the testing team provides reports for the current project status.

Report number 1: Before the integration with other systems.

The report for all reported and still not resolved defects is given in the Fig. 4. It can be noticed that the majority of the defects are related to the user interface. There are 21 defects in total and 13 of them are critical.

Two Dimensional Filter Statistics: All open defects					
Components	Critical	Major	Minor	Moderate	T:
User Interface	13	6	1	1	21
Backend implementation	0	0	2	0	2
Backend implementation	0	0	1	0	1
Backend implementation	2	5	0	0	7
Backend implementation - database	4	6	0	0	10
Total Unique Issues:	19	17	4	1	41
Grouped by: Severity					Showing 5 of 5 statistics

Fig. 4. Report for all open defects before integration

Report number 2: After the integration with other systems.

In the Fig. 5, it can be noticed that after the integration is done, most of the critical defects linked with the user interface are resolved and now there is only one critical defect. That means that the user interface depended on the integration with the other systems. Generally, as it may be expected, the integration leads to increasing the number of defects. The total number of defects is increased for almost 4 times, summing up to 161 defects, with 3 out of them being blockers.

Two Dimensional Filter Statistics: All open defects						
Components	Blocker	Critical	Major	Minor	Moderate	T:
User Interface	0	1	33	22	59	115
Backend implementation	1	2	9	6	15	33
Backend implementation	2	1	0	0	2	5
Backend implementation - database	0	0	0	0	3	3
Integration with external system A	0	0	2	0	1	3
Integration with external system B	0	0	0	0	2	2
Integration with external system C	0	1	0	0	0	1
Integration with external system D	0	0	0	0	1	1
Total Unique Issues:	3	5	44	28	81	161
Grouped by: Severity						Showing 8 of 8 statistics

Fig. 5. Report for all open defects after the integration

Summary: The number of open defects within the first report points out that the user interface state is critical. But, in that moment it is necessary to be emphasized that the user interface behavior depends on the integration with the other systems. Having the metrics collected regularly (after an iteration, phase, or on regular demand) helps identifying the real source of the issues. In this case, with focus on the user interface, it is evident that even though there is a big number of defects after the integration, there are no blocking issues for the UI.

4 Results from the Survey

For the purpose of this research, a survey was conducted among the testers in an international IT company. The main emphasis is put on the perception for the testing metrics and their influences over the development process.

4.1 Testing Metrics

The Fig. 6 shows the results on the question "If the testing metrics are:" (a) Confirmed by both client and testing team in the IT company; (b) Provided by the IT company without the requirement from the client; (c) Specifically required from the client. It can be seen that 17% of the clients, which is a percentage that should not be ignored, do not require metrics as evidence. In these cases, the testing team submits testing metrics as a proof for their work and the quality they deliver.

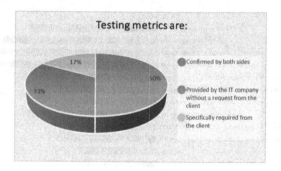

Fig. 6. Demand for testing metrics

4.2 Entry/Exit Criteria and Definition of Done

Every methodology with its implementations brings rules and procedures, including ones for testing and defining a completed task. Start and completion criteria need to be clearly set up. In the analyzed projects, in 17 instances this process is clear and well defined. However, in one instance the lack of criteria causes problems like delayed delivery, redundant testing, suboptimal time consumption, etc., (Fig. 7). The testing team should avoid this approach as much as possible.

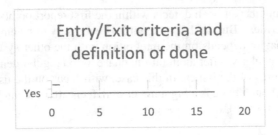

Fig. 7. Entry/Exit criteria and definition of done

4.3 Testing Metrics Required for the Testing Processes

The following three figures represent the answers on the question "Which testing metrics are required from the testing team".

The results from the survey regarding the testing metrics for kanban are distinguished as most critical. In comparison to scrum and planned iterative model (Figs. 8 and 10), projects developed by kanban hardly provide metrics for the testing process (Fig. 9). One of the reasons for this situation is the lack of sequences within kanban and lack of visualization dashboard updated by the team and the clients. However, the case study for kanban which is analyzed above should be taken as a good example for all the teams. There are plenty of ways to succeed in easy and on-time metrics delivery and therefore, many ways to show the quality of the work done and the software developed.

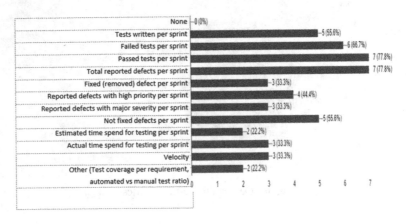

Fig. 8. Most represented testing metrics for scrum

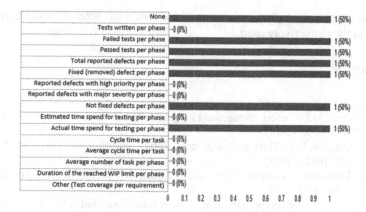

Fig. 9. Most represented testing metrics for kanban

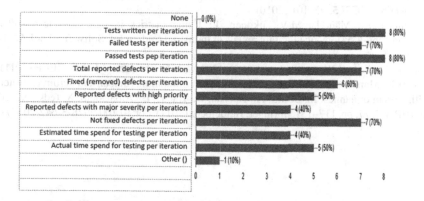

Fig. 10. Most represented testing metrics for planned iterative methodology

5   Conclusion

The development of each project in the IT industry is different and there is no general approach for testing metrics which would be suitable for all of them. However, each development methodology imposes its own principles and rules which are beneficial to be respected. Testing metrics within each of the methodologies have a great importance and it is necessary to pay attention. They are the main measurements not only for the software quality, but also for the quality of the testing team. The survey results show that there are clients who do not respect the idea of spending quality time on analysis and metrics evaluation since they do not see its effective value.

There is a wide range of metrics and visualization charts, which contribute for clear evidence and control over the software quality. The metrics and graphs which are used for the case studies are specifically chosen as representatives. The software testing metrics are essential for proper assessment whether any modules are critical and to point out the real reason for that. The test management tools which provide easy monitoring for defects, development progress and trend for removing defects should carefully be selected and then effectively used.

References

1. Heikkinen, E.S.: Testing metrics in a software development company (2015)
2. Noor, R., Khan, M.F.: Defect management in agile software development. Int. J. Mod. Educ. Comput. Sci. **6**(3), 55 (2014)
3. Hooda, I., Chhillar, R.S.: Software test process, testing types and techniques. Int. J. Comput. Appl. **111**(13), 10–14 (2015)
4. Lee, R.C.: The success factors of running scrum: a qualitative perspective. J. Softw. Eng. Appl. **5**(06), 367–374 (2012)
5. Bin Noor, T., et al.: A novel approach to implement Burndown chart in SCRUM methodology. Int. J. Adv. Res. Comput. Sci. Softw. Eng. (2012)
6. Munassar, N.M.A., Govardhan, A.: A comparison between five models of software engineering. IJCSI **5**, 95–101 (2010)
7. Kupiainen, E., Mäntylä, M.V., Itkonen, J.: Using metrics in agile and lean software development–a systematic literature review of industrial studies. Inf. Softw. Technol. **62**, 143–163 (2015)
8. Polk, R.: Agile and Kanban in coordination. In: Agile Conference (AGILE). IEEE (2011)
9. Cocco, L., Mannaro, K., Concas, G., Marchesi, M.: Simulating Kanban and Scrum vs. waterfall with system dynamics. In: Sillitti, A., Hazzan, O., Bache, E., Albaladejo, X. (eds.) XP 2011. LNBIP, vol. 77, pp. 117–131. Springer, Heidelberg (2011). doi:10.1007/978-3-642-20677-1_9

Reflections on Data-Driven Risk Valuation Models for MSMEs Based on Field Research

Jasmina Trajkovski[1(✉)] and Ljupcho Antovski[2]

[1] Trajkovski and Partners Management Consulting, Sveti Kliment Ohridski 24/2/1,
1000 Skopje, Macedonia
jasminat@tpconsulting.com.mk
[2] Faculty of Computer Science and Engineering, University Ss. Cyril and Methodius,
Rugjer Boshkovikj 16, 1000 Skopje, Macedonia
ljupcho.antovski@finki.ukim.mk

Abstract. There are many approaches to risk management, and the practice shows that they are not suitable for IT-centric Micro Small and Medium Enterprises (MSME), but more targeted to large and complex organizations. At the same time, the existing approaches in isolation are aimed at a particular type of risk, not taking into account that MSMEs need a more integrated approach, constraint on time and resources and availability of data. Based on the field research of over 150 organizations, the initially proposed risk management framework was revised generally in the area of scope of the risk, duration, risk management team and risk valuation model. Various risk models were reviewed for appropriateness. The development of IT and its use in organizations allows for preference to data-driven models, but the limitation of MSMEs with resources, and understanding of complex data-driven model limits their use. The field research showed that MSMEs prefer a hybrid method for assessment of risks, as they couldn't sustain a fully quantitative approach and as managers feel more confident with qualitative estimates.

Keywords: ISO31000 · IT-centric companies · Micro Small and Medium Enterprises (MSME) · Data driven risk models · Risk assessment · Risk management framework

1 Introduction

Over the past decades we are witnessing the development of a knowledge society where information has enormous significance, and as well extremely rapid development of information technologies and their penetration into every aspect of our life generating large amounts of data for the operations of the organizations. On one side, the new technology creates new risks that directly threaten the work of organizations [1], but on the other hand, the generation and collection of vast amounts of data open opportunity for analysis of these risks based on facts as big data enables organizations to better understand its business, market and make timely decisions [2]. But are the micro, small and medium IT-centric companies (MSMEs) using this opportunity?

© Springer International Publishing AG 2017
D. Trajanov and V. Bakeva (Eds.): ICT Innovations 2017, CCIS 778, pp. 251–264, 2017.
DOI: 10.1007/978-3-319-67597-8_24

While there are many approaches to risk management, the practice from the experience in the 20+ organization where we have implemented risk management shows that they are not suitable for these MSMEs, but more targeted to large and complex organizations. At the same time, the existing approaches in isolation are aimed at a particular type of risk, not taking into account that MSMEs need a more integrated approach that they could apply to various types of risks they are exposed and in order to obtain adequate information for making quality decisions for managing the companies. Identified problems include lack of adequate integrated and tailored methodology for managing risks related and lack of adequate model for the assessment and valuation of the risks.

In this paper we review the risk management concepts, analyze the findings from field research on perception of risk among IT-centric organization, propose a revised risk management framework for MSMEs and give an opinion based on these experiences on the feasibility of using data-driven risk models among MSMEs. The revised framework for risk management is based on our published framework [3], but the elements are revised as a result of the findings from the field research. The research is based on the direct experience of the leading author in the last 7 years with over 20 micro and small companies that are heavily IT-centric in their operations, and on survey conducted in early 2017.

2   Overview of Risk, Risk Management and Risk Models

The main concepts of risks management in IT-centric micro and small companies are divided into 2 groups: (i) definition of risk, types of risks and risk management, and (ii) risk management frameworks and standards.

Based on the International standard for Risk Management – ISO31000 [4], risk is defined as: "effect of uncertainty on objectives", where the uncertainties include events (which may or not happen) and uncertainties caused by ambiguity or a lack of information, while the objectives can have different aspects (health and safety, financial, IT, environmental) and can apply at different levels (such as strategic, organizational, project, process). It also includes both negative and positive impacts on objectives. The risk is often expresses as a combination of the consequences of an event and the associated likelihood of occurrence. As we discuss risks management frameworks for IT-centric micro and small companies, the main focus are the organizational risks. There are various types of organizational risks such as program management risk, investment risk, budgetary risk, legal liability risk, safety risk, inventory risk, supply chain risk, and security risk [5].

For the needs of the management of the IT-centric micro and small companies, all these risks could not be approached independently, and an integrated approach is necessary. This approach should be focused on the main drivers in the company, like the continual operations thru IT operation and known business processes so that the employees can understand what they should do. The reliance on IT as well puts the information security risks among the top as well. For the purposes of the research questions, we make the assumption that the management of these IT-centric micro and small companies deals with the legal and financial risks intuitively, and that they are not necessary to be included in the integrated risk management framework and approach of the company.

Having said that, we will look into the IT risk, information security risk and operational risk, which are respectively defined as:

- IT risk—that is the business risk associated with the use, ownership, operation, involvement, influence and adoption of IT within an enterprise [6];
- Information security risk—that is, the risk associated with the operation and use of information systems that support the missions and business functions of their organizations [5];
- Operational risk - The most common definition, first published in The Next Frontier and also adopted in recent operational risk documents issued by the Basel Committee [7], is that "Operational risk is the direct or indirect loss resulting from inadequate or failed internal processes, people and systems, or from external events".

The next concept to be introduced is the risk management. ISO31000 defines the risk management very broadly as the coordinated activities to direct and control an organization with regards to risk. Other institutions have a more precise definition, as described for example in NIST special publication SP800-39 [5], where risk management is defined as a comprehensive process that requires organizations to:

- frame risk (i.e. establish the context for risk-based decisions);
- assess risk;
- respond to risk once determined; and
- monitor risk on an ongoing basis using effective organizational communications and a feedback loop for continuous improvement in the risk-related activities of organizations.

Risk management is carried out as a holistic, organization-wide activity that addresses risk from the strategic level to the tactical level, ensuring that risk based decision making is integrated into every aspect of the organization.

With the development of risk management as an organizational discipline, a more defined concept evolved, named Enterprise Risk Management (ERM). There are many definitions of ERM, but a representative one is from the COSO framework [8, 9]: "Enterprise risk management is a process, effected by an entity's board of directors, management and other personnel, applied in strategy setting and across the enterprise, designed to identify potential events that may affect the entity, and manage risk to be within its risk appetite, to provide reasonable assurance regarding the achievement of entity objectives".

After setting the stage with the definition of the concept, lets look at the available risk management frameworks. Nowadays, there are several types of risk management methodologies, some of them issued by national and international organizations such as ISO, NIST, AS/NZS, BSI, others issued by professional organizations such as ISACA or COSO, and the rest presented by research projects. Each of these methods has been developed to meet a particular need so they have a vast scope of application, structure and steps. The common goal of these methods is to enable organizations to conduct risk assessment exercises and then effectively manage the risks by minimizing them to an acceptable level [10].

Vorster and Labuschagne [11] go even deeper in the analysis focusing solely on the methodologies for information security risk analysis and define a framework for comparing them. The objective of their framework is to assist the organization in the selection process o the most suitable methodology and/or framework. The elements that they are taking into consideration include:

- Whether risk analysis is done on single assets or groups of assets;
- Where in the methodology risk analysis is done;
- The people involved in the risk analysis;
- The main formulas used;
- Whether the results of the methodology are relative or absolute.

Some of these criteria are tightly related to the risk management considerations we have identified in the following section for the IT-centric micro and small companies.

A summary overview of elements taken into consideration in various risk management frameworks is given in Table 1.

Table 1. Overview of elements in risk management frameworks and methodologies

Type of framework	Main elements	Resource
Generic risk management frameworks	11 Principles for managing risks 5 segment framework: mandate and commitment; design framework; implement risk management; monitor and review the framework; continual improvement 5 step process: establish the context; risk assessment; risk treatment; monitoring and review; communication and consultation	ISO31000:2009 Risk Management Standard [4]
	It has 4 sub-processes: Risk assessment process; Risk treatment process; Risk communication process; Risk review and monitoring process.	Corpuz and Barnes in their 2010 paper on integration information security policy into corporate risk management [12]
Information Security Risk Management Frameworks	The tiers are: Organization, Mission/business processes and Information Systems, while the phases are Frame, Assess, Respond and Monitor	NIST SP800-39: Managing Information Security Risk [5]
	6 step process: context establishment; risk assessment; risk treatment; risk acceptance; monitoring and review; risk communication	ISO27005:2008 Information Security Risk management [13, p. 27]
	Views: STROPE - strategy, technology, organization, people, and environment Phases: DMAIC - define, measure, analyze, improve, and control cyclic phases	Information security risk management (ISRM) framework for enterprises using IT [10]
IT Risk management frameworks	Domains: Risk governance, Risk evaluation and Risk response	RiskIT framework [6]
Operational Risk Management Framework	Components: identify, assess, respond to and control risk	COSO Enterprise risk management integrated framework [8]
	Elements: 1. leadership, 2. management, 3. risk, and 4. tools	RMA Operational risk management framework [14]

Risk assessments/analysis are the first step in determining how to safeguard enterprise assets and reduce the probability that those assets will be compromised. There are several different approaches to risk analysis, but they can be broken down into two essential types: quantitative and qualitative. As described in Behnia's A survey of information security risk analysis methods [15], the following is the short overview of both types of risk analysis and assessment.

- Quantitative Risk Analysis: This approach uses two basic elements: the probability of an event occurring and the losses that may be incurred. Quantitative risk analysis uses one number produced from these elements. This is called the Expected Annual Loss (ALE) or Estimated Annual Cost (EAC). This is calculated for an event by simply multiplying by the probability of potential losses. Therefore, in theory, one may rank events in order of risk (ALE) and make decisions based on that risk. The problem with this type of risk analysis is usually associated with the unreliability and inaccuracy of data. Probability can rarely be accurate and can, in some cases, promote complacency. In addition, control and action steps that often deal with a number of potential events and the events themselves are often inter-related.
- Qualitative Risk Analysis: The qualitative method rates the magnitude of the potential impact of a threat as high, medium, or low. Qualitative methods are the most common measures of the impact of risks. This method allows covered entities to assess all potential impacts, whether they are touchable or untouchable. The qualitative risk analysis methodology uses several elements that are interconnected: Threats, Vulnerabilities, Controls.

Based on Lo Chen's work on hybrid risk assessment models [16], both types of risk assessment previously explained have deficiencies. Specifically, Quantitative risk assessment methods, using mathematical and statistical tools, attempt to assign specific numbers to the costs of safeguards and the amount of damage that can take place. These methods require a large amount of preliminary work to collect precise values of all elements, including asset values, threat frequency, safeguard effectiveness, and safeguard costs. The lack of good quality data for estimating probabilities of occurrence or loss expectancies, such as the probability of rare threats, is be a problem when the assessments are performed.

3   Field Research on the Perception for Risk Management

The research was conducted in March 2017 and 157 responses were collected with respondents matching the target sample. For the research, an online survey tool was used to create and validate the questions, to distribute and collect responses and to analyze the results. The survey was targeted at various individuals that represent/work in organizations that are IT intensive such as organizations in fields like IT services, IT outsourcing, Business Process Outsourcing, Telecoms, government institutions. The industry sector spread was wide to allow to compare the experiences from organization in different sub-sectors. Even though the spread is significant, with a few exceptions, the respondents do use IT on regular basis for performing their daily operations, and cannot operate without IT, thus can be included on the margins of the IT-intensive organizations. The size of the organization was a well varied, from micro organizations with less than 10 people to large enterprises with over 250 employees. This again was designed in such a way to be able to compare the practices across various sizes of organizations, and to be able to isolate the specifics of the small and micro organizations. Based on the results, there is an equal representation from the categories, coming to a 74% of respondents within the main target group of the research i.e. micro, small and medium (MSME) companies.

The first content specific questions was about the past experience with performing risk assessment i.e. to identify which of the respondents have consider performing a formal risk assessment for its operations. 23% have not considered it, while 24% have considered but are not doing it due to knowledge or time constraints, and 54% are doing it with varying dynamics.

When we look at the same questions but segmented by size of organization, we notice that in large enterprises the predominant practice is to have regular risk reviews, while in the MSMEs, some are doing it regularly or in key situations (44–47%), but the majority is not doing them either because they don't know how, they don't have time or it simple was not considered at all. Going into the details of what was the reason for not considering or pardoning risk assessments, the following diagram shows the answers, and they are not surprising: the most represented reason is that they organization did not see any benefit from it or that the tool used was asking for too much data, that it is overly complex or that it doesn't cover the scope needed.

When asked which methodology or framework they used for their risk assessments, 24% were not aware which methodology (if any) was used for the risk assessment, while 22,9% did not use any formal methodology. Of those who used some formal methodology, most represented is ISO27001/ISO27005 i.e. the information security risk assessment methodology. 17% of the responses named the international standard for risk management ISO31000 as the used methodology. Among the identified methodologies outside of the provided choices, the responses identified Prince2, AIRIMC, SWOT, OWASP, FMEA, CMMI Risk management SPI, and internally developed methodologies.

The duration of the risk assessment exercises varies and depends on various factors. The majority finish their risk assessment within 4 weeks, but an significant percentage of respondents did not know how long the overall risk assessment took. Looking per size of organization, in micro companies they finish in less than 1 week, while in the larger organizations the majority have answered 2–4 weeks. In majority of the surveyed organization, the risk assessment teams were small i.e. 2–4 team members, while in micro companies, sometime it is a one-man activity.

When asked to compare the experiences with risk assessment and risk management, the organizations had a slightly more positive opinion on the risk management process, i.e. in all sizes of companies the experience with the risk management is more positively perceived than the one with risk assessment. The used qualitative scale was useless, bad, so-so, ok, good, great. Only 5% of the respondents think that the risk assessment and risk management were useless or a bad experience, while a bit more than 45% believe it was a good or a great experience, and the remaining are in the range between.

When asked about the challenges in the implementation, again a qualitative scale was used, this time with gradations such as impossible (value = 1), difficult (value = 2), doable (value = 3), fairly easy (value = 4), easy (value = 5). The assessment covers the following 8 elements: identification of assets, identification of risks, evaluation of risks, identification of mitigation actions and their effectiveness, estimation of residual risks, monitoring of risks and revision of risks assessment. The following graph gives a summary comparison across all elements, with indication of opinions of organization per their size (Fig. 1).

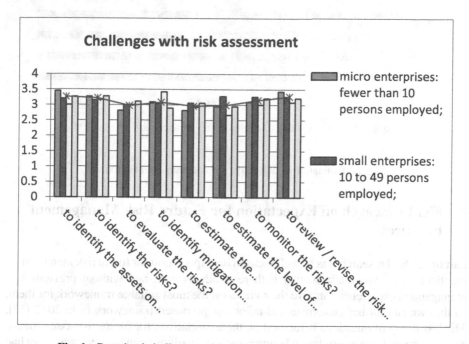

Fig. 1. Perceived challenges with various elements of risk management.

Putting the focus on the micro and small organizations, the following diagram shows the responses per element of the risk management activity. It can be seen that there are several elements that respondents found impossible or difficult to do, thus implying that they need external assistance for them. They include: evaluation of risks, identification of mitigation actions, estimation of effectiveness of mitigation action, estimation of residual risks and risk review. Nevertheless, the majority of the respondents are under the perception that risk management activities are at least doable (Fig. 2).

Fig. 2. Evaluation of difficulty of risk management activities

4 Field Research on Expectation for Future Risk Management Practices

Part of the field research was as well focused on expectation for future risk management practices. This was made so that both respondents with and without previous risk management experience can give their views on the most suitable framework for them, so that we can further customize and tailor our proposed framework from 2012 [17]. The last group of questions is focused on the expectations for future risk assessments. As there might have a correlation between the expectations from the ones who have had previous experience with risk assessments from those who have not, the analysis was done separately for each of the sub-sets of respondents.

Scope of Risk Assessment and Management
The scope options that the respondents could choose from were identified based on the previous experience with implementation of risk assessment in IT-centric MSMEs but as well on the prevailing good practice as described in Chapter 2. The provided options to select for scope of risk assessment and management in the survey included: IT and security risks, operation risks, compliance risks, financial risks, strategic risks, other to be specified. Multiple selections were allowed for this question. Looking at the entire population of respondents, the top identified options for the scope of risk management were: operational risks, strategic risks and IT/security risks.

Size and Composition of Risk Assessment Teams

As people are one of the main elements of the new proposed Risk management framework for IT-centric MSMEs, is was as well included in the survey for expectation of future risk assessments. Based on the responses about their past experiences, micro and small companies identified that 2–4 person team are ideal, while for the medium and large it was up to 10 people. On the composition of the team, the most selected options included: representatives from all departments, top management and IT staff. 40% of the organizations that had previous experience in risk assessment and management indicated that risk professionals or dedicated risk personnel is needed in the team.

Modality for Performing the Assessment

As important as the framework for risk assessment is, so is the modality or the way how it will be implemented in practice, as that is where the organizations struggle and get lost in the depth and complexity of the topic. When asked about the preferred modality, the sample population chose the following three as the most preferred: facilitated group workshop, delegated to various individuals using a computer cool and risk team with subsequent consultation.

Valuation models for risks

Maybe the biggest difference between academic and practical risk assessment is in the valuation models [18]. Current research work ventures in the direction of fuzzy models [19], complex quantitative models [20], economic models [21]. These models that are based on extensive scientific research provide a very in-depth approach and analysis of the value of the specific risks depending on multitude of factors. But, at the same time, these models are very difficult and time consuming for managers to implement in risk assessments in organizations. Another obstacle for implementation of these models is the need for large amounts of data and stronger mathematical skills. Based on the experiences in implementation of risk assessments in MSMEs, we have found that they prefer the quick and simple model. During the survey, the questions was posed to the sample population and their priority was a mix of qualitative and quantitative model, using value ranges from low to high or from 1 to 10 (Fig. 3).

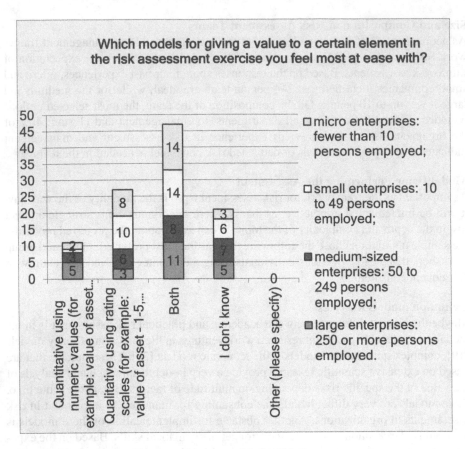

Fig. 3. Valuation models for risk elements

Duration of risk management exercises

The existing risk management frameworks do not give estimates or guidance on duration of risk assessment exercises [4, 6, 8, 13]. On the other hand, that is a deciding factor for managers in organizations – when will that have the result as it is needed for making decisions. The surveyed population, comprising of over 150 organizations was with equal distribution of micro, small, medium and large organization as shown on the following diagram. Regardless of their size, the predominant answer was that the duration of the risk assessment exercise should be 2–4 weeks. Other well represented answers were: as long as it takes and 1–2 months.

Frequency of review

The risk management is a continual process, where risk assessments are done on regular basis [4]. Even though the international standard on risk management ISO31000 recommends "regular review at least annually" [4, p. 3] it is up to the organization to decide its frequency. Based on the survey results, it can be seen that this recommendation is taken seriously as it was the most dominant response. Other significant responses included review on quarterly basis as well as review per project or significant change.

5   Impact on Previously Proposed Risk Management Framework

The previously proposed Risk management framework for IT-centric MSMEs was supported by the findings from the field research, but some areas for further fine-tuning were identified. The customized framework for risk management in IT-centric SMEs consists of: people, policy, methodology and process, and tools. Table 2 represents the changes which need to be implemented based on the field research into the proposed risk management framework for MSMEs.

Table 2. Needed changes to proposed RM Framework

	Initially proposed framework in 2012	Findings from field research
People - Risk management team	5–7 people, to include the representatives from the main processes or units in the company, as well as the management team	2–4 people, representatives from all departments, top management and IT staff
Policy	Annual review, generic scope	Annual review, scope –: operational risks, strategic risks and IT/security risks
Methodology and Process	Clear steps for risk management	Need for guidance on evaluation of risks, identification of mitigation actions, estimation of effectiveness of mitigation action, estimation of residual risks and risk review
Tools	Simple tool presented	Need for guidance on evaluation of risks i.e. risk models

6   Use of Data-Driven Risk Models

Based on the findings from the field survey, the proposed framework will be extended with guidance on the risk evaluation specifically on the risk models. The respondents' perception is that the model should be a mix of qualitative and quantitative, which opens the opportunity for using the vast amounts of data available nowadays to companies from the IT-centric operations.

Based on the international standard ISO31010 [22] for risk assessment techniques, methods used in analyzing risks can be qualitative, semi-quantitative or quantitative. The degree of detail required will depend upon the particular application, the availability of reliable data and the decision-making needs of the organization. Some methods and the degree of detail of the analysis may be prescribed by legislation.

- Qualitative assessment defines consequence, probability and level of risk by significance levels such as "high", "medium" and "low", may combine consequence and probability, and evaluates the resultant level of risk against qualitative criteria;
- Semi-quantitative methods use numerical rating scales for consequence and probability and combine them to produce a level of risk using a formula. Scales may be linear or logarithmic, or have some other relationship; formulae used can also vary;

- Quantitative analysis estimates practical values for consequences and their probabilities, and produces values of the level of risk in specific units defined when developing the context. Full quantitative analysis may not always be possible or desirable due to insufficient information about the system or activity being analyzed, lack of data, influence of human factors, etc. or because the effort of quantitative analysis is not warranted or required. In such circumstances, a comparative semi-quantitative or qualitative ranking of risks by specialists, knowledgeable in their respective field, may still be effective.

Even where full quantification has been carried out, it needs to be recognized that the levels of risk calculated are estimates. Care should be taken to ensure that they are not attributed with a level of accuracy and precision inconsistent with the accuracy of the data and methods employed.

The selection of risk assessment techniques should be done with respect to several factors such as depth or details required, possibility to be replicated, traceable and verifiable, to be appropriate for the organization and finally to provide understandable results for the management.

Some of the quantitative techniques include: Human reliability analysis, Toxicological risk assessment, Fault tree analysis, Event three analysis, Cause/consequence analysis, FMA/FMECA, Reliability centered maintenance, LOPA, Bowtie analysis, Markov analysis, Monte-Carlo analysis, Bayesian analysis, etc. [22]. Unfortunately, none of these is suitable for MSMEs within their expected timeframe for doing the risk assessment as identified by the field survey. Following the recommendation from Muntenau et al. [23], there has to be a balance between qualitative and quantitative assessment. For the short duration of 2–4 weeks, qualitative or semi-qualitative models have a higher chance for completion.

7   Conclusion

The literature review of various risk management frameworks as presented in Chapter 2 describes the current development in this field. It as well references our previously developed specific framework for IT-centric MSMEs. Based on that, a field research was conducted with over 150 organization on their perceptions and experiences about risk management as a process and as a model for evaluation of risk. The key findings are presented in this paper, focusing on the need for a comprehensive simple framework, to be implemented by a mixed team over a short period of time, utilizing a mixed risk evaluation methods. The paper gives focus on the feasibility for use of data-driven risk models. The conclusion based on the field research, is that even though there is precondition for having sufficient data for use in a data-driven model, the MSMEs perceive it as overly complex, so more simple models need to be defined. These findings contribute to a refinement of our proposed risk management framework for MSMEs and its fine-tuning as they come from research conducted to the specific target group addressed by our initial framework.

Further work will be focused on defining the criteria for the risk evaluation model taking in consideration the findings from the field research on one side and the academic developed models on the other side. This, together with the refined risk management framework will then be tested in several IT-centric organization to check its validity and usefulness.

References

1. Zhang, X., Wuwong, N., Li, H., Zhang, X.: Information security risk management framework for the cloud computing environments, pp. 1328–1334 (2010)
2. Chen, H., Chiang, R.H., Storey, V.C.: Business intelligence and analytics: From big data to big impact. MIS Q. **36**(4), 1165–1188 (2012)
3. Trajkovski, J., Antovski, L.: Risk management framework that meets the implementation challenges for IT centric SMEs. IJHCITP 3 (2013)
4. ISO 31000 risk management - principles and guidelines.pdf. ISO (2009)
5. Joint Task Force: NIST SP800-39- Managing information security risk.pdf. NIST Special Publication
6. The RiskIT framework -Excerpt.pdf. ISACA (2009)
7. Haubenstock, M.: The operational risk management framework. Oper. Risk Regul. Anal. Manag. Prentice Hall-Financ. Times. **84**(4) (2003)
8. COSO: Internal Control-Integrated Framework. Committee of Sponsoring Organizations of the Tread way Commission (COSO), AICPA/COSO (1992)
9. Aguilar, M.K.: COSO releases a risk management framework.pdf. Account. Today 18(19), 1 (2004)
10. Saleh, M.S., Alfantookh, A.: A new comprehensive framework for enterprise information security risk management. Appl. Comput. Inform. **9**(2), 107–118 (2011)
11. Vorster, A., Labuschagne, L.: A framework for comparing different information security risk analysis methodologies. In: Presented at the SAICSIT 2005 (2005)
12. Corpuz, M., Barnes, P.H.: Integrating information security policy management with corporate risk management for strategic alignment. In: Proceedings of the 14th World Multi-Conference on Systemics, Cybernetics and Informatics (WMSCI 2010) (2010)
13. ISO/IEC JTC1: ISO 27005-2008 Information security risk management.pdf. ISO (2008)
14. Taylor, C.: The RMA operational risk management framework.pdf. RMA J. **88**(5), 4–7 (2006)
15. Behnia, A.: A survey of information security risk analysis methods. Smart Comput. Rev. (2012)
16. Lo, C.-C., Chen, W.-J.: A hybrid information security risk assessment procedure considering interdependences between controls. Expert Syst. Appl. **39**(1), 247–257 (2012)
17. Trajkovski, J., Antovski, L.: Risk management framework for IT centric SMEs. In: Proceedings from ICT Innovation (2012). Ohrid
18. Eloff, J.H.P., Labuschagne, L., Badenhorst, K.P.: A comparative framework for risk analysis methods. Comput. Secur. **12**(6), 597–603 (1993)
19. Lee, M.-C.: Information security risk analysis methods and research trends: AHP and fuzzy comprehensive method. Int. J. Comput. Sci. Inf. Technol. **6**(1), 29–45 (2014)
20. Bojanc, R.: A quantitative model for information-security risk management. ResearchGate 25(2) (2012)

21. Bojanc, R., Jerman-Blažič, B.: An economic modelling approach to information security risk management. Int. J. Inf. Manag. **28**(5), 413–422 (2008)
22. Editor ed., ISO 31010 - Risk Management - Risk Assessment Techniques. ISO (2009)
23. Munteanu, A.: Information security risk assessment: The qualitative versus quantitative dilemma. In: Managing Information in the Digital Economy: Issues & Solutions-Proceedings of the 6th International Business Information Management Association (IBIMA) Conference, pp. 227–232 (2006)

The Model for Gamification of E-learning in Higher Education Based on Learning Styles

Nadja Zaric[1], Snezana Scepanović[1(✉)], Tijana Vujicic[1],
Jelena Ljucovic[1], and Danco Davcev[2]

[1] Faculty of Information Technologies,
University "Mediterranean", Podgorica, Montenegro
`snezana.scepanovic@unimediteran.net`
[2] Faculty of Computer Science and Engineering,
University "Ss. Cyril and Methodius", Skopje, Macedonia

Abstract. This paper describes the results of first phase of our original research about implementing gamification in the university e-courses based on the learning styles. Main research aims were to conduct comprehensive review of literature and current practice in implementation of gamification concept in higher education and to identify game elements that can make a positive impact on a specific learning style (LS). Result of first research phase is conceptual model for including game elements in e-learning courses in higher education based on Felder-Silverman Learning Style Model (FSLSM). The implementation and evaluation of the proposed model is planned as a future work. Our model presents extension of FSLSM model providing not only information about the learning styles, but also about their semantic groups in game-based learning contexts. The proposed model can be used as theoretical framework for further research on relationships between student's learning style, achievements and behaviors in Virtual Learning Environments (VLE).

Keywords: Gamification · Learning style · E-learning · Instructional design · Felder Silverman Learning Style Model · Adaptive hypermedia systems

1 Introduction

We all advance and learn throughout our lives. First, we learn about ourselves and the world around us through playing games. As we grow up, learning leaves the form of entertainment, and it gradually takes on the outlines of formal education. Formal education, with its strict, predefined rules and goals, leaves no room for individual creativity and play. In a standardized environment, such as formal education, we often find ourselves demotivated and limited, which can lead us to poor learning outcomes or even to giving up on learning [1]. This issue is more often seen in e-learning systems, which, by their nature, require additional educational measures to compensate physical distance, as well as the diversity of individuals who participate in the process of e-learning [2].

For the purpose of increasing the motivation and engagement of students and, therefore, improving learning outcomes, researchers are more often considering the

© Springer International Publishing AG 2017
D. Trajanov and V. Bakeva (Eds.): ICT Innovations 2017, CCIS 778, pp. 265–273, 2017.
DOI: 10.1007/978-3-319-67597-8_25

inclusion of games and game elements in the learning process [3–6]. Incorporating game elements in non-game context is called gamification [7]. Gamification is still fairly new in higher education, but it is built on the success of the gaming industry, social media and decades of research on human psychology.

On the other side, we cannot ignore the fact that students belong to heterogeneous groups and that they learn, perceive, absorb and understand information in different ways i.e. they have different learning styles. Furthermore, learning styles are considered more and more in technology enhanced learning systems as an important factor in improving the learning progress of students and to make learning easier for them [8].

This brings us to the idea that application of game elements that are consistent with learning styles can be used in Virtual Learning Environment (VLE) to evoke student motivation, improve engagement and performance of learning.

2 Theoretical Background

2.1 Learning Styles Theories

Learning style is an ongoing issue of great importance to educational research in technology enhanced learning. LS refers to the preferential way in which the student absorbs the processes, comprehends and retains information. Learning styles are the psychological and cognitive behaviors that serve as indicators of how students learn and how they interact with learning environment. The roots of learning styles originate from the classification of psychological types and depend on cognitive, emotional and environmental factors, as well as one's prior experience [9]. There are many theories and classification of learning styles. Most current in psycho-logical and educational research are: Myers-Briggs Type Indicator, Kolb theory and the theory based on the Felder-Silverman Learning Styles Model [10].

Myers-Briggs Type Indicator model is created on the basis of Jung's theory of psychological types. According to Jung, there are four criteria (intuiting, sensing, thinking and feeling) which define the eight types of personality with different learning styles [11]. Based on intuition criteria, two types are noted: extraversion type who learns from the phenomenon and has expressed interpersonal communication and introversion type who prefers learning on personal experience. Sensing criteria defines sensing type and intuition type. According to feeling criteria, judgment and perceiving types are defined. Judgment type learns and explore through agendas, while perceiving type needs a lot of information and rich content to adopt a fact.

Kolb's learning theory sets out four distinct learning styles [12]. In his model, Kolb differentiates students according to their ability for receiving (concrete or abstract) and processing information (active or reflective) [13]. Based on this theory, Kolb defined four types of learning: diverging, assimilating, converging, and accommodating type.

Felder-Silverman LS model describes learning styles in very much detail, distinguishing between preferences on four dimensions. FSLSM is one of the most often used learning style model in technology enhance learning and some researchers even argue that it is the most appropriate model for the use in adaptive learning systems [14]. In the following, the four dimensions of FSLSM are described.

The first dimension distinguishes between an active and a reflective way of processing information. Active learners learn best by working actively and trying things. In contrast, reflective learners prefer to think about and reflect on the material. The second dimension covers sensing versus intuitive learning. Learners who prefer a sensing LS like to learn facts and concrete learning material. In contrast, intuitive learners prefer conceptual thinking, concerned with theories and meanings. The third, visual-verbal continuum determines how learners prefer information to be presented. This dimension differentiates learners who remember best what they have seen, and learners who get more out of textual representations - written or spoken. In the fourth dimension, the learners are characterized according to their understanding. Sequential learners learn in small incremental steps and therefore have a linear learning progress. In contrast, global learners use a holistic thinking process and learn in large leaps. They tend to absorb learning material almost randomly without seeing connections but after they have learned enough material they suddenly get the whole picture.

2.2 Games and Gamification

Game can be defined as sequence of activities that involves one or more players. Each game has goal, consequences and rewards. Game rules are defined in advance and game involves competition against another player or yourself [15]. Games are a part of our daily life and they are research subject of many scientific fields such as psychology, philosophy, anthropology, pedagogy and others [16]. Nowadays, game concepts are being increasingly incorporated in different areas. Gamification is using game elements in non-game contexts like marketing, business, e-commerce, education, work environment, social media, etc. Regardless of the domain in which it is applied, gamification is mainly used to strength users/customers motivation and engagement [17].

In order to understand gamification, it is necessary to understand the core concepts of games and games design. Based on the so-called MDA framework there are three equally important game design elements [18]:

1. **Mechanics** – refers to behavior and control mechanisms. It defines actions that player has to take in order to achieve a specific goal.
2. **Dynamics** – describes the run-time behavior of the mechanics which depends on player actions and inputs.
3. **Aesthetics** – design, visibility and usability of user interface. Aesthetics describes users' expressions, feelings and emotional responses evoked in the player.

Ten most used game elements are: avatars, 3D environment, narrative context, feedback, leaderboards, rankings and levels, economy, competition, team play, time restrictions and systems for real-time communication [19].

2.3 Gamification and E-learning

Game elements in education are mainly used in order to increase motivation and engagement in the learning process [20]. Learning problems arise when a student is not motivated enough to learn or does not have enough knowledge to solve a particular problem. Scientists believe that incorporating elements of a game can make learning fun and easier, and students more motivated and persistent [21].

There are many overlaps between games and learning that justify the idea of incorporating game elements in e-learning curriculum. The principal of MDA framework for game design can be applicable in process of designing an online learning course. Mechanics would refer to actions (task, assignment, lessons) that student has to finish in order to complete the course. Dynamics refers to behavior of VLE caused by students' actions and moves. Finally, aesthetics represents pupils' emotions and reactions during the learning process.

Main similarity between learning process and game is that both of them are goal-focused activities. Players/students are being directed to undertake tasks in order to achieve a desired outcome, moving to the next level/mission (in the case of a game), or understanding a complex topic (in the case of learning process). We can draw a parallel between course activities and game activities, while course goals can be defined as game goals [14].

Playing a game to "win" is very similar to student learning harder in order to get a better grade. Games usually have levels, while learning process has lessons. When you complete a level, you get an award – when you learn a lesson you get points. Most common game reward mechanics are points, leaderboards and achievements [3–6]. In learning, student's grades (or points) can be used to show their success and to form the leaderboard, so students could compete to accomplish higher ranking.

Furthermore, method of progress tracking within a game is analogous to the provision of feedback within education. An important part of providing feedback to users is to let them know how much progress they've made, either in game or learning. Also, the parallel between time factor in games and time factor in learning is also evident. The game can determine the exact time within which player can perform a task. Analogously students can have a time-limited quiz, or test, or assignment.

2.4   Current Studies on Using Gamification in E-learning in Higher Education

In order to identify game elements that were implemented in e-learning so far and which of them had the biggest impact on participants' cognitive and emotional behavior relevant studies on gamification in higher education are analyzed in our previous research. Concrete models and experiments results are discussed and presented in [22]. Generally, all of the studies considered the learning outcomes of gamification as mostly positive in terms of increased motivation and engagement in the learning tasks as well as enjoyment over them. The most successfully implemented elements were experience points (XPs), badges, scoreboards and feedback that had positive impact. Opposite of which, negative or poor impact on students' motivation had avatars, time track and competition.

However, during the analysis of the previous studies on gamification, a significant number of limitations were found. For example, there is no study that implements more than five game elements at once, which questions the scope of implementation of their model. Further, in most examples practical assignments and tasks were gamified, with lack of examples on gamifying theoretical material. Related to examine results, we noticed that researches mainly measured the impact of gamification on students' motivation and engagement with few measurements on learning outcomes. The most important limitation that made us question the relevance of studies' results was that no one took into account the fact that pupils learn on different ways i.e. that they have different learning styles. Learning style as an important part of learning process was not taken into account during the design phase of gamified learning material, nor during evaluation.

Regard to this, model presented in this paper is a step forward in terms of gamification, putting different student profiles to the forefront. If it has already been established that the learning outcome will depend on the learning style to which student belongs [23], then it is justifiable to assume that the effect will be analogous in the gamification.

3 The Model for Gamification E-learning in Higher Education Based on Learning Styles

The basic idea of the proposed conceptual model is to incorporate game elements in online education based on students' learning styles. The subject of this paper is gamification of e-courses within Modular Object-Oriented Dynamic Learning Environment (Moodle LMS) according to learning styles defined in FSLSM. Proposed model aims to define specific game elements which can improve the learning process and outcomes for each individual LS defined in FSLSM.

3.1 Conceptual Design and Architecture

The model proposes architecture of VLE for creating gamified e-learning courses in which learning styles are the base for the creation of teaching material and course content. Furthermore, gamified e-learning courses should increase student motivation, engagement and provide an atmosphere in which students are proactive, motivated and express a positive attitude toward learning. The model is presented in Fig. 1 and it consists of the following main elements: management of e-learning, management of FSLSM, ADDIE model (analysis, design, development, implementation and evaluation) for e-courses development, gamification elements in e-learning and their effects on students.

Moodle LMS stands out as one of the most common systems to support e-learning. Moodle is widely used in higher levels of education, with over 50,000 registered sites [24]. As a service that has been successfully used in formal and informal teaching for many years, Moodle LMS 3.2v is used as the basis for the development of gamified e-learning model. Tools for managing user accounts and roles, creating and monitoring online courses, developing and updating learning materials and collaborative tools are just some of the features of Moodle that are used in this model. During the course, Moodle stores data about a variety of students' actions and activities such as: time spent on the course, time spent in the development of tasks, the list of materials examined, the results of tests and questionnaires and etc. Those data are used to determine the typical behavior of students in the system for managing the learning process and the ways in which individual resources and activities of the course are used.

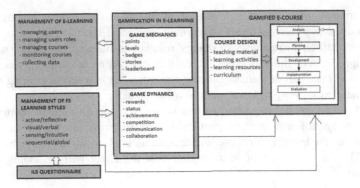

Fig. 1. Conceptual model of gamification e-learning based on students learning style

3.2   Development of Gamified E-learning Course

Analysis

As the proposed model is focused on introducing game elements based on students' learning style, the first step in development process is to identify the LS that student belongs to. Questionnaire that is used for the purpose of identification is Index of Learning Styles (ILS). The ILS [25], developed by Felder and Silverman, is a 44-item questionnaire for identifying the learning style according to FSLSM. This standardized questionnaire has to be extended to include questions about student personal information (personal data, gender, year of study and average grade) in order to be able to identify the respondents in further steps. Based on the results, tested sample is divided in two groups - experimental and control. Students from first group will enroll the course with gamified elements, while those from control group will learn from non-game learning materials. Table 1 provides suggestions for structuring of the teaching material, so that it corresponds the FSLSM.

Table 1. Methods of structuring of the teaching material, based on the FSLSM

	Materials and resources	Activities and assessments
Active	Concrete examples, case studies	Practical assignments, experiments
Reflective	Presentations with topics to think about	Student gives his opinion
Visual	Material with pictures, videos and graphs	Question with visual elements
Verbal	Textual documents, books and lectures	Essay questions
Sequential	Materials must be read in specific order	Task with multiple steps
Global	Students creates their own learning plan	Conceptual solutions
Sensor	Concrete examples and strict information	Practical tasks
Intuitive	Concepts and theoretical frames	Conceptual issues

Design

Aim of the design phase is to formalize the game elements that can be incorporated into the online course in a way that they contribute to increasing the motivation and participation of students, as well as to the expression of positive behavior towards learning. Table 2 provides an overview of the game elements that can make a positive impact on a specific LS. Proposals is based on the examined theoretical literature and researches [3–6, 9–12].

Table 2. Game elements that can make a positive impact on a specific LS(elements with impact are marked by "x" sign)

	Active	Reflective	Visual	Verbal	Sequential	Global	Sensor	Intuitive
Badges	x	x	x		x	x	x	x
Leaderboards	x		x		x	x		
XP's	x		x		x	x	x	
Mystery			x	x				x
Levels	x	x	x	x	x		x	
Stories		x		x			x	x
Feedback	x	x	x	x	x		x	
Time track	x		x	x				
Progress bar	x	x	x		x	x	x	x
Competition	x	x	x		x			

Development

For the purposes of course development, existing Moodle LMS services were integrated and modified to support gamification. We used: Open badges to incorporate award game element, Level up! Module for XP's points, Ranking block for showing scoreboards, User tours for implementing story game element, Progress bar for tracking students' progress and Game mod for developing educational games for assignments and practice [26].

Implementation

During implementation, we introduced the system to its users and general public. Before the evaluation starts, all the users (professors and pupils) should be familiarized with gamification concept, assessment methods and curriculum. It is also necessary to monitor system's usage, performance and collect feedback.

Evaluation

Evaluation results are supposed to show whether the gamified material made a positive impact on the learners, increased their learning motivation, and if it did, how the game elements and learning styles are connected. Data collected from e-learning system serves to analyze learning outcomes and should be used to compare results of two groups. Since we know which students belongs to which LS and his score on each individual e-test, we can easily draw conclusions about success and impact of game elements on the level of FSLSM. In the process of evaluation, we use *learning style* as an independent variable. With Pearson's chi-squared test [27] we can relatively easy find possible correlations and conditionals between LS and their impact on success of each game element implemented in e-course. For determining students' level of satisfaction and impressions on using gamification the opinion questionnaire is used.

4 Conclusion

An effective teaching strategy is a proven way of transferring knowledge and can make students more productive. If it is pointless to consider tailoring instruction to each individual student, it is equally misguided to imagine that a single one-size-fits-all approach to teaching can meet the needs of every student. There is definitely a need for new techniques in e-learning to address the issues of boredom and loneliness in online platforms. Gamification has brought forward an opportunity for educators to provide a comprehensive framework by which playful learning in the context of serious adult level content can be realized. It does so with no disruption to effective pedagogical practice and provides the means to engage students in learning. Adaptation to learning styles and gamification seem to be promising concepts in e-learning worth further exploration, even if neither of the techniques will cure all problems or reach all students. To determine the usefulness of gamification across academic disciplines and learning styles a more thorough research is required.

The research presented in this paper contributes to the fact that the use of game elements in developing a model for e-learning and teaching can significantly contribute to the success of learning, if they comply with the learning styles of students. The proposed model will contribute to the further development of adaptive systems for e-learning and will serve as the basis for future theoretical and empirical research in the field of gamification in e-learning. Future work will deal, on one hand, with implementation and evaluation of proposed model with a larger number of students, considering more detailed variables about students' behavior in the gamified course as well as the interactions between different LS preferences and additivity features in VLE.

References

1. Keller, J., Suzuki, K.: Learner motivation and E-learning design: a multinationally validated process. J. Educ. Media **29**(3), 229–239 (2004). doi:10.1080/1358165042000283084
2. Ćamilović, D.: Trendovi u obrazovanju na daljinu. Globalisation Challenges and the Social Economic, pp. 125–133. Faculty of Business and Management Sciences, Novo Mesto (2012)
3. Barata, G., Gama, S., Jorge, J., Gonçalves, D.: Engaging engineering students with gamification. In: Proceedings of the 5th International Conference Games and Virtual Worlds for Serious Applications (VS-GAMES) (2013)
4. Kosmadoudi, Z., Lim, T., Ritchie, J., Louchart, S., Liu, Y., Sung, R.: Engineering design using game enhanced CAD: the potential to augment the user experience with game elements. Comput. Aided Des. **45**, 777–795 (2013)
5. Codish, D., Ravid, G.: Personality based gamification – educational gamification for extroverts and introverts. In: Proceedings of the IX CHAIS Conference for the Study of Innovation and Learning Technologies: Learning in the Technological Era, pp. 36–44 (2014)
6. Dickey, M.D.: Engaging by design: how engagement strategies in popular computer and video games can inform instructional design. Educ. Train. Res. Dev. **53**, 67–83 (2005). http://medicina. iztacala.unam.mx/medicina/Engaging%20by%20design.pdf. Accessed 16 June 2015
7. Glover, I.: Play as you learn: gamification as a technique for motivating learners. In: Proceedings of the World Conference on Educational Multimedia, Hypermedia and Telecommunications, pp. 1999–2008. AACE, Chesapeake (2013)

8. Šćepanović, S., Debevc, M.: Adaptation of learning objects in virtual learning environment to learning styles of students. In: Proceeding of the 5th International Conference of Education, Research and Innovation, Madrid, Spain, pp. 5061–5066 (2012)
9. Sillaots, M.: Gamification of higher education by the example of computer games course. In: Proceedings of the Seventh International Conference on Mobile, Hybrid, and On-line Learning – eLmL, Lisbon, Portugal (2015). http://www.thinkmind.org/index.php?view= article&articleid=elml_2015_4_20_50048. Accessed 08 July 2015
10. Open Badges for Higher education. https://www.pearsoned.com/wp-content/uploads/Open-Badges-for-Higher-Education.pdf. Accessed 09 July 2015
11. Keefe, J.W.: Learning Style: Cognitive and Thinking Skills, pp. 22–26. National Association of Secondary School Principals, Reston (1991)
12. Cassidy, S.: Learning styles: an overview of theories, models and measures, Directorate of Psychology, pp. 2–4. University of Salford (2004). ISSN 0144-3410/ISSN 1469-046X
13. Catherine, B., Wheeler, D.: The Myers-Briggs personality type and its relationship to computer programming. J. Res. Comput. Educ. **26**(3), 358–370 (1994). doi:10.1080/08886504.1994.10782096
14. Stash, N., Bra, P.D.: Incorporating cognitive styles in AHA! (The Adaptive Hypermedia Architecture). In: IASTED International Conference Web-Based Education, Innsbruck, pp. 378–383 (2004)
15. Schäfer, A., Holz, J., Leonhardt, T., Schroeder, U., Brauner, P., Ziefle, M.: From boring to scoring – a collaborative serious game for learning and practicing mathematical logic for computer science education. Comput. Sci. Educ. **23**(2), 87–111 (2013)
16. Simões, J.: A brief history of gamification: part I - the origin. EduLearning: http://edulearning2.blogspot.pt/2014/03/a-brief-history-of-gamification-part-i.html. Accessed 5 Mar 2017
17. Žarić, N., Šćepanović, S.: Edukativne igre u online marketingu visokoškolskih ustanova - fikcija ili stvanost? (2015)
18. Hunicke, R., LeBlanc, M., Zubek, R.: MDA: a formal approach to game design and game research. In: Proceedings of the AAAI Workshop on Challenges in Game AI. vol. 4, no. 1, pp. 17–22 (2004)
19. Kim, B., Park, H., Baek, Y.: Not just fun, but serious strategies: Using meta-cognitive strategies in game-based learning. Comput. Educ. **52**(4), 800–810 (2009)
20. Adams, E.: Fundamentals of Game Design, 2nd edn. New Riders, Berkeley (2009). p. 700
21. Schäfer, A., Holz, J., Leonhardt, T., Schroeder, U., Brauner, P., Ziefle, M.: From boring to scoring – a collaborative serious game for learning and practicing mathematical logic for computer science education. Computer Science Education (2013). doi:10.1080/08993408.2013.778040
22. Scepanovic, S., Zaric, N., Matijevic, T.: Gamification in higher education learning – state of the art, challenges and opportunities. In: Proceedings of the VI International Conference of E-learning, Belgrade, Serbia (2015)
23. Bjekić, D.: Psihološki faktori e-učenja i e-nastave. Čačak: Tehnički fakultet. Accessed Apr 2017
24. Moodle homepage. https://moodle.org/. Accessed 10 July 2017
25. Learning styles and strategies. http://www4.ncsu.edu/unity/lockers/users/f/felder/public/ILSdir/styles.htm. Accessed 10 July 2017
26. Moodle Plugins for Gamification. https://www.paradisosolutions.com/blog/moodle-plugins-for-gamification/. Accessed 10 July 2017
27. Živković, S.: Analysis of data in SPSS, Handbook for Statistics. De Facto Consultancy, Podgorica (2015)

Behavioral Targeted vs Non-targeted Online Campaign in Macedonia

Borce Dzurovski and Smilka Janeska-Sarkanjac[✉]

Faculty of Computer Science and Engineering,
Ss Cyril and Methodius University in Skopje, Skopje, Macedonia
borce.dzurovski@gmail.com,
smilka.janeska.sarkanjac@finki.ukim.mk

Abstract. Behavioral Targeting (BT) is a technique used by online advertisers to increase the effectiveness of their campaigns, and is playing an increasingly important role in the online advertising market. Although it's a technique that has been used for more than 10 years in online advertising, and there are surveys of its effects on developed markets, there is little evidence for the results of behavioral targeted campaigns in the developing countries as Macedonia. In this paper we give a short review of what behavioral targeting is, we list the forms of behavioral targeting, its advantages and disadvantages, and we present a comparative analysis of two advertising campaigns for a Macedonian food company - using behavioral targeting vs. using only geo-targeting (only users in Macedonia) to show the difference in results. From this analysis we may conclude that the targeted advertising provides far more than non-targeted at similar conditions. The difference in clicks is huge, even with slightly fewer impressions, and if we consider that the cost of the campaign are usually dependent on impressions (more impressions - more expensive campaign), then the cost of the campaign in targeted advertising would be lower, and finally the overall CTR is almost 5 times higher for targeted advertising. This paper tries to contribute to the scientific-research field and offer a motivation for the digital marketing agencies in developing countries and their clients to start using more behavioral targeting in their campaigns and accomplish higher results.

Keywords: Online advertising · Behavioral targeting · Click-Through Rate (CTR) · Google AdWords · User tracking

1 Introduction

Behavioral targeting or behavioral advertising is a form of targeted advertising which tries to offer appropriate ad content for each internet user based on collected user profiles. This allows advertisers to use their marketing budget more efficiently by only reaching people who are likely to become customers. These user profiles may contain information such as sex, age group, location, estimated income and interests and are, to a large extent, built from search queries and browsing history of the users [1]. This data may be augmented with public information from social networks, resulting in highly detailed profiles. Behavioral targeting platforms take into account some of the aspects of the consumer profiles and their behavior over time. There is evidence that behavioral

© Springer International Publishing AG 2017
D. Trajanov and V. Bakeva (Eds.): ICT Innovations 2017, CCIS 778, pp. 274–283, 2017.
DOI: 10.1007/978-3-319-67597-8_26

targeting significantly increases the effectiveness of online advertisements, making individuals more likely to buy an advertised product [2].

In this paper we will examine the effects of two campaigns of a Macedonian food company, one non-targeted and one with implemented behavioral targeting by interests, in order to contribute to the evidence of the effectiveness of behavioral advertising with a Macedonian case study.

The structure of the paper is as follows. First, forms of behavioral targeting are presented in Sect. 2. The advantages and disadvantages of the behavioral targeting are sketched in Sect. 3. The discussion of the comparative analysis of two advertising campaigns for the Macedonian food company is given in Sect. 4 and finally a conclusion is given in Sect. 5.

2 Forms of Behavioral Targeting

Behavior targeting has two major categorizations: one which helps with targeting all users according to behavior for the specific website, also well known as network behavioral targeting and the other which helps with targeting users within the website with different offers and promotions, is called On-site behavioral targeting [3].

- On-site behavioral targeting – it is performed on one particular web site, and the content of the web site is customized according to the preferences of a particular group of visitors. There are web sites that are personalized according to the preferences of each visitor, offering unique appearance for each of them.
- Network behavioral targeting – it uses the visitor's browsing history to collect the information about that person - marital status, sex, age, interests, etc. to target the user with the relevant advertising on different sites in Internet. The data is collected using cookies, web beacons and similar technologies, and/or third-party ad serving software, to automatically collect information about site users and site activity [4].

Alongside these two main categorizations, there are 8 different forms of behavioral targeting [5]:

- Retargeting – Based on user's previous visits on an advertiser's website, relevant ads will be served on sites where the advertiser has banner placements.
- Contextual Targeting – Ads are specifically targeted by content, such as an ad for sporting goods shown on sport web site.
- Audience Targeting – This approach matches ads to specific audiences that share the same demographics, behavior, devices, geography, day, time, etc.
- Demographic Targeting – Demographic features that are usually used are gender, age, income, ethnicity, etc.
- Geographic Targeting – Targeting users based on location information including country, city, postal code, and other locale information.
- Keyword Targeting - Targeting users based on keywords including search terms, hash tags or keywords used on social media sites.
- Time-Based Targeting – There are ads that are targeted based on daytime or weekdays.

- Emotional Targeting – This form of behavioral targeting guesses the emotional state of users using information from web cams, and analyzes facial expressions, speech patterns and gestures, or semantic recognition on users' communication such as IMs, Texts, Tweets, Facebook posts, etc.

3   Advantages and Disadvantages

Some of the advantages of behavioral targeting, subject to a number of studies, including this are [6]:

- Higher Click-Through Rate – Behavioral targeting places relevant ads to internet users, and therefore, they are far more likely to click through and see the offer.
- Higher Conversion Rate – Although CTR is important for marketers, conversions, such as sign up for the weekly newsletter, or download a specific eBook, or buy a product or service, are essential for any business. Behavioral targeting selects the right audience and helps marketers to design a creative that appeals to the targeted customers.
- Higher Return on Investment – Behavioral targeting reduces the marketing budget, by offering the advertisements only to the people that are potentially interested in the product or service; it can convince more people to become sales for the company and the amount earned through adequate leads is improved.

Aside from the advantages, there are several disadvantages of behavioral targeting that should be considered [7]:

- Identification of Users Tracked – to create profiles of information, advertisers use unique identifiers to identify users to provide contextual ads. When the unique identifier can be tied back to the user, it becomes threat to the privacy of users.
- Collection of Sensitive Data – such as personally identifiable information, personal traits or medical records.
- Accountability of Advertisers – it should be transparent which advertisers are serving the ads, including malicious or inappropriate ads.
- Unreliable Opt-out Mechanism – the efficacy of the opt-out mechanism depends on the type of cookies used (HTTP, flash, etc.). Opt-out should mean that all tracking activities of the user should be stopped, which usually is not the case.
- Multiple User System – behavioral targeting may not be completely successful in cases when one computer is used by several users (such as members of a family, adults and children, or when a computer is in a public place).
- Cookie Security – cookies have many potential security threats, and could be used to steal information of users.
- Use of outdated information from cookies – the period of interest may be very important variable in the targeted campaign, due to the fact that people's interests vary over time.

4 Comparative Analysis of Two Advertising Campaigns for Macedonian Food Company

To contribute to the evidence of the effectiveness of behavioral advertising with a Macedonian case study, a comparative analysis of two campaigns for advertising products of a large Macedonian company from the food industry will be presented - one without a defined behavioral targeting, with the exception of geographical targeting (all users in the territory of Macedonia, because it is the market in which the company performs) and the other using targeting based on interests, (a segmentation of users into separate groups is made previously). Both campaigns are performed by the Macedonian digital marketing agency Media Solutions, conducted through the platform Google AdWords.

Non-targeted campaign was conducted in the period between 24.09.2016 and 03.10.2016, while the targeted campaign between 03.09.2015 and 10.26.2016. Approximately the same number of impressions (1,468,493 impressions during the non-targeted campaign, and 1,477,062 impressions for the targeted one) has been achieved, so both campaigns are comparable according to the results of online metrics. Also, the analysis is conducted on two campaigns from the same company, which are conducted in almost same time, therefore the preferences of buyers towards the company in terms of the products or the brand are on the same level, and the attractiveness of the product doesn't have influence on the results, but the choice of internet users who were shown the banner. Therefore, this comparative analysis shows the effects of targeting, opposed to a traditional, non-targeted campaign.

4.1 Non-targeted Campaign

During the non-targeted campaign, a total of 1.468.493 impressions were placed to Macedonian internet user to a number of web sites that belong to Google Display Network. Click Through Rate was 0.2, since there were 2.875 clicks on the banners. Table 1 shows the distribution of impressions, clicks and CTR by days, and Fig. 1 shows the elementary demographic characteristics of the impressions and clicks (Figs. 2, 3 and 4).

Table 1. Number of clicks, impressions and CTR in the period from 10.01–28.11.2016.

Date	Clicks	Impressions	CTR (Click-through rate)
10.01.2016	194	102656	0.19%
12.01.2016	117	45157	0.26%
10.02.2016	198	116109	0.17%
10.03.2016	72	44741	0.16%
10.04.2016	83	45363	0.18%
10.05.2016	14	8304	0.17%
10.06.2016	43	26295	0.16%
10.07.2016	39	25187	0.15%
10.08.2016	40	25191	0.16%

(continued)

Table 1. (*continued*)

Date	Clicks	Impressions	CTR (Click-through rate)
10.09.2016	17	8413	0.20%
11.09.2016	64	33286	0.19%
23.09.2016	77	28778	0.27%
24.09.2016	97	30122	0.32%
25.09.2016	112	41481	0.27%
26.09.2016	8	7818	0.10%
10.10.2016	43	14654	0.29%
11.10.2016	109	41886	0.26%
10.11.2016	38	17752	0.21%
11.11.2016	69	41191	0.17%
10.12.2016	40	11551	0.35%
11.12.2016	73	38089	0.19%
10.13.2016	36	23662	0.15%
10.14.2016	40	23203	0.17%
10.15.2016	46	21206	0.22%
10.16.2016	37	24878	0.15%
10.17.2016	42	24440	0.17%
10.18.2016	37	26313	0.14%
10.19.2016	40	25835	0.15%
10.20.2016	40	20253	0.20%
10.21.2016	32	16180	0.20%
10.22.2016	38	14499	0.26%
10.23.2016	39	17284	0.23%
10.24.2016	76	33433	0.23%
10.25.2016	36	20477	0.18%
10.26.2016	35	16730	0.21%
10.27.2016	23	15638	0.15%
11.13.2016	72	34450	0.21%
11.14.2016	66	32397	0.20%
11.15.2016	32	20013	0.16%
11.16.2016	32	19439	0.16%
11.17.2016	38	21579	0.18%
11.18.2016	31	10585	0.29%
11.19.2016	34	15120	0.22%
11.20.2016	36	26677	0.13%
11.21.2016	31	21778	0.14%
11.22.2016	35	24064	0.15%
11.23.2016	34	14533	0.23%
11.24.2016	34	17819	0.19%
11.25.2016	30	18630	0.16%
11.26.2016	33	20610	0.16%
11.27.2016	36	20463	0.18%
11.28.2016	129	57616	0.22%
11.29.2016	28	14665	0.19%
Total	2875	1468493	0.20%

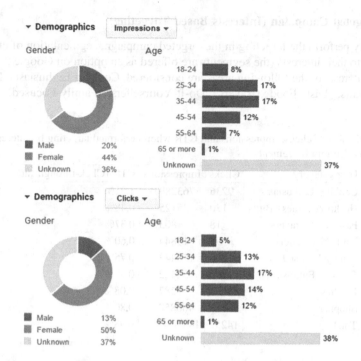

Fig. 1. Demographic segmentation of customers in non-targeted campaign. The criteria taken into are clicks and impressions

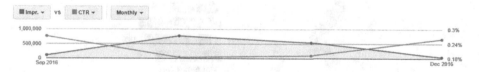

Fig. 2. Impressions vs. CTR (Click-through rate) in the non-targeted campaign

Fig. 3. Clicks vs. impressions in the non-targeted campaign

Fig. 4. Clicks vs. CTR in the non-targeted campaign

4.2   Targeted Campaign (Interests Based Targeting)

To properly perform the targeting in the targeted campaign, segmentation of customers according to their interests (the segments are offered as an option on Google AdWords) is made. Interests of the following areas are considered: Cooking Enthusiasts, Health & Fitness Buffs, Fast Food Cravers, Do-It-Yourselfers, Family-Focused, Outdoor

Table 2. Number of clicks, impressions and CTR when performed targeting by interests, where users are divided into 8 segments.

User category	Clicks	Impressions	CTR (Click-through rate)
Cooking Enthusiasts	9236	763289	1.21%
Health & Fitness Buffs	170	35125	0.48%
Fast Food Cravers	18	4803	0.37%
Do-It-Yourselfers	95	15844	0.60%
Family-Focused	849	113442	0.75%
Outdoor Enthusiasts	196	52422	0.37%
Foodies	1	1287	0.08%
Shoppers	3687	460850	0.80%
Total	14252	1447062	0.98%

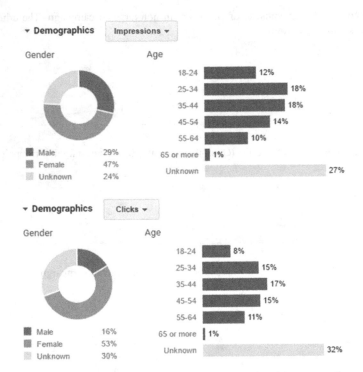

Fig. 5. Demographic segmentation of users in targeted advertising according to interests. Impressions and clicks.

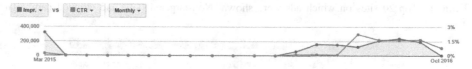

Fig. 6. Impressions compared to CTR in targeted campaign (monthly)

Fig. 7. Clicks vs. CTR in the targeted campaign (monthly)

Fig. 8. Clicks vs. Impressions in the targeted campaign (monthly)

Enthusiasts, FoodiesShoppers. The results obtained from targeting by interest shown in the Table 2 below.

During the targeted campaign, a total of 1.447.062 impressions were placed to

Table 3. Difference between the number of clicks, impressions and CTR between non-targeted and targeted advertising based on interests.

	Non-targeted	Targeted	Difference
Clicks	2.875	14.252	11.377
Impressions	1.468.493	1.447.062	−21.431
Total CTR in percent	0.20%	0.98%	4.9 times more interest

Macedonian internet user to the same Google Display Network. In this case, the number of clicks was 14.252, and therefore Click Through Rate was 0.98. Figure 5 shows the elementary demographic characteristics of the impressions and clicks (Figs. 6, 7 and 8).

We compared the above-mentioned results of the two campaigns, and the results are shown on Table 3. On the Table 4 there is a list of web sites on which the advertisements were shown, for the non-targeted and the targeted campaign.

From the results, we can conclude that the targeted advertising provides far better than non-targeted at relatively same conditions. The difference in clicks is immense, even with slightly fewer impressions, and if we consider that the cost of the campaign

Table 4. Top 20 sites on which ads were shown. Non-targeted (left) and targeted campaign (right).

Site	Clicks	Impressio	CTR	Site	Clicks	Impressio	CTR
blic.rs	30	97361	0,03%	blogspot.mk	680	35778	1,90%
kurir.rs	34	64685	0,05%	blic.rs	11	27181	0,04%
sportske.net	20	44871	0,04%	ekskluziva.ba	19	25781	0,07%
ekskluziva.ba	11	35308	0,03%	kurir.rs	10	24896	0,04%
devetmeseci.net	10	32430	0,03%	fokuzz.com	109	16848	0,65%
story.rs	1	28010	0,00%	receptizasve.work	104	14483	0,72%
heroquizz.com	10	27864	0,04%	kuhinjaideje.com	112	13667	0,82%
accuweather.com	15	24578	0,06%	domacikuhar.info	263	11804	2,23%
atraktiva.in	24	21679	0,11%	index.hr	8	10543	0,08%
tekstovi.net	1	20001	0,00%	web-tribune.com	73	9577	0,76%
receptidomacekuhii	107	19393	0,55%	telegraf.rs	1	9059	0,01%
beograd.in	68	17677	0,38%	torteikolacirecepti.(	354	8366	4,23%
webtribune.rs	45	16312	0,28%	igrezadecu.com	22	8329	0,26%
espreso.rs	23	16051	0,14%	malakuhinjica.com	72	7996	0,90%
blogspot.mk	113	15406	0,73%	heroquizz.com	9	6915	0,13%
srbijajavlja.rs	4	13990	0,03%	receptibezmane.inf	266	6798	3,91%
fokuzz.com	47	12317	0,38%	tagged.com	2	6690	0,03%
lijekizprirode.com	24	11703	0,21%	webtribune.rs	37	6632	0,56%
minimagazin.info	14	11465	0,12%	biljnaapoteka.info	109	6515	1,67%
youtube.com	6	11359	0,05%	zenakraljica.work	66	6491	1,02%

is usually dependent on impressions (more impressions - more expensive campaign), then the cost of the campaign in targeted advertising would be lower, and finally the overall CTR is almost 5 times higher for targeted advertising. From diagrams by comparison, we can conclude the following: first, seems that the non-targeted campaign although shows increasing in impressions, the CTR line is diametrically opposite (not increased), while in the targeted, the values are almost on the same line after a certain period. Second, although the number of clicks is increased there is no increase in CTR for non-targeted campaigns, while for targeted campaigns there are small difference (between clicks and CTR). Unfortunately, we can`t provide information from the costs and profits of the campaigns, and we cannot calculate Conversion rate and Return on investment rate, because they are sensitive data, but according to marketing campaign performances, we can predict with high possibility that targeted advertising based on interests, gave far better conversion and financial results in comparison with non-targeted campaign (using only geo-targeting - users from Macedonia).

5   Conclusion

From what is written above, we may conclude that the behavioral targeting may appear in many forms, and usually combining them will get the desired result. The process is not simple, but now, with online platforms that offer this kind of targeting and tracking results in real time it's quite easier (intuitive) and more advanced for marketers. Marketers just review the reports of user behavior and decide which criteria will be used in targeting, and the same procedure can be repeated several times until the desired result come in a manner of "test - results - test - results - test - a successful campaign". From the comparative analysis for the Macedonian food company, it is evident that behavioral targeting campaign (compared to non-targeted campaign) gives almost 5 times higher CTR with less impressions (less impressions - lower costs).

Finally, we may say that it will be interesting to observe how behavioral targeting will develop in the future, if we consider the latest trends like intelligent IP and the rise of account-based marketing. As the chief revenue officer of eXelate (company for behavioral data), Mark Zakorski said [8]: "Behavioral marketing is not going away any time soon, so it pays to watch its growth and integration into the marketing mainstream."

References

1. Jaworska, J., Sydow, M.: Behavioural targeting in on-line advertising: an empirical study. In: Bailey, J., Maier, D., Schewe, K.-D., Thalheim, B., Wang, X.S. (eds.) WISE 2008. LNCS, vol. 5175, pp. 62–76. Springer, Heidelberg (2008). doi:10.1007/978-3-540-85481-4_7
2. Behavioral Targeting, Advertising, and the Web: An Evaluation of Today's "Don't Ask, Don't Tell" Regime of Consumer Tracking Online, p. 9. Harvard Law, May 2008
3. Stratigent.   http://www.stratigent.com/community/websight-newsletters/onsite-behavioral-targeting. Accessed 15 May 2017
4. Know Online Advertising.   http://www.knowonlineadvertising.com/targeting/behavioral-targeting. Accessed 15 May 2017
5. AdPeople Worldwide Austin: Behavioral Targeting and Dynamic Content Creation, p. 4, April 2015
6. ExactDrive: exactdrive,   http://www.exactdrive.com/news/benefits-of-behavioral-targeting. Accessed 15 May 2017
7. Behavioral Targeting, Advertising, and the Web: An Evaluation of Today's "Don't Ask, Don't Tell" Regime of Consumer Tracking Online, pp. 11–17. Harvard Law, May 2008
8. Propel Marketing. http://responsive.propelmarketing.com/media/site_assets/57fdb2627e670c d09dd6823d9930dfeb/assets/whatisbehavioraltarg.html. Accessed 15 May 2017

Author Index

Printed in the United States
By Bookmasters

Printed in the United States
By Bookmasters